L'Amérique Méridionale: the Map That Shaped Brazil in the 18th Century

Mapping the Past

VOLUME 4

The titles published in this series are listed at *brill.com/mtp*

L'Amérique Méridionale: the Map That Shaped Brazil in the 18th Century

By

Junia Ferreira Furtado

BRILL

LEIDEN | BOSTON

Cover illustration: Detail of the front page of d'Anville, *Atlas, ou Recueil de cartes géographiques, Collection de cartes géographiques en trois volumes que son Excellence l'Ambassadeur a fait choisir pour son usage par le Sieur d'Anville, géographe du roy*. FBN, AT 015,02,001/003.

This book was translated by Sabrina Gledhill, Flora Thomson-DeVeaux, and Eoin O'Neill.

Library of Congress Cataloging-in-Publication Data
Names: Furtado, Junia Ferreira, author.
Title: L'Amérique méridionale : the map that shaped Brazil in the 18th century / By Junia Ferreira Furtado.
Description: Leiden ; Boston : Brill, [2025] | Series: Mapping the past, 2589-9945 ; Volume 4 | Includes index. | Summary: "This book explores how the origins of Brazil's modern borders can be traced to the cartography of the Americas produced by the eighteenth-century French cartographer J.B.B. d'Anville. It argues that this map reflects the geopolitical policies of the Portuguese diplomat D. Luis da Cunha, who was involved in Portugal's negotiations with the Spanish to formally establish Brazil's frontiers, and highlights how and why these policies were adopted in the Treaty of Madrid in 1750"– Provided by publisher.
Identifiers: LCCN 2024036800 (print) | LCCN 2024036801 (ebook) | ISBN 9789004708310 (hardback) | ISBN 9789004710764 (ebook)
Subjects: LCSH: Cartography–Brazil–History–18th century. | Geopolitics–Brazil–History–18th century. | Cunha, Luiz da, 1662-1749. | Brazil–Politics and government–History–18th century.
Classification: LCC GA673.6.A1 F87 2025 (print) | LCC GA673.6.A1 (ebook) | DDC 320.1/20981–dc23/eng/20240815
LC record available at https://lccn.loc.gov/2024036800
LC ebook record available at https://lccn.loc.gov/2024036801

Typeface for the Latin, Greek, and Cyrillic scripts: "Brill". See and download: brill.com/brill-typeface.

ISSN 2589-9945
ISBN 978-90-04-70831-0 (hardback)
ISBN 978-90-04-71076-4 (e-book)
DOI 10.1163/9789004710764

This book is printed on acid-free paper and produced in a sustainable manner.

To baby Otto, newly arrived
and already fulfilling my world

Contents

Acknowledgments

Firstly, I would like to thank the Bosch Company for having granted me access to their collection, in their headquarters in Stuttgart, where I found the first documents that made me started this research. The institutional support I received in Brazil was crucial in bringing the research to fruition. I would especially like to thank Universidade Federal de Minas Gerais (UFMG) for all the academic and funding support, especially the PhD Program in History, the Instituto de Estudos Avançados/IEAT, and the Reitoria. I received public funds at different moments to support my research, I would especially thank CNPq, CAPES, Fapemig, and Faperj.

I would like to thank the Odebrecht Group and the Jury of the 2011 edition of the Odebrecht Historical Research Prize Clarival do Prado Valadares for having embraced and funded this project; and the Jury of Human Science of the 2013 Jabuti Award, sponsored by Câmara Brasileira do Livro, Elias Thomé Saliba, and Luiz Mott, to grant the award to the Brazilian first edition of *O Mapa que inventou o Brasil*. Also the translators Sabrina Gledhill, Flora Thomson-DeVeaux, and Eoin O'Neill, who in different moments helped me to translate the book into English.

Abroad, I was able to count on some institutions and colleagues who welcomed me, allowing the research to reach the international scope necessary to achieve the goals I set myself, as I ventured to cross the destinies of a Portuguese ambassador and a French cartographer, in a European world much more globalized than one might expect. In 2007, I obtained a visiting research fellowship at the Newberry Library, granted by the Hermon Dunlap Smith Center for the Study of Cartography, where I would especially like to thank James Ackerman. In 2009 and 2017, I was welcomed as a visiting researcher at the École des Hautes Études en Sciences Sociales (EHESS). There, I was able to count on the invaluable welcome and friendship of Serge Gruzinski, Cláudia Damasceno Fonseca, Carmen Bernand, Louise Benat-Tachot, Carmen Salazar-Soler, Catarina Madeira Santos, and Eloi Fiquet. In Lisbon, where I was a visiting professor at the Institute of Social Sciences of the University of Lisbon, I thank Nuno Gonçalo Monteiro. In Stanford University where I received the Cátedra Joaquim Nabuco de Estudos Brasileiros in 2012, I would like to thank the Center for Latin American Studies, and the historians Herbert Klein and Karen Wigën and Barry Ruderman and Katherine Parker who invited me to present my work at the Barry Ruderman Conferences.

I would like to thank all the staff at the libraries and archives where I went looking for clues about D. Luis da Cunha and d'Anville, in the persons of ambassador Álvaro da Costa Franco, from the Arquivo Histórico do Itamaraty; Maria Dulce de Faria, from the Fundação Biblioteca National; Helène Richard and Catherine Hofmann, from the Bibliothèque Nationale de France; Margarida Lages and Manuel Côrte-Real, from the Arquivo do Ministério Estrangeiro de Portugal; Maria Joaquina Esteves Feijáo and Conceição Chambel, from the Biblioteca Nacional de Portugal; and Helena Greco, from the Sociedade Geographica de Lisboa; also the Museum of Versalhes, Bosch Company, Biblioteca Guita e José Mindlin/USP, for authorizing the reproduction of the images and maps that illustrate this book.

The colleagues and friends who have contributed in some way with suggestions, documents, support, invitations to discuss this research in seminars and publications, or friendship are precious and countless. At the risk of forgetting some, I would like to thank Neil Safier and Iris Kantor, special friends and map lovers, Eddy Stols, Luís Felipe de Alencastro, André Ferrand de Almeida, Jorge Pimentel Cintra, Mary Pedley, Hal Langfur, Matthew Edney, João Carlos Garcia, Diogo Ramada Curto, Jean-Marc Besse, Chet Van Duzer, Zsolt G. Török, Sergueï Karp, Carla Lois, Mauricio Onetto Pavez, Catherine Delano-Smith, Carme Montaner, Wolbert Smith, Marie-Noelle Bourguet, Charlotte de Castelnau-L'Estoile, Lucien Bély, Andrea Doré, Jordana Dym, Mirela Altic, Stuart Schwartz, J.B. Shank, Maria Fernanda Bicalho, Maria de Fátima Gouvêa (in memory), João Fragoso, Gabriel Paquette, Fernando Bouza, Pedro Cardim, Mauricio Nieto Olarte, Sebastián Díaz Ángel, Marie-Thérèse Mandroux-França, Roberta Stumpf, Laurent Vidal, Ronaldo Vainfas, Rodrigo Bentes, Márcia Menendes Motta, Heloisa Meireles Gesteira, Luís Miguel Carolino, Pedro Marinho, Thomas Werner, Eddy Stols, Pedro Cardim, Antônio Gilberto Costa, Friedrich E. Renger, Márcia Maria Duarte Santos, Katia Queirós Matoso, Rákóczi István, Maria Emilia Madeira Santos, Abilio Diniz Silva, Anna Canas Delgado Martins, Isabel Ferreira da Mota, Ângela Domingues, Eliana Dutra, Regina Horta Duarte, Luiz Carlos Villalta, Mauro Condé, José Newton Coelho Meneses, Andre Luis Pereira Miatello, Heloísa Maria Murguel Starling, Lígia Beatriz de Paula Germano, Eduardo França Paiva, Vanicléia Silva Santos, Valquíria Ferreira da Silva, Carmem Marques Rodrigues Maria Leônia Chaves, Francisco Eduardo de Andrade, Ronaldo Pereira de Jesus, Andréa Lisly Gonçalves, Álvaro Antunes Rodrigues, and Patricia Kauark Leite also for all her care.

Last but not least, I would like to thank my family, who in one way or another shared in the construction of this book, my husband Lucas, my daughters Clara and Alice, my son-in-law, Lucas Henrique, and my grandson Otto.

Figures and Tables

Figures

Tables

Abbreviations

AAIBL	Archives de la Académie des Inscriptions et Belles-Lettres		
AASP	Archives de la Académie Royale des Sciences de Paris		
	PV	Process Verbaux	
ADO	Archives Départamentales de l'Orne		
	SHAO	Fonds de la Société Historique et Archéologique de l'Orne	
AAE	Archives des Affaires Étrangères		
	CP	Correspondance Politic	
	E	Espagne	
AGS	Arquivo Geral de Simancas		
	E	Estado	
AHI	Arquivo Histórico do Itamaraty		
	DEL	Documentos Encadernados: limites	
AHU	Arquivo Histórico Ultramarino		
	MAG	Manuscritos Avulsos de Goiás	
ANTT	Arquivos Nacionais da Torre do Tombo		
	IL	Inquisição de Lisboa	
	ML	Manuscritos da Livraria	
	MR	Ministério do Reino	
	MNE	Ministério dos Negócios Estrangeiros	
		LM	Legação de Madrid
		LPB	Legação dos Países Baixos
AUC	Arquivo da Universidade de Coimbra		
	CLC	Cartório de dom Luís da Cunha	
BNP	Biblioteca Nacional de Portugal		
	R	Reservados	
	AT	Arquivos do Conde de Tarouca	
BNF	Bibliothèque Nationale de France—Paris		
	M	Manuscrits	
		MF	Manuscrits Français
		NAF	Nouvelles Acquisitions Françaises
		MP	Manuscrits Portugais
	DCP	Département des Cartes et Plans	
FBN	Fundação Biblioteca Nacional, Brazil		
	M	Manuscritos	
NL	Newberry Library		
	VB	Vault Box	
RACL	Real Academia das Ciências de Lisboa		
RBC	Robert Bosch Collection		
SGL	Sociedade Geográfica de Lisboa		

Introduction

Night was falling in Madrid on August 18, 1747. A traveler passed quickly through the city walls.[1] He was dusty and tired after coming hastily from Paris in a hurried journey of almost ten days.[2] The cities of Blois, Poitiers, Bordeaux, and Bayonne, all on French soil, Vitoria, Miranda de Ebro, Burgos, Aranda de Duero, Boceguillas, and Alcobendas,[3] all of which were in Spain, merely served as brief rest stops, giving him just enough time to recover from the aches and pains of the day's journey, to replace or rest the horses, and eat a scanty meal—barely sufficient to restore the energy spent on his journey. The pain that scourged his body was already part of his daily life, common to everyone who was charged with swiftly and safely carrying correspondence between Portuguese ambassadors scattered throughout the courts of Europe, as well as between them and the authorities of their own kingdom. He was the courier tasked with swift delivery—although there were those who said the opposite and called those in charge turtles[4]—of the correspondence of D. Luis da Cunha (1662–1749), then the Portuguese ambassador in Paris.

Once inside Madrid, the courier[5] moved quickly, because the winding streets and narrow alleys, still warm with the heat of the day, were old friends. Without delay, he found the door of the Portuguese ambassador, Tomás da Silva Teles, 13th Viscount of Vila Nova de Cerveira (1683–1762).[6] His bag contained correspondence from D. Luis da Cunha, and together with those letters, fitted snugly against his back, carried with the greatest care, as D. Luis had instructed him to do before setting off, a map was safely tucked away. Late at night, the courier knocked on the viscount's door, but he was not at home. Therefore, he delivered the precious package to the diplomat's servants.[7] The Viscount of

1 AUC, CLC, 994, August 21, 1747, 1.

2 AUC, CLC, 1017, October 30, 1747; Didier Ozanan, *La diplomacia de Fernando VI: correspondencia reservada entre D. Jose de Carvajal y el Duque de Huescar, 1746–1749* (Madrid: C.S.I.C., 1975), p. 80.

3 AAE, CP II, E, livro 501.

4 Didier Ozanan, *La diplomacia de Fernando VI*, p. 150; Jaime Cortesão, *Alexandre de Gusmão e o Tratado de Madrid*, vol. 4, pt. 1 (Rio de Janeiro: Instituto Rio Branco, 1957), pp. 56–57.

5 At the time, D. Luis's couriers in Paris were António Rodrigues Cavalheiro and Vicente or João Baptista Mannikin. See Isabel Cluny, *D. Luis da Cunha e a ideia de diplomacia em Portugal* (Lisbon: Livros Horizontes, 1999), p. 234; ANTT, MNE, caixa 2, maço 2, f. 117, November 28, 1746.

6 AUC, CLC, doc. 894, September 1746.

7 Probably to António Lopes, his "most esteemed servant." Nuno Gonçalo Monteiro, ed., *Meu pai e meu senhor muito do meu coração: correspondência do Conde de Assumar, para seu pai, o Marquês de Alorna* (Lisbon: ICS/Quetzal Editores, 2000), p. 58.

 | DOI:10.1163/9789004710764_002

Vila Nova de Cerveira was immediately informed of his arrival, bringing news from Paris.[8] Delivered along with the letters from D. Luis, was a map of South America,[9] probably a manuscript or one of the first printed proofs of *Amérique méridionale* by Jean-Baptiste Bourguignon d'Anville (1697–1782), geographer to the King of France, accompanied by a letter from him.

The starting point of this book and its subject is this map, entitled *Amérique méridionale*,[10] of which three variations were produced: a preliminary, more or less detailed, manuscript version dated 1737 (Fig. 1), made for the private use of the Duke of Orleans, Philippe (1674–1723), First Prince of the Blood and Regent of France; another very similar one without the dedication, perhaps dated 1742 (Fig. 2); and a larger version (Fig. 3) that is more elaborate and detailed, printed in 1748.[11] After that year, the map was reprinted several times in France—the first six reprintings were under the direction of d'Anville and aimed to introduce small changes to reflect greater knowledge of the interior of the continent, such as the introduction of Parima Lake on the page showing the north of the continent (Fig. 4).[12] The map was produced in close collaboration with D. Luis da Cunha. The focus of this book centers on the investigation of this collaboration and the impact of the map on the Spanish–Portuguese negotiations over the borders of their empires in South America during the first half of the eighteenth century.

This research commenced in 2003, when I became aware of the existence of a letter, written in 1711 in the captaincy of Minas Gerais, Brazil, by the so-called "Nunes brothers." This is held in the *Brasiliana* Library of the Robert Bosch Collection (GmbH), in the head office of Bosch, in Stuttgart, Germany. I managed to go there the following year and discovered that the document was held in a blue folder entitled *J.B.B. D'Anville—Collection of Eight Manuscripts concerning Brazil*. All of these were related to *Amérique méridionale*, and consisted

8 "When I received news, away from home, it was very late." AUC, CLC, doc. 994, 1, August 21, 1747.

9 "The same Portuguese courier delivered me the map, and a letter from M. d'Anville." AUC, CLC, doc. 994, 1, August 21, 1747.

10 Junia F. Furtado, *Oráculos da geografia iluminista: Dom Luís da Cunha e D'Anville na construção da cartografia sobre o Brasil* (Belo Horizonte: Ed. UFMG, 2012); and Furtado, *O mapa que inventou o Brasil* (Rio de Janeiro: Odebrecht/Versal, 2013).

11 Cortesão, *Alexandre de Gusmão e o Tratado de Madrid*, vol. 3, pt. 5, p. 787; vol. 4, pt. 5, p. 908.

12 Louis Charles Joseph de Manne, Jean-Denis Barbié du Bocage and Bon-Joseph Dacier, *Notice des ouvrages de M. D'Anville précédée de son éloge* (Paris: Imprimerie de Delance, 1802), pp. 98–99.

FIGURE 1 *Amérique méridionale*, J.B.B. d'Anville (Paris, by the author, manuscript, January 1737). Bibliothèque Nationale de France (BNF), DCP, GE D-11795 (RES)
REPRODUCED COURTESY OF BNF

FIGURE 2 *Amérique méridionale*, J.B.B. d'Anville (Paris, by the author, manuscript, ink on paper, *c.*1742?). BNF, DCP, GE D-10659
REPRODUCED COURTESY OF BNF

FIGURE 3 *Amérique méridionale*, J.B.B. d'Anville (Paris, printed by Gravelot, 1748). BNF, DCP, GE DD-2987 (9168B)
REPRODUCED COURTESY OF BNF

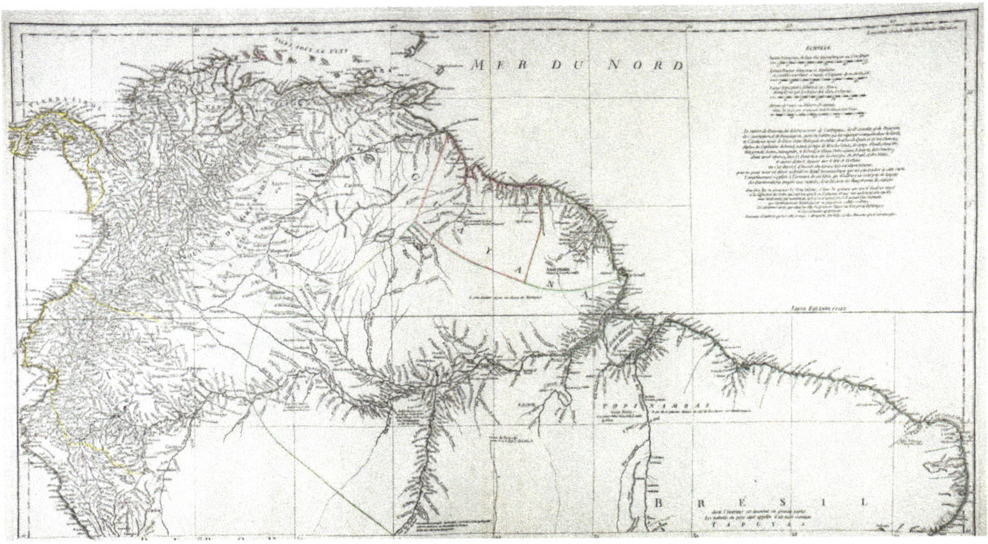

FIGURE 4 Page 1 (north) of *Amérique méridionale*, J.B.B. d'Anville (Paris, printed by Gravelot, 1779). Furtado Lins Collection
REPRODUCED COURTESY OF FURTADO LINS COLLECTION

of the letter from the Nunes brothers, another from a certain Antonio Blem about the diamonds in Minas Gerais, a *Description of Minas Gerais* by an unknown author, a description of the mouth of the Tocantins and Pará Rivers,[13] a geographic discussion about the measurement of longitude that better suits the Portuguese in South America by d'Anville himself, and two geographic memoirs written by d'Anville in 1779 about his map of South America in which he describes in detail the process by which he drew up his map and makes reference to the maps and documents he used for this, and also two manuscript maps representing the territory between Minas Gerais and São Paulo captaincies (Figs. 31 and 32).[14] Except for the two geographic memoirs all documents and the maps were given to d'Anville by D. Luis da Cunha.

D. Luis da Cunha and d'Anville met in Paris at the beginning of the 1720s, when the ambassador was given the task by the Portuguese king, D. João V (1689–1750), of purchasing all the engravings produced in Europe until that moment.[15] What the king wanted was to go on a Grand Tour of the continent

13 The Pará River was the name given by the Portuguese to the Amazon close to its mouth in the Atlantic Ocean.

14 RBC, 229, nos. 537, 539 (1), 539 (2), 555 (1), 555 (2), 555 (3), 558, 562.

15 Marie-Thérèse Mandroux-França, "La Collection royale portugaise," in Mandroux-França and Préaud, eds., *Catalogues de la collection d'estampes de Jean V, Roi de Portugal, par*

without leaving his throne in Lisbon. The enormous volume of engravings, which included maps and atlases, demanded the organization and indexation of items and it was for this reason that d'Anville, then a young geographer at the service of the House of Orleans, was initially contracted. Following this, in compliance with the diplomatic interests of the Portuguese crown, d'Anville produced geographic studies commissioned by the ambassador, discussing the new methods of measuring longitude, the Tordesillas Meridian, the crossing of southern Africa, and the frontiers of South America. He also produced various manuscript and handwritten maps of the two continents, notably the 1748 *Amérique méridionale* and the 1749 *Afrique*, which marked the beginning of his series of large printed mural maps of the four continents, which were fundamental in establishing his reputation as an armchair geographer.

In 2012, I published the book *Oráculos da geografia iluminista: Dom Luís da Cunha e D'Anville na construção da cartografia sobre o Brasil*. This centered on how the cartography of the Americas produced by d'Anville reflected, during the first half of the eighteenth century, the geopolitical conceptions which D. Luis argued Portugal should adopt in the diplomatic negotiations with the Spanish to establish the frontiers with Brazil, highlighting how and why they were used—if at all—in the Treaty of Madrid, signed by Portugal and Spain in 1750.[16] It also examines, albeit not very deeply, the production process of the map *Amérique méridionale*, which D. Luis advocated as the base map for the negotiations. This justified it being the central theme of another book, published the following year, entitled *O mapa que inventou o Brasil*.[17] Since then, I have verticalized the analysis of some of the topics which appeared in the two books in articles and book chapters. These have dealt with the central influence of the geopolitical thought of D. Luis da Cunha in the importance which Brazil acquired in Portuguese politics in the first half of the eighteenth century;[18] the

---

Pierre-Jean Mariette (Lisbon/Paris: Fundação Calouste Gulbenkian/Bibliothèque Nationale de France/Fundação da Casa de Bragança, 2003), p. 55; ANTT, MNE, LL, livro 14, ff. 122v–23, July 12, 1724.

16 Furtado, *Oráculos da geografia iluminista*.

17 Furtado, *O mapa que inventou o Brasil*. The book received the 2012 Prado Valadares Clarival Prize the Jabuti Prize for Best Book in Human Sciences in 2013, granted by the Brazilian Book Council.

18 Furtado, "O oráculo que Sua Majestade foi buscar: Dom Luís da Cunha e a geopolítica do novo império Luso-Brasileiro," in João Fragoso and Maria de Fátima Gouvêa, eds., *Na trama das redes: política e negócios no império português, séculos XVI–XVIII* (Rio de Janeiro: Ed. Civilização Brasileira, 2010), pp. 373–400; Furtado, "Dom João V e a década de 1720: novas perspectivas na ordenação do espaço mundial," in Fragoso and Gouvêa, eds., *O Brasil colonial (1720–1821)*, vol. 3 (Rio de Janeiro: Editora Civilização Brasileira, 2014), pp. 61–110.

role played by D. Luis and d'Anville in art[19] and scientific collections during the reign of D. João V;[20] the role of maps as an instrument of diplomatic negotiation of frontiers, through the Treaty of Utrecht (1712–1713), as the ambassador advocated;[21] the central role of D. Luis's thought in the formulation of innovative diplomatic conceptions used by Portugal for the first time in the Treaty of Madrid;[22] the geographic conceptions which d'Anville, instigated by D. Luis, suggested that Portugal use in its favor in the negotiations with Spain;[23] the role of the Jesuits in Enlightenment cartography and the way the first peoples of the various continents were represented in the maps of the time.[24] I also analyzed in greater detail the relationship existing between the documents which D. Luis gave to d'Anville and the representation of some geographic elements of Brazilian territory in *Amérique méridionale*, the case of the road

---

19 Furtado, "Collectionisme en Smaak. De Portugese Aankopen van Boeken en Prenten in de Zuidelijke Nederlanden," in Werner Thomas and Eddy Stols, eds., *Een Wereld op papier: Zuid-Nederlandse Boeken, Prenten en Kaarten in het Spaanse en Portugese Wereldrijk* (*16de–18de Eeuw*) (Leuven: Acco, 2009), pp. 411–25 or the Brazilian version: "Colecionismo e gosto. As compras portuguesas de livros e estampas nos Países Baixos Meridionais," in Werner Thomas, Eddy Stols, Iris Kantor, and Junia F. Furtado, eds., *Um mundo sobre papel: livros, gravuras e impressos flamengos nos impérios português e espanhol (séculos XVI–XVIII)* (São Paulo/Belo Horizonte: Edusp/Ed.UFMG, 2014), pp. 351–66.

20 Furtado, "Bosque de Minerva: artefatos científicos no Colecionismo Joanino," in Heloisa Meireles Gesteira, Luís Miguel Carolino, and Pedro Marinho, eds., *Formas do império: ciência, tecnologia e política em Portugal e no Brasil; séculos XVI ao XIX* (Rio de Janeiro: Paz e Terra, 2014), pp. 229–73; Furtado, "El embajador portugués, el cartógrafo francés y la Biblioteca Real de Don João V en la construcción de la cartografía de la América," in Mauricio Nieto Olarte and Sebastián Díaz Ángel, eds., *Dibujar y pintar el mundo: arte, cartografía y política* (Bogotá: Uniandes, 2016), pp. 131–36.

21 Furtado, "O Tratado de Utrecht e a construção da cartografia como instrumento da diplomacia portuguesa," in José Damião Rodrigues and Sónia Borges, eds., *Da restauração às revoluções Atlânticas: estratégias políticas e práticas culturais da diplomacia na época moderna (sécs. XVII–XIX); estudos oferecidos a Ana Leal de Faria* (Lisbon: Editora do Centro de História da Universidade de Lisboa, forthcoming).

22 Furtado, "Portuguese America under Foreign Threat and the Creation of the Concept of *Uti possidetis* in the First Half of the 18th Century," *Espacio, tiempo y forma: série IV, Historia moderna*, 34 (2021), pp. 45–74.

23 Furtado, "Embates na terra de papel: as longitudes e a cartografia na negociação das fronteiras da América portuguesa na primeira metade do século XVIII," in Heloisa Meireles Gesteira, Fernando Figueiredo, Luís Miguel Carolino, and Thomás Haddad, eds., *As matemáticas e as ciências afins na construção dos territórios* (Coimbra: Editora da Universidade de Coimbra, forthcoming).

24 Furtado, "Society of Jesus (Rome)," in Mathew H. Edney and Mary S. Pedley, eds., *The History of Cartography*, vol. 4, *Cartography in the European Enlightenment* (Chicago: Chicago University Press, 2020), pp. 1311–19; Furtado and Neil Safier, "Indigenous Peoples and European Cartography," in the same volume, pp. 663–72.

between the captaincies of Bahia and Minas Gerais, based on information provided by the Nunes brothers;[25] the Amazonas River, drawn in page 1 of the map;[26] and the Tocantins River, in the center-west region.[27] I also look in great detail at the project from the crossing of southern Africa, which resulted in the 1749 map *Afrique*, another collaboration between the diplomat and the geographer which benefited from the geographic methodology d'Anville had used to draw *Amérique méridionale* the previous year.[28] All the perspectives of verticalized analysis in these studies were incorporated into this book which is an updated sum of all these studies, particularly my two books: *Oráculos da geografia iluminista* and *O mapa que inventou o Brasil*.

One aspect which runs through this study is Cunha's view of Portuguese geopolitics, which, over the course of the first half of the eighteenth century, had to be formulated for Portugal's possessions in the Americas, and how that outlook was reflected in the construction of the *Amérique méridionale* map. The seminal study on geopolitics and diplomacy used by the Portuguese crown to negotiate the frontiers of Brazil in the first half of the eighteenth century is Jaime Cortesão's (1884–1960) monumental work, entitled *Alexandre de*

25 Furtado and Neil Safier, "O sertão das Minas como espaço vivido: Luís da Cunha e Jean-Baptiste Bourguignon D'Anville na construção da cartografia européia sobre o Brasil," in Eduardo França Paiva, ed., *Brasil-Portugal: sociedades, culturas e formas de governar no mundo português (séculos XVI–XVIII)* (São Paulo: Editora Annablume, 2006), pp. 263–77; Furtado, "Trajetórias carto-geográficas de uma família de Cristãos-Novos dos sertões das Gerais aos cárceres da Inquisição: o caso dos Irmãos Nunes," in Furtado and Maria Leônia Chaves de Resende, eds., *Travessias inquisitoriais das Minas Gerais aos cárceres do Santo Ofício: diálogos e trânsitos religiosos no império luso-brasileiro (sécs. XVI–XVIII)* (Belo Horizonte: Fino Traço, 2013), pp. 189–238.

26 Furtado and Jorge Pimentel Cintra, "Bourguignon D'Anville's *Carte de l'Amérique méridionale*: A Comparative Amazonian Cartography in Perspective," *Revista brasileira de história*, 31, 62 (2011), pp. 273–316.

27 Furtado, "Metamorphoses of Colonization: The Tocantins River and the Expansion to the West in Maps and Reports (18th Century)," *Tempo*, 22, 40 (May–August 2016), pp. 367–99; Furtado, "A expansão para o oeste: a *Carte manuscrite de la navigation de la Rivière des Tocantins*, c. 1742–3," in Furtado and Andréa Doré, eds., *História do Brasil em 25 Mapas* (São Paulo: Companhia das Letras, 2022), pp. 119–33.

28 Furtado, "Between Angola and Mozambique: The Portuguese Ambassador Dom Luís da Cunha, the French Geographer Jean-Baptiste Bourguignon D'Anville and the Discovery of the South Africa's Geography (XVIII Century)," in Sergueï Karp, dir., *Век Просвещения, Le siècle des lumières: la géographie à l'époque des lumières; entre imaginaire et réalité*, vol. 5 (Moscow: Naouka, 2015), pp. 168–85; Furtado, "Evolving Ideas: J. B. D'Anville's Maps of Southern Africa, 1725/1749," *Imago mundi: The International Journal for the History of Cartography*, 69 (2017), pp. 202–15; Furtado, *Quebra-Cabeça africano: como um embaixador português, um geógrafo francês, um escritor Inglês e um pirata imaginário transformaram a cartografia da África no século XVIII* (Belo Horizonte: Odisseia/Miguilim, 2021).

Gusmão e o Tratado de Madrid,[29] which continues to influence in the present studies on the process of the formation of Brazilian territory. On the one hand, as the title indicates, Cortesão highlights the central role played by D. João V's private secretary, the Brazilian Alexandre de Gusmão (1695–1753), in the formulation of diplomatic strategies used by Portugal in this treaty. On the other, he defended the idea that the formation of the Brazilian frontiers was a natural, linear, and progressive occurrence resulting from the westward advance of Luso-Brazilian settlement from the coast. He especially highlights the role of the expeditions of the *Paulistas*, people from the captaincy of São Paulo, called *Bandeirantes*, as the main drivers of this territorial expansion beginning in the seventeenth century.[30] His two interpretations are marked by a nationalist bias and had a strong historiographic impact, even being used diplomatically in all Brazilian frontier negotiations, which lasted until the twentieth century.[31] This is not surprising since from the 1940s onwards Cortesão gave history classes with this content, with an emphasis on cartography, to Brazilian diplomats in the Rio Branco Institute, a body linked to the Ministry of Foreign Affairs of Brazil (Itamaraty), influencing successive generations of historians/diplomats with his interpretations.[32]

This study opposes these two main lines of interpretation. First, it reveals that the main formulator of diplomatic conceptions used in the Treaty of Madrid was actually D. Luis da Cunha. Furthermore, although Alexandre de Gusmão, responsible in Portugal for contributing to the formulation of Portuguese strategies, had put into practice most of his ideas, they disagreed on central points. Included here was the case of the introduction of the *Uti possidetis*, which determined as the main diplomatic strategy that each crown would keep the land it had colonized. This was the great diplomatic novelty used in the Treaty of Madrid as until then legitimacy had rested on the concepts of conquest and

29 Jaime Cortesão, *Alexandre de Gusmão e o Tratado de Madrid*, 10 vols. (Coimbra: INIC/ Universidade de Coimbra, 1990).

30 Cortesão, *Raposo Tavares e a Formação Territorial do Brasil* (Rio de Janeiro: MEC, 1958).

31 Barão do Rio Branco, *Questão de Limites: Exposições de Motivos* (Brasília: Fundação Alexandre de Gusmão, 2012); Íris Kantor, "L'appropriation des cartes de d'Anville dans le monde luso-brésilien: mémoire toponymique et stratégie diplomatique dans la region amazonienne, 1798 et 1904," in Haguet and Hofmann, eds., *Une carrière de géographe au siècle des Lumières*, pp. 287–303.

32 By way of example, see, Luis Ferrand de Almeida, *Alexandre de Gusmão, o Brasil e o Tratado de Madrid (1735–1750)*; Synesio Sampaio Goes Filho, *Navegantes, bandeirantes, diplomatas: um ensaio sobre a formação das fronteiras do Brasil* (Brasília: FUNAG, 2015); Sérgio Eduardo Moreira Lima, "Apresentação," in the same volume, pp. 5–8; and Rubens Ricupero, *A diplomacia na construção do Brasil, 1750–2016* (Rio de Janeiro: Versal Editores, 2016).

discovery.[33] Cortesão believed this was because Gusmão "found in *Uti possidetis* from Roman law a conservative form of sanctioning a new and revolutionary principle of the law of peoples."[34] It was actually D. Luis who conceived this concept and advocated that it be used as the primary argument in the frontier negotiation leading to the Treaty of Madrid. Gusmão actually resisted the use of this strategy until the last moment.[35] Another point of disagreement between them was related to how maps should be used as a diplomatic strategy in the same treaty. D. Luis da Cunha had already advised Portugal to adopt an official policy of surveying and mapping Brazil. For this purpose, he provided Portuguese maps and documents so that d'Anville could produce *Amérique méridionale* in order to present a geographic configuration of Brazil that was as mathematically accurate as possible. He also defended that the map be presented in Madrid and used in the frontier negotiations and for this reason ordered it to be sent to the Viscount of Vila Nova de Cerveira in Madrid as soon as possible. Gusmão, in turn, was in charge of producing in Lisbon the manuscript *Mapa dos confins do Brazil com as terras da Coroa de Espanha na América meridional* (Map of the boundaries of Brazil with the lands of the Spanish crown in South America), better known as the *Mapa das cortes* (Map of the Courts, Fig. 51),[36] ordering Portugal's representatives to present it in the negotiations and to ensure that it should be used to guide the drafting of the treaty that established the new borders between the two crowns in South America. Based on a comparison of these two maps, I intend to identify the similarities but most especially the differences that respectively reflect the geopolitical outlooks of D. Luis da Cunha and Alexandre de Gusmão in relation to Portuguese America. I stress that, for the ambassador, the establishment of a policy for that region was inseparable from in-depth knowledge of its geography to be displayed in maps and, in an enlightened vision, cartography for him needed to be a real mirror of the geography of the territory represented. For Gusmão, a map distorted mathematically—the territory of Brazil was moved

33 In relation to the transition of these two forms of negotiation over borders, see Furtado, "Da conquista à dicotomia colônia-metrópole: uma cartografia dos vocábulos políticos no contexto da independência," in Furtado and Andréa Slemian, eds., *Uma cartografia dos Brasis: poderes, disputas e sociabilidades na independência* (Belo Horizonte: Fino Traço, 2022), pp. 41–68.

34 Cortesão, *Alexandre de Gusmão e o Tratado de Madrid*, vol. 2, p. 390.

35 Furtado, "Portuguese America under Foreign Threat and the Creation of the Concept of *Uti possidetis*."

36 Mário Clemente Ferreira, *O Tratado de Madrid e o Brasil Meridional* (Lisbon: Comissão Nacional para Celebração dos Descobrimentos Portugueses, 2001); Ferreira, "O *Mapa das cortes* e o Tratado de Madrid: a cartografia a serviço da diplomacia", *Varia Historia*, 23, 37 (2007), pp. 53–74.

to the east in relation to longitude—as is the case of the *Mapa das cortes*, could be justifiably used as a legitimate diplomatic strategy.

In second place and this is the central thesis of this book, it was d'Anville's *Amérique méridionale* map that invented Brazil and not, as is usually thought, the other way around. In other words, the map was not simply a faithful mirror of the Portuguese territories in America, which the ambassador believed it was. This book also proposes, contrary to the idea that Brazilian frontiers were naturally formed, that it was an artificial process conducted by Luso-Brazilian diplomacy over several centuries. D'Anville's map had a key role in this of constructing a new image of Portuguese America, drawing imaginary border lines which had nothing to do with the reality of the time and much to do with the new frontiers proposed by D. Luis da Cunha, frontiers soon to be negotiated in Madrid. As the ambassador predicted, he was the oracle of the geopolitics in the negotiations of Brazil's frontiers, responsible for tracing the diplomatic strategy which Portugal and, afterwards, independent Brazil, would use in the following centuries. It consisted of inventing a gigantic territory, which did not exist, whose legitimacy was expressed in a cartographic invention: d'Anville's *Amérique méridionale*. It was the first map to configure Brazil as a massive territory extending from north to south, from the present-day states of Amapá and Rondônia to the mouth of the Rio de la Plata (River Plate). From this extreme southern point, Brazil swelled outwards as its northwestern border advanced, taking on a triangular shape that encompassed the present-day states of Mato Grosso, Goiás, and a large part of the Amazonas basin, far overstepping the imaginary line of the Tordesillas Meridian, valid until then. In general terms, this representation corresponds to the current frontiers of Brazil, and, contrary to what Jaime Cortesão defended, this configuration was painstakingly built in the centuries following the Treaty of Madrid, hereafter guiding the official expansionist colonization policy and Luso-Brazilian frontier negotiations.

Another aspect analyzed is the process of producing the map itself, scrutinizing the sources d'Anville used (most supplied by D. Luis da Cunha), inserting the map and the geopolitical image it expresses into a broader context of the building up of cartographic knowledge characteristic of the Enlightenment. In this sense also examined here is how the map reflects the scientific debates among European *savants* about the actual outlines of South America, particularly under the aegis of the *Académie Royale des Sciences de Paris* and the *Academia Real da História Portuguesa*, created in 1720 by D. João V. At that time, European geographers were also engaged in an intense debate on the best system to measure longitude and about the shape of the Earth, which would result in different portrayals of the continents, particularly South America, due to its extensive north–south dimensions. These debates impacted on d'Anville's *Amérique méridionale*.[37]

This study is located in the field of the history of cartography. Its great contribution in this area of study is an in-depth examination of the biography, cartographic methodology, and the role of the modern geographic maps of Jean Baptiste Bourguignon d'Anville. An important armchair geographer, for a long time he has been neglected by the historians of cartography, more focused on Guillaume Delisle (1675–1725), First Geographer to the King of France; and in the Cassini Family—Jean Dominique (1625–1712), Jacques (1677–1757), and Cesar François (1714–1784)—and the construction of their map of France.[38] Until my 2012 and 2013 books, d'Anville had enjoyed only occasional mentions by various authors. By way of example, Numa Broc stated that he "prolonged and perfected the work of Delisle." José Aguilar argued that he was "more productive than Delisle," publishing more than 200 maps and various editions of atlases, among them the 1737 China atlas, in which he adopted the surveys carried out by Jesuits missionaries. Crone reports that "his talent lay in the critical assessment and correlation of older topographical sources, and their reconciliation with contemporary observations"; and Jeremy Black that his "*Carte de L'Inde* (1752) (...) labelled most of east-central India as a large extent of country of which there is little accurate knowledge."[39]

More recently, Lucile Haguet has written about d'Anville's methodology of work as an armchair geographer and the use of white in his maps as a cartographic tool. She also cowrote with Stéphane Bond an entry about him in volume 4 of the *History of Cartography*.[40] In 2012, the Bibliothéque Nationale da France (BNF) held a commemorative seminar on Jean Baptiste Bourguignon

37 Furtado, "Embates na terra de papel."

38 See for example, Nelson Martin Dawson, *L'atelier Delisle: l'Amérique du Nord sur la table à dessin* (Québec: Editions du Septentrion, 2000); Monique Pelletier, *Les cartes des Cassini: la science au service de l'état et des régions* (Paris: Éditions du CTHS, 2002).

39 José Aguilar, *Historia de la cartografia: la tierra de papel*, vol. 1 (Buenos Aires: Codex, 1968), pp. 218–20, 228; Numa Broc, *La geographie des philosophes: géographes et voyages français au XVIIIe siècle* (Lille: University Paul Valéry de Montpellier, 1972), p. 33; C.R. Crone, *Maps and Their Makers: An Introduction to the History of Cartography* (Folkestone: Dawson/Archon Books, 1978), pp. 89–90; Jeremy Black, *Visions of the World: A History of Maps* (London: Octopus Publishing Group, 2003), p. 74.

40 Haguet, "J.-B. D'Anville as Armchair Mapmaker: The Impact of Production Contexts on His Work," *Imago mundi*, 63, 1 (2011), pp. 88–105; Haguet, "Specifying Ignorance in Eighteenth-Century Cartography, a Powerful Way to Promote the Geographer's Work: The Example of Jean-Baptiste d'Anville," in Cornel Zwierlein, ed., *The Dark Side of Knowledge* (Leiden/Boston: Brill, 2016), pp. 358–84; Haguet and Stéphane Bond, "Anville,

d'Anville resulting in the publication of the collective work *Une carrière de géographe au siècle des Lumières: Jean-Baptiste D'Anville*, edited by Lucile Haguet and Catherine Hofmann. Written by different authors, the chapters of this work detailed his biography, his role in the education of Louis XV (1710–1774), his work as an armchair geographer, his intellectual exchange with other intellectuals from his time, how he incorporated geographic sources into his maps, the dissemination and critique of his cartography in French gazettes, the reception of his maps among the public, and the paths which his cartographic collection followed in French institutions after its purchase by the king.[41] I contributed a chapter to the collective book which emphasized the importance of the patronage of the Portuguese crown, intermediated by D. Luis da Cunha, for d'Anville, then a still young but promising geographer, to venture into modern cartography, which exercised a determinant impact on his career.[42] This collaboration, which extended throughout the ambassador's life and lasted until his death in 1749, is little known to readers who do not speak Portuguese, and can only be unveiled through the consultation of the Luso-Brazilian archives, which hold the vast documentation produced by D. Luis da Cunha, as well as about the kingdom of D. João V. Many of the chapters of this collective work return to and update information and analyses I used in my previous studies, unfortunately, most of them did not take advantage of my studies as these were mostly written in Portuguese. With the exception of one or two points which will be discussed at an opportune moment, the interpretations advanced by various authors were incorporated into the analyses made in this book, which allows English-speaking readers have access to the first up-to-date vertical study of d'Anville and his cartographic work, centering on *Amérique méridionale*, from 1748, which was a turning point in his career.

Portuguese efforts to occupy territories beyond the Tordesillas Line Meridian, the numerous negotiations on the question of borders in the Americas, the major transformation of the art of cartography during that time, and the need to establish new frontiers for its territory in the Americas sparked a veritable cartographic fever in the Portuguese crown regarding Brazil in the first half of the eighteenth century. This was most intense in relation to the three major border regions that were disputed (the Amazonas River basin, the center-west mining regions, and the Rio de la Plata estuary where the

---

Jean-Baptiste Bourguignon d'", in Mathew H. Edney and Mary S. Pedley, eds., *The History of Cartography*, vol. 4, pp. 111–15.

41 Haguet and Hofmann, eds., *Une carrière de géographe au siècle des Lumières*.

42 Furtado, "Elargir ses réseaux, diversifier ses commandes: les travaux de D'Anville pour la Couronne Portugaise," in *Une carrière de géographe au siècle des Lumières*, pp. 53–83.

Portuguese established Colônia do Sacramento on the north bank in 1680).[43] It was intended to provide its diplomatic representatives, as in the case of D. Luis da Cunha, with accurate maps and geographical information about the disputed areas.

The process of producing the *Amérique méridionale* is part of this broader context of renegotiating the borders of Portuguese and Spanish possessions set in 1494 by the Treaty of Tordesillas. However, during the two centuries that followed, the occupation of those territories did not take place as uniformly as the treaty had required. Thus, in the eighteenth century, the boundaries between the two crowns were being contested in South America and the South Sea (Pacific Ocean), as in the case of the Moluccas. Over the course of that century, as Portuguese settlements expanded into the Brazilian hinterland, the Treaty of Tordesillas presented a growing impediment to Portugal's aim of establishing sovereignty in the mid-west of South America. By establishing the division of the new lands in the Americas on the basis of an imaginary line drawn 370 leagues west of an unspecified island in the Cape Verde Archipelago, the treaty had allotted to Spain the mid-west region of South America, which was being systematically explored, occupied, and claimed by Portugal. The unspecified location of the Tordesillas Meridian also cast doubt on the exact boundaries between the two crowns in northern Brazil—in the basins of the Amazonas and Tocantins Rivers and their tributaries. In the far south, the dispute focused on the Rio de la Plata, where the Portuguese had established Colônia do Sacramento in the late seventeenth century. In the early eighteenth century, in 1705, in the context of the War of the Spanish Succession (1700–1713)[44] Colônia do Sacramento, which was still a Portuguese possession, was invaded by Spain.[45] "The problem was discussed extensively by representatives of Portugal in Madrid and in other royal capitals, such as Vienna and Paris, but they failed to change Spain's position."[46]

The inexact location of the Tordesillas Meridian Line, the actual settlements that the two Iberian powers had established in the territory, the Portuguese imperial project of territorial expansion which intended to take possession of the mid-west of South America where gold and diamond mines had recently

43 Almeida, "A Colónia do Sacramento e a formação do Sul do Brasil," in *Páginas dispersas: estudos de história moderna de Portugal* (Coimbra: Instituto de História Econômica e Social/Faculdade de Letras da Universidade de Coimbra, 1995), pp. 163–82.

44 Portugal entered this conflict in 1704.

45 Luís Ferrand de Almeida, *A Colônia do Sacramento na época da Sucessão de Espanha* (Coimbra: Instituto Histórico Dr. António de Vasconcelos, Faculdade de Letras da Universidade de Coimbra, 1957).

46 Almeida, *Alexandre de Gusmão, o Brasil e o Tratado de Madrid*, p. 9.

been discovered, and disagreements regarding the border regions with Spain produced a line of argument in Portugal that favored jointly negotiating the boundaries between the two crowns encompassing territories in the Pacific Ocean. A position that was suggested by d'Anville and that D. Luis da Cunha systematically upheld in all the negotiations in which he represented Portugal after the War of the Spanish Succession (particularly for the treaties of Utrecht, Cambrai, Breda, and Aachen). According to D. Luis da Cunha, the Portuguese Colônia do Sacramento and the Moluccas[47] in the Pacific were the two main bargaining chips for both nations, an opinion that d'Anville shared in a report he wrote on that subject.[48] This process of renegotiating the boundaries of Portuguese and Spanish possessions culminated in 1750 with the signing of the Treaty of Madrid. Utilizing the concept of *Uti possidetis*, as developed by D. Luis, gave to Portugal's share of the Americas a triangular shape that is very close to Brazil's present-day borders.[49]

It is in this context of renegotiating the borders of Portugal and Spain in the first half of the eighteenth century that these two iconic figures—D. Luis da Cunha and d'Anville—met and established a profitable collaboration. This happened because, in the course of the negotiations in which he took part, the former became aware of the need to build up a solid cartographic basis to enable Portugal in that century to take the lead over Spain in the process of negotiating the borders of the Americas. This is the context of production of d'Anville's *Amérique méridionale*. Commissioned by the celebrated ambassador, the map was intended to serve the Portuguese crown's purposes in the negotiations in Madrid in 1750 regarding its borders with Spanish America,[50] but Cunha's plans were thwarted. The desire to see the map play the role for which it was produced led the ambassador in 1747, with great precautions, to send a copy of the manuscript or a printed proof to the Viscount of Vila Nova de Cerveira, entrusted with initiating negotiations between the two nations.

Jaime Cortesão was the first to posit the theory that the collaboration to produce the *Amérique méridionale* began in 1742 and that D. Luis da Cunha had supplied "supplemental information and maps (...) to which (d'Anville)

---

47 According to the new measurements of longitude, the Moluccas were in the Portuguese part of the world as agreed in the Treaty of Tordesillas. This made the purchase of the Moluccas by the Portuguese from the Spanish in the 1529 Treaty of Zaragoza illegitimate. D. Luis suggested that the Portuguese be compensated with Spanish territory west of Brazil.

48 NL, 552, J8, K9.

49 Furtado, "Portuguese America under Foreign Threat and the Creation of the Concept of *Uti possidetis*."

50 Cortesão, *Alexandre de Gusmão e o Tratado de Madrid*, vol. 4, pp. 820, 875.

gave cartographic shape." According to Cortesão, D. Luis da Cunha decided to make a direct contribution to the production of a map of the Americas from the moment he received instructions from Alexandre de Gusmão in December 1741, asking him "to indicate a real equivalent of lands (in the Americas) so His Majesty can determine what he deems most convenient to his service," with regard to the exchange of Colônia do Sacramento and the payment for the Moluccas for Spanish territory in the Americas, as he had always advocated. Cortesão, who identified only one map of southern Brazil, produced by José da Silva Paes (1679–1760), as one of the sources of the map, believes D. Luis embarked on a close collaboration with d'Anville, supplying him with the information required to produce the map,[51] but the collaboration had in fact started much earlier in the 1720s.[52]

The starting date for this study is 1697, the year when D. Luis da Cunha began his career as a Portuguese ambassador, as well as also being the year of d'Anville's birth. D. Luis built up his vision of the Portuguese empire, and particularly of the geopolitics of the Americas, in the course of his diplomatic career in several European courts that started around the same year. As of 1724, this led to his collaboration with d'Anville, which would result in the creation of the 1748 *Amérique méridionale* map with a view to improving cartographic knowledge of South America while providing Portuguese ambassadors with a solid basis for negotiations with the Spanish monarchy. This book is not intended to be, nor will it be, a study of either D. Luis da Cunha's life or his career as an ambassador. The same applies to d'Anville. Instead, it will focus on his diplomatic residences in several European capitals to understand the context and process of constructing the *Amérique méridionale* and the collaboration he formed with d'Anville whose career will be examined. The end date is 1750, the year that Portugal and Spain signed a new frontier treaty in Madrid related to their borders in South America. During this period, we also see the end of Pedro II's reign and that of D. João V, during which Portugal engaged in important negotiations and treaties with Spain with a view to resolving issues that originated from the War of the Spanish Succession and which, among many other matters, dealt with the boundaries between the two crowns' territories in the Americas. In addition to the Treaty of Utrecht and preparations for the Treaty of Madrid (1750), D. Luis was also involved in European negotiations in Cambrai, between 1720 and 1725, Breda (1747), and the Treaty of Aix-la-Chapelle (1748). In the course of his many diplomatic missions, D. Luis da Cunha insistently requested more accurate maps as a guide for territorial

51 Cortesão, *Alexandre de Gusmão e o Tratado de Madrid*, vol. 4, p. 754.
52 Furtado, "Elargir ses réseaux, diversifier ses commandes."

negotiations. In 1715, when the Treaty of Utrecht was in its final stages, he complained disconsolately about the lack of maps suitable for establishing the new borders in South America in an irrefutable and enduring manner. His interest in cartography was not unusual in the context of diplomacy.

The geographic scope of this study is broad: it covers Meridional[53] America, which is portrayed in d'Anville's map; Brazil, the site where most of the documents used as sources for the cartographic work on the map were produced; D. Luis da Cunha's diplomatic residences in several locations in Europe; and Paris, where d'Anville lived, worked as a cartographer, established a collaboration with D. Luis da Cunha, and engaged in a debate on his map with the scholarly community of his time.

Part 1—*Oracles*, divided into Chapters 1 and 2—discusses how D. Luis da Cunha and Jean-Baptiste Bourguignon d'Anville came to be raised to the position of oracles, by themselves as much as anyone else, their contemporaries, and by future generations, in Portugal and France, respectively. These chapters seek to understand the ambiguous meanings of the term "oracle," highlighting the prophetic character of both men's work, but also the often-controversial nature of their ideas. The objective here is also to analyze the performance of the former's diplomatic activity as well as the latter's cartographic work. For this purpose, these chapters trace D. Luis da Cunha's personal trajectory as a diplomat and d'Anville's education trajectory and his conception of geography and methodology of work that I call a cartographical puzzle.[54] Heir to the tradition of great court armchair geographers, he developed a keen methodology of cartographical criticism and, in order to implement this, amassed an impressive collection of cartographic and geographic documents.

Part 2—*Encounters*—starts with Chapter 3. It looks at the two men's collaboration, starting in the 1720s, when the ambassador was serving in Paris. Their meeting came about principally in the context of the formation of D. João V's library; this was not simply a matter of collecting books but a broader project of intellectual transformation and renovation, with an Enlightenment bent, meant to give Portugal the most modern cultural and scientific facilities possible. The art of cartography was given particular emphasis, in large part due to the insistence of D. Luis da Cunha, since the process of negotiating the American borders demanded that Portugal be as familiar with its territories' geography

53 The same as South America. The *Equator* "divides the globe into two equal parts, and the two equal parts of the Earth's surface separated by the Equator are called *septentrional* and *meridional*." François de Dainville, *Le langage des géographes: termes, signes, couleurs des cartes anciennes, 1500–1800* (Paris: Éditions Picard, 1964), p. 11.

54 Furtado, *Quebra-Cabeça africano*, pp. 35, 149–51.

as possible. D. Luis's association with d'Anville was designed to improve this knowledge. Chapter 4 analyzes the geopolitical concept that D. Luis da Cunha developed over the course of his diplomatic career as Portugal's representative in Europe's main royal capitals in the first half of the eighteenth century. From there he carefully observed the disputes over Brazil's territory. Thus, he observed the shift of the empire's economic center to Brazil in the wake of the expansion of mining activities and began to advocate the transfer of the political axis in the same direction. Next, I discuss the process of constructing the idea of *Uti possidetis*, which he began to advocate as an important negotiating tool. It is crucial to understand this because it is reflected in the geographic concept of the Americas presented in *Amérique méridionale*.

Part 3—*Maps*—focuses on the process of producing *Amérique méridionale* itself. This involves the investigation of several aspects: how is the map situated in the context of Enlightenment geography, how d'Anville selected his sources, and how this data was subjected to the standards of validation that determined whether they were reliable. Once the sources were selected, how the geographer consolidated them in a map with all the hallmarks of the Enlightenment. So, Chapters 5 and 6 thus focus on the sources used for and the geographical configuration given to specific regions of the map, such as Minas Gerais and the Amazonas, de la Plata, and Tocantins Rivers, and how they were used in d'Anville's *Amérique méridionale* map. I also investigate the different meanings acquired by the white empty spaces that extend within the interior of the Americas, as well as the persistence of some myths, such as Lake Parima, and the warlike Amazonas women.

Finally, the Epilogue covers the political context of the Luso-Spanish negotiations regarding the frontiers in the Americas at the time when the *Amérique méridionale* was sent to Madrid, so that, according to D. Luis da Cunha's wishes, it could visually represent the geopolitical concept that he believed to be best for Portugal. It begins by discussing D. Luis da Cunha's geopolitical project as expressed in the different versions of d'Anville's *Amérique méridionale* map (1737, c.1742, and 1748), and follows the contrast regarding the differences between D. Luis da Cunha's and Alexandre de Gusmão's geopolitical view, with the latter entrusted by João v with handling the negotiations from Lisbon. Finally, I seek to understand why d'Anville's *Amérique méridionale*, which reflected D. Luis da Cunha's geopolitical vision and portrayed Portugal's territory in the Americas as being mathematically and geometrically closer to reality, did not become the key cartographic element used to set the boundaries established in the Treaty of Madrid, becoming in that moment what I describe as a temporarily "lost treasure."

PART 1

Oracles

CHAPTER 1

D. Luis da Cunha: Oracle of Geopolitics

1 Mirror for a Prince

In 1736, D. Luis da Cunha,[1] while serving as ambassador in The Hague, despite having been appointed the previous year to a new position in Paris, was informed of the death in Portugal of the minister Diogo de Mendonça Corte Real (1658–1736). The latter had been responsible for running the Portuguese State Secretariat of Foreign Affairs and his longtime friend. At this time, Cunha had just completed his 39th year as a Portuguese diplomat, having spent that period in several royal capitals of Europe. When he was informed that D. João v was about to appoint Corte Real's successor, and aware of the reputation he had built up over the years, D. Luis was sure that he was the oracle His Majesty sought.[2] He was flattered that the king would "remove him from so far away to serve [him] more closely, preferring me to so many subjects he must have in our land who are much more qualified for a similar task." However, he was aware that his long stay outside Portugal prevented him from participating in palace politics and intrigue, which were essential when royal appointments were made, and concluded that "I could be a less bad Secretary of State for Foreign Affairs because experience may have given me more knowledge of them than any other who has never dealt with them, nor heard of them, except very superficially."[3]

By referring to himself in his missive as an oracle of the king, D. Luis da Cunha seems to have been very much aware of his role in establishing policy for the empire. He attributed to himself a prophetic nature, since "oracle" could mean the answer that angels gave with a human voice, or the words of God in

---

1 In relation to his diplomatic career and biography see Cluny, *D. Luis da Cunha e a ideia de diplomacia em Portugal*; Furtado, "Dom Luis da Cunha: oráculo da geografia portuguesa," in *O mapa que inventou o Brasil*, pp. 28–63; and Abílio Diniz Silva, "Introdução," in Silva, ed., *Instruções políticas de D. Luís da Cunha* (Lisbon: Comissão Nacional para as Comemorações dos Descobrimentos Portugueses, 2001), pp. 9–117.

2 ANTT, MNE, LPB, caixa 789, June 5, 1736; Cluny, *D. Luis da Cunha e a ideia de diplomacia em Portugal*, p. 165; Furtado, "O oráculo que Sua Majestade foi buscar: Dom Luís da Cunha e a geopolítica do novo império Luso-Brasileiro," pp. 373–400.

3 ANTT, MNE, LPB, caixa 789; Cluny, *D. Luis da Cunha e a ideia de diplomacia em Portugal*, p. 165.

 | DOI:10.1163/9789004710764_003

Holy Scripture.[4] Enlightenment *savants* or intellectuals like him believed that theirs was a civilizing mission to be placed at the service of the state. Therefore, the life of the ambassador, who was always far removed from Portugal when serving in distant royal European capitals, was marked by the intense production of written correspondence, since it was primarily through his writings that he sought to influence the major decisions of his time, seeking to ensure that his ideas reached D. João V's ears. His counsel must have served as a mirror for princes, because, as an Enlightenment philosopher, the ambassador shared the "conviction of working to advance the reign of universal reason."[5]

Confirming his own conviction that he was an oracle, there is abundant evidence to attest that D. Luis da Cunha was a respected figure in the Portuguese court, and the monarch himself heeded his opinions and held them in high regard when it came to affairs of state. In 1746, in his letter in response to Marco António de Azevedo Coutinho (1688–1750), then serving as Secretary of Foreign Affairs and War, we can see that he was assigned the role of royal adviser. Here D. Luis da Cunha was responding to a request from Azevedo Coutinho and Cardinal da Mota, João da Mota e Silva (1685–1747), another member of D. João V's inner circle, and perhaps at that time, the king's most important adviser.[6] The two men had asked him to write directly to the monarch on an important subject the three had discussed privately and on which they shared the same opinion. The matter in question is not directly referred to in the letter, but it was probably the French proposal that Portugal serve as mediator at the peace conference that was about to begin in Breda. We can see that although the latter two resided in Portugal, and were thus much closer to the king, they turned to D. Luis, who lived in far-off Paris, to advise the monarch. The elderly ambassador, then in his 84th year,[7] replied with apparent modesty that "I do not understand why it is necessary [for the monarch to] use other lights than his own, which are the ones that will give more clarity and greater power, because objects from afar never seem as big as those seen close at hand."[8] It was mere wordplay, since D. Luis's vigorous epistolary output leaves no doubt that, even from afar, he sought to advise the monarch and influence the high circles of the ruling nobility that shaped policy during D. João's reign.

4 Raphael Bluteau, "Oráculo," in *Vocabulário Português e Latino*, vol. 6 (Coimbra: No Collegio das Artes da Companhia de Jesus, 1728), p. 97.

5 Didier Masseau, *L'invention de l'intellectuel dans l'Europe du XVIIIe siècle* (Paris: PUF, 1994), p. 15.

6 AUC, CLC, 828, f. 1–1v, May 16, 1746.

7 ANTT, MNE, caixa 562, November 3, 1749; Cluny, *D. Luis da Cunha e a ideia de diplomacia em Portugal*, p. 20.

8 AUC, CLC, 828, f. 1–1v, May 16, 1746; ANTT, MNE, caixa 562, f. 3, 1745–1748.

The ambassador was born on January 23 or 25, 1662 into a high-ranking noble family, one of the most ancient of Portugal. His father served the kings of Portugal and D. Luis benefited from a very good education in court. He graduated in canon law from Coimbra University in 1685, when he was 23 years old. He started his career as a judge of the higher chamber, appointed knight of the Order of Christ, and in 1697 became a diplomat with London being his first post.[9] Diplomacy was the career that he would follow until the end, never returning to Portugal.

In the last years of D. Luis da Cunha's life, D. João V recognized him as his most experienced ambassador. In a letter to his daughter, Princess Maria Barbara de Bragança (1711–1758), who became Queen of Spain in 1746, D. João refers to him as follows: "In Paris, as you know, there is D. Luis da Cunha, who is now the dean of ministers, and for his renowned ability and extensive experience, he is greatly respected by the others (…) so that in this way I usually receive the most pure and accurate news."[10] He "read all of (his) missives regularly (…) and personally resolved all the matters with which they dealt."[11]

As an oracle, D. Luis was constantly called upon by members of his circle for his opinions on Portuguese politics, providing them with his advice. Viscount of Vila Nova de Cerveira also attests to his influence as a councilor, fed by long experience as a diplomat. When Cerveira returned to assume his post as ambassador in Madrid in 1746, he asked D. Luis, with all his vast knowledge, to instruct him as to how to best carry out his new role, asking him to "occupy your time with giving me such detailed instructions as you can in order to enable me to perceive Portugal's true interests in the present context."[12] A few months later, when asking further advice about establishing peace between Spain and Portugal, Vila Nova de Cerveira insisted on D. Luis's wisdom and prudence as one of his majesty's oracles. He asked D. Luis to advise him "about everything I must share with this court [Madrid], even in the form of advice, because your [counsel] is not suspect and is always full of prudence and based on your tremendous knowledge and experience of European affairs."[13]

Two of the extant portraits of D. Luis da Cunha show him in all his glory (Figs. 5 and 6). His clothing is both in the style worn by the nobility but also reflects the luxury that should be characteristic of diplomats representing

---

9 Cluny, *D. Luis da Cunha e a ideia de diplomacia em Portugal*, pp. 19–34.

10 A. Pinto Ferreiro, ed., *Correspondência de Dom João V e D. Bárbara de Bragança, Rainha de Espanha (1746–1747)* (Coimbra: Livraria Gonçalves, 1945), p. 195.

11 Ferreiro, ed., *Correspondência de Dom João V*, p. 172.

12 AUC, CLC, 894, f. 1–1v, September 1746.

13 AUC, CLC, 917, f. 1, November 10, 1746.

FIGURE 5 Portrait of D. Luis da Cunha, by unknown author, in *Memórias da Paz de Utrecht*, by D. Luis da Cunha (manuscript, miniature, *c.*1715). BNP, R, PBA 450A
REPRODUCED COURTESY OF BNP

FIGURE 6 Portrait of D. Luis da Cunha, unknown author (oil on canvas). MNE, Palácio das Necessidades, 1745
REPRODUCED COURTESY OF MNE

their monarchs in foreign courts. The vast wig, embroidered coat, lace bibs, silk waistcoat, red cloak, medals, Cross of the Order of Christ, and powdered face are elements of dignity, and at the same time, typical of the sociability and conventions of the noble court and diplomacy. While the first portrait shows an ambitious young man, the second reveals an old, seasoned, and shrewd one, just as he presented himself to his countrymen. In another aspect, this portrayal of the ambassador reveals the status of an oracle of Portugal with which he believed himself to be imbued.

Frequently, as the years went by and his foreign experience grew, orders were issued in Portugal advising newly appointed ambassadors to seek D. Luis da Cunha's advice. In 1748, when Manuel Freire de Andrade e Castro (1698–?)

was appointed ambassador to the States General (or the Netherlands), where D. Luis had served for many years, he was ordered "to carefully communicate all news you obtain regarding peace negotiations to D. Luis da Cunha, and in the affairs of your ambassadorship, to seek his prudent counsel as far in advance as possible."[16] Towards the end of his life, also in the French court and amidst other foreign ministers serving in Paris, D. Luis enjoyed particular respect and deference, including that of Louis xv. The Duke of Luynes, Charles Philippe d'Albert (1695–1758), held him in good esteem and confirmed that all of Parisian society recognized his intelligence and sagacity. He wrote that the ambassador was "a man of great spirit and very well known," and that "all the foreign ministers viewed him as a father and gathered every week in his home. To them, those visits seemed to be a duty and his friendship a favor."[14] The 2nd Marquis D'Argenson, René Louis de Voyer de Paulmy (1694–1757), who in 1737 was awaiting his appointment as French ambassador in Lisbon, spent a great deal of time with D. Luis during that period, and had this to say about him:

> he was 70 years old. He was a man who was consumed with negotiations, in which he has been employed for 45 years. He has a great deal of spirit; I believe it is a just spirit, but age may have weighed upon his views and perspicacity. However, this experienced old man still retains an abundance of good hits and mysteries.[15]

Francisco Xavier de Oliveira (1702–1783), the Cavaleiro de Oliveira (or Knight of Oliveira), also gave a favorable account of D. Luis da Cunha. Oliveira was active in the diplomatic service as secretary of the 4th Count of Tarouca, João Gomes da Silva (1671–1738), at the congress leading to the Treaty of Utrecht (1713–1715), where he came into contact with D. Luis. The Cavaleiro de Oliveira's radical ideas, particularly against the Holy Inquisition and the persecution of Jews, earned him banishment from Portugal and a condemnation from the Inquisition. Years later, in 1741, when he wrote his memoirs, he described the elderly ambassador as follows: "The integrity of D. Luis da Cunha, together with his portly stature, advanced age, and well-known body of knowledge, are attributes which are well suited to making him respectable."[16]

---

14 Abílio Diniz Silva, "Dom Luis da Cunha e o Brasil," in Maria Beatriz Nizza da Silva, ed., *De Cabral a Pedro I: aspectos da colonização Portuguesa no Brasil* (Porto: Universidade Portucalense, 2001) p. 268, note 15.

15 René-Louis de Voyer d'Argenson, *Journal et mémoires du Marquis d'Argenson*, vol. 1 (Paris: Vve. de J. Renouard, 1859), p. 248.

16 Francisco Xavier de Oliveira, *Memórias das viagens de Francisco Xavier de Oliveira*, vol. 1 (Amsterdam, 1741), p. 137.

The youthful 4th Count of Assumar, D. João de Almeida e Portugal (1663–1733), who spent time with D. Luis in Paris between 1744 and 1745, observes that during a visit to the Louvre Palace in the company of D. Luis, who had gone there to speak to the king, Louis XV, he perceived that "everyone generally at court venerates him and gives him attention that is given to no one else. Even the king, who is highly reserved, spoke to him very kindly, asking him about his health and showing interest in him."[17] This deference to the elderly ambassador seems to have been shared by the queen, who displayed similar interest, and when the young man was presented to her when the ambassador was not present, "she asked after D. Luis da Cunha's health."[18]

Several contemporaries mentioned his wit and sagacity. The 4th Count of Assumar wrote that he "was a man of tremendous *sang froid*."[19] When his death was announced, the *Gazeta de Lisboa* (Lisbon Gazette) reported that he had served in several diplomatic posts, "confirming to everyone their high regard of his capacity."[24] Friar Gaspar da Encarnação (1685–1752) was also amazed by the ambassador's spirit and tenacity. When commenting on his death and referring to the last period of his life, which was marked by confrontation with the king, he commented that "what human and royal forces failed to do in such a long time, the divine power accomplished in an instant."[20]

2 The Fifth Evangelist

For all these reasons, his influence was not limited to his lifetime, but rather transcended it. Emphasizing his prophetic character, and fulfilling the 4th Count's of Assumar's expectations that "only the end of the world could make him leave it,"[21] the French historian Alphonse de Beauchamp (1767–1832), who wrote a history of Brazil in 1815, affirmed that, for the Portuguese, D. Luis da Cunha was the fifth evangelist.[22] This did not occur by chance—it was a strategy devised in life by D. Luis himself, because it was part of his mission, like those of the other Enlightenment philosophers, to situate his work within the vast field of historical development. Therefore, the most significant part of his writings was focused on the future and was intended as his legacy for posterity,

17 Cf. Monteiro, ed., *Meu pai e meu senhor muito do meu coração*, p. 43.
18 Monteiro, ed., *Meu pai e meu senhor muito do meu coração*, p. 30.
19 Monteiro, ed., *Meu pai e meu senhor muito do meu coração*, p. 126.
20 ANTT, MNE, LM, caixa 612–555/11, January 4, 1750.
21 Cf. Monteiro, ed. *Meu pai e meu senhor muito do meu coração*, p. 126.
22 Alphonse de Beauchamp, *Histoire du Brésil: depuis sa découverte en 1500 jusqu'en 1810* (Paris: A La Librairie D'Éducation et de Jurisprudence D'Alexis Eymery, 1815).

thereby enabling him to transcend his own time and space. Some of his ideas were disseminated through his vigorous literary output, expressed to a great extent in his vast correspondence with various Portuguese authorities, and arising principally, but not exclusively, from his diplomatic role. Although he spent most of his adult life working outside Portugal, D. Luis da Cunha maintained an intense and uninterrupted correspondence with the main architects and pillars of Portuguese politics of his time, furnishing them with his advice and worldview. Even at a distance, he sought to influence the most important decisions of his time through these missives.

Although most of his letters were primarily intended for private circulation, some of them had a major public impact, and a few were even produced with the aim of reaching a wider audience, going beyond their recipients. Created with the same aim and intent, D. Luis's main political writings sought directly to advise the kings, not just D. João V, but also his successors, such as D. José I (1714–1777), as well as more than one generation of Portugal's governing elite.[23] In addition to his massive correspondence, most of which has survived to this day, the ambassador produced some texts in which he sought to express the lessons learned from his participation in several diplomatic missions, more particularly the negotiations of the major European peace treaties to establish diplomatic policy in Portugal.[24]

Besides the documents of a diplomatic nature, we will focus on three of his political texts. All were written with the intention of being made public, although some were maintained in secret during a certain period due to the heterodox nature of the suggestions they contained. The first, dedicated to his nephew, D. Luis da Cunha Manuel (1703–1775), is known as the *Carta de instruções* (Letter of Instructions).[25] The second, entitled *Instruções políticas* (Political Instructions),[26] was probably begun around 1736. The third came to be known as the *Testamento político* (Political Testament),[27] as it was supposedly addressed to the future King D. José I, establishing the agenda that D. Luis wanted Portugal to follow in the years to come. Not by chance, these documents in which he sketched his vision of Portuguese geopolitics were related to both his nephew and two of his pupils, Marco António de Azevedo and Sebastião José de Carvalho e Melo (1699–1782), later Count of Oeiras and finally Marquis

23 Masseau, *L'invention de l'intellectuel dans l'Europe du XVIIIe siècle*, p. 16.
24 Cluny, *D. Luis da Cunha e a ideia de diplomacia em Portugal*, p. 89.
25 Luis da Cunha, "Carta de instruções a Dom Luis da Cunha Manuel," in Silva, ed., *Instruções políticas*, pp. 173–79.
26 Cunha, "Carta de instruções a Marco António de Azevedo Coutinho," pp. 181–373.
27 Cunha, *Testamento político (1748)*, edited by Abílio Diniz Silva (São Paulo: Alfa-Omega, 1976), pp. 81–114.

of Pombal, personalities of their time, whom D. Luis hoped might come to be mirrors for princes in turn. As a whole, these documents pointed to the present and future of the Portuguese empire.

D. Luis began writing the *Instruções políticas* at the behest of Marco António de Azevedo Coutinho, his favorite protégé, who was appointed (instead of him) as Secretary of Foreign Affairs in 1736. Azevedo Coutinho expected the ambassador to write an essay on the "system he judges best suited to the interests of Portugal so that it can be governed in its true interests."[28] In fact, this missive already makes clear D. Luis's intention to consolidate an extensive political agenda for Portugal, although in the last lines of the document the ambassador stated that it "greatly exceeds the limits of the letter you requested or ordered me to write."[29] However, we do know that the document was not sent to Portugal at that time, and later on the ambassador had a more complete version bound and sent to his nephew, D. Luis da Cunha Manuel, marking the beginning of the public trajectory of that famous document, which sought to establish an entire political and economic agenda to be followed by Portugal in the years to come. Although they were not published at the time, the *Instruções políticas*, along with some of D. Luis's letters, began to circulate in the form of manuscript copies, which indicates the importance and impact of his political opinions, not only in his closest circle, but in other larger circles and successive generations of the Portuguese administrative elite, who occasionally referred to the ambassador's thinking.

The *Testamento político* recommended Sebastião José de Carvalho e Melo (future Marquis of Pombal) to the future king as the man who was best qualified to become Secretary of State, a position he would later in fact achieve.[34] D. Luis's influence on policy during Pombal's period (1755–1777) as D. José I's First Minister is so clear that historian Joaquim Veríssimo Serrão (1925–2020) considered D. Luis da Cunha "a kind of oracle for the future Marquis of Pombal."[30] Pombal's main directives were subsequently identified as having come from ideas that D. Luis da Cunha recorded in his policy documents. The historian Camilo Castelo Branco (1825–1890) refers to D. Luis as "the dean of ambassadors" and states that Pombal was the executor of the ideas "of his teacher" and,

28 RACL, maço 175A. D. Luis da Cunha himself attested that his initial aim was to instruct Marco Antônio for the post of Secretary of State, to which he was appointed in 1736.

29 Cunha, "Carta de instruções a Marco António de Azevedo Coutinho," pp. 372–73.

30 Joaquim Veríssimo Serrão, *História de Portugal (1640–1750): a restauração e a monarquia absoluta*, vol. 5 (Lisbon: Verbo, 1982), p. 330.

from that perspective, he considered the ambassador "the oracle of Sebastião de Carvalho."[31]

If at first sight it seems that the impact of his writings on the administrative elites of Portugal made him the oracle of Portuguese policy in the coming years, the opposite may well have been true. In other words, it was those leaders who sought to base or justify some of their actions on D. Luis da Cunha's political writings, which continued to circulate within the ruling circles of Portugal long after his death and his ideas were reappropriated for different purposes. Thus, it is posterity that characterizes him as an oracle, making it appear that his predictions had come true. The transfer of the Portuguese royal court to Brazil in 1808 seems to have been the loudest echo, in the early nineteenth century, of the impact of D. Luis da Cunha's vision of the empire. Not coincidentally, during the context of the court's removal to Brazil, there was renewed interest in the latter's ideas, especially by those who sought in his thought the embryo of the reforms made at the dawn of the nineteenth century to change the face of the Portuguese empire. By suggesting the removal of the seat of the empire to Brazil, *Instruções políticas* seemed to direct Portuguese policy from the first half of the eighteenth century until well after the time they were written. However, the fact is that some nineteenth-century Portuguese intellectuals, such as D. Rodrigo de Sousa Coutinho (1755–1812), then Secretary of the Navy and Overseas Dominions, or the Prince Regent himself, later D. João VI (1767–1826), sought identification between their own ideas and D. Luis da Cunha's thought, based on the presence of common elements that were being put into practice, such as moving the seat of government to Brazil, which ended up enhancing the old ambassador's status as an oracle of his time, and not the other way around. This strategy intensified the feeling that his political influence transcended his own lifetime. Furthermore, although one cannot say that moving the court to Brazil was a natural development or the implementation of the pure ideals outlined by D. Luis, it is symptomatic that, in this regard, a manuscript copy of the *Instruções políticas* became

> part of the Collection of Manuscripts of the crown, subject to special care, which, kept in a separate archive, accompanied the Prince Regent to Brazil, remaining at his behest "with his Person" and (being) one of the few manuscripts of said Collection to return to Lisbon in 1822.[32]

---

31 Serrão, *História de Portugal*, pp. 99–100.

32 Silva, "Dom Luis da Cunha e o Brasil," p. 261.

This strategy of keeping the manuscript close at hand makes it clear that it was Prince D. João who was assigning the status of oracle to D. Luis da Cunha, as the ambassador had so ardently wished. Thus, the apparently visionary nature of the plan to transfer the court's location led to the rebirth in the early nineteenth century of an interest in his writings and helped reinforce the idea of the importance and influence of his ideas on the development of Portuguese policy beyond his own time. In fact, over the next two centuries, interest in his thought reappeared cyclically, always reborn when the governing generation or that which aspired to govern wanted to justify its policy on the basis of his ideas, as if they constituted an agenda for the future.

It is interesting to note, however, that when he wrote suggesting the transfer of the seat of the Portuguese empire to Brazil, D. Luis da Cunha himself employed a similar strategy and sought to ground his ideas on a long family tradition that began when his "great-great-grandfather, D. Pedro da Cunha [15?–?], gave to Antonio, Grand Prior of Crato [1531–1595]," the same advice that "His Highness could embark with those who would go, and (remove to) Brazil, whose vastness and riches were well known at that time."[33] If his thinking was focused on the future, because as he said: "I want to be seen by you as a visionary," the idea is anchored in the past, in family traditions, as lineage was part of the power structure in *ancien régime* monarchies, because it was first set forth by his great-great-grandfather, and it was the invocation of this tradition that gave the legitimacy to change the future.

3 Academies

Though living far from Portugal, D. Luis da Cunha played an important role in the Enlightenment-tinged Republic of Letters forming in Portugal around its principal patron, D. João v. This group of scholars sought to build a new body of knowledge genuinely focused on the issues faced by the Portuguese empire. D. Luis da Cunha was highly conscious of this pedagogical role assumed by elite thinkers in the modernization and transformation of Portuguese culture, since "it is not enough to instill knowledge in people, one must instruct them,"[34] as well as of the importance of a new understanding turned toward the service of the nation.

In their participation in Portuguese and European academies, which emerged everywhere as loci of discussion and interchange between these

33 Cunha, "Carta de instruções a Marco António de Azevedo Coutinho," p. 363.

34 BNP, R, cód. 10.484, f. 88.

savants; in meetings in various European capitals, around which revolved a new intellectual sociability;[35] on travels, since the Grand Tour formed part of the education of all the networks of Enlightenment court intellectuals, not only of the Portuguese; in the shared practice of diplomacy in Portuguese embassies and in the great peace congresses held in Europe during the first half of the eighteenth century; in indirect contact sparked by the exchange of letters or by reading the same books and gazettes, this group shared visions of the world and grew ever closer.

In Portugal, France, and even Brazil in the late seventeenth century and throughout the eighteenth century, literary academies were important centers of convergence for these educated men, places for them to exchange and spread their ideas.[36] As in France, "the man of letters does not exist outside the institutions apparent in a social space. His activity depends on the cultural apparatus of the state and all the networks of civil society (salons, circles, and academies)."[37] Not by chance, D. Luis da Cunha participated in both of the most important academies functioning in Portugal at the time: the *Academia dos Generosos* (Academy of the Generous) and the *Academia Real da História Portuguesa* (Portuguese Royal Academy of History), having emerged out of the former. In France and England, he attended or followed the proceedings of the local scientific academies.

The *Academia dos Generosos* was founded in 1647 by, among others, Luis da Cunha's father—António Álvares da Cunha (1626–1690). While it existed, the *Academia dos Generosos* was based in D. António's house in Lisbon and had an important influence on young D. Luis,[38] carrying on its activities until 1668, when it was temporarily suspended. In its second phase, from 1693 to 1696, D. Luis da Cunha played a crucial role in ensuring the institution's success over the next three years. In 1696, upon leaving for London as ambassador, it came to an end permanently.[39] This first academy laid the foundation for the Portuguese Republic of Letters in the first half of the eighteenth century, and the acquaintances and friendships forged there would be invoked by

---

35 Jeremy Black, *Italy and the Grand Tour* (New Haven/London: Yale University Press, 2003), pp. 2–3.

36 In Brazil, notable academies included the *Academia Brasílica dos Esquecidos* and that of the *Renascidos*. See Iris Kantor, *Esquecidos e renascidos: historiografia acadêmica Luso-Americana* (São Paulo: Hucitec, 2004).

37 Masseau, *L'invention de l'intellectuel dans l'Europe du XVIIIe siècle*, p. 6.

38 João Palma-Ferreira, *Academias literárias dos séculos XVII e XVIII* (Lisbon: Biblioteca Nacional de Portugal, 1982), pp. 31–08.

39 Silva, "Introduction," pp. 27–29; Cluny, *D. Luis da Cunha e a ideia de diplomacia em Portugal*, pp. 23–24; Kantor, *Esquecidos e renascidos*, pp. 45–57.

participants throughout their lives. As was the case with so many other members, it was in the *Academia dos Generosos* that Luis da Cunha wove the first social and intellectual connections that he would forever maintain in Portugal, though he lived abroad for the rest of his life.[40]

The *Academia Real da História Portuguesa* was founded by D. João V in 1720, drawing together the principal representatives of Portuguese administration and intellectual life of that time.[41] Upon the founding of this new academy, the former members of the *Academia dos Generosos* also joined. In 1723, D. Luis da Cunha was made a supernumerary member by royal appointment.[42] The supernumerary members were those who did not live in Lisbon; in this way, the intellectual elite of the capital could connect with residents living elsewhere in the country and outside it, as in the case of its diplomats. The creation of this academy served the purposes of the educated elite in terms of formulating new knowledge, constructed according to the principles of the Cartesian method. Strict submission to methodological rules would confer the status of science on this knowledge. According to this method, the historical texts produced by members of the institution ought to be preceded by rigorous investigation, and sources would first be critiqued according to this new method.[43] This was history tied to power, and, for that reason, it was primarily a history of administration. These two considerations—a new care taken with sourcing and an administrative angle—gave rise to a need to collect and organize documents, official ones in particular. It was not by chance that D. João V initially entrusted the administration of the Torre do Tombo archives to António Álvares da Cunha, who set the institution on a new path. D. António hoped that being his son D. Luis might succeed him; and so he raised the boy with all the intellectual preparation—informed by the new methodology of the academies—that the position demanded. According to the ambassador, "my father wanted me to accompany him and educated me in the hopes that I might succeed him and carry on his work,"[44] but the succession never came to pass as he became a diplomat.

Portuguese ambassadors, as in his case, had a privileged viewpoint over this theater of the world; and, for that very reason, they constantly produced texts reflecting on the events of the day, which they debated amongst their

40 Silva, "Introduction," p. 34.

41 Isabel Ferreira da Mota, *A Academia Real da História: os intelectuais, o poder cultural e o poder monárquico no século XVIII* (Coimbra: Minerva, 2003).

42 Cluny, *D. Luis da Cunha e a ideia de diplomacia em Portugal*, pp. 107–08.

43 Mota, *A Academia Real da História*, p. 60.

44 ANTT, MNE, caixa 561, f. 42, February 18, 1742.

correspondents. This was an eminently political history, one which sought to reconstitute the events of the past but also instruct the crown in terms of its future action. José da Cunha Brochado (1651–1733), writing about the time he served as ambassador in France between 1696 and 1702, states that one of an ambassador's greatest virtues—besides "great grace, keen attention, great wisdom and great dissimulation, a countenance of many faces and an apparatus with enough artifices to serve all types"—was a "great knowledge of recent history."[45] Brochado, as was to be expected, was also a member of the *Academia Real da História Portuguesa*, having been its first director.[46] D. Luis writes about diplomats in similar terms. As he saw them, "ambassadors (if they are as they should be, not as I am) have precisely the obligation of being journalist historians of present events, being obliged to combine them with those of the past in order to form an idea of those of the future."[47] One can see here the messianic mission of history, that of projecting the action of present-day men into the future, enlightened all the while by past knowledge.

Through many historical-political writings, among which his *Memórias sobre a Paz de Utrecht* holds a special place, D. Luis da Cunha fulfilled the role assigned to the Republic of Letters and specifically to the members of the *Academia Real da História Portuguesa* in the construction of a modern body of knowledge at the service of the state. "The power of these intellectuals ultimately rests on their belief in producing history,"[48] not merely in order to reveal the past, but to instruct princes as to their place in historical development. In his speech upon joining the *Academia*, D. Luis da Cunha emphasized the role of the academic historians in the education of enlightened monarchs:

> The study they make of antiquity, giving them experience of all times, makes them fit so that Your Majesty might choose from among them one worthy of the task of educating our lord the prince, since, being so well-versed in the art of praising heroes, it would seem that they better than others would know how to form them.[49]

---

45 BNF, M, MP 18, *Memórias particulares ou anedotas da corte de França, apontadas por José da Cunha Brochado no tempo que serviu como enviado naquela corte*, 1696–1702, p. 41.

46 Mota, *A Academia Real da História*, p. 47.

47 Cunha, "Carta de D. Luis da Cunha em resposta do aviso que o Secretario da Academia lhe Fez de Estar Nomeado Académico Supranumerário," in Marquez de Alegrete Manoel Telles da Sylva, eds., *Collecçam dos documentos estatutos e memorias da Academia Real da Historia Portugueza* (Lisbon Occidental: Officina de Pascoal da Silva, Impressor de Sua Magestade, e da Academia Real, 1723), p. 87.

48 Masseau, *L'invention de l'intellectuel dans l'Europe du XVIIIe siècle*, p. 15.

49 Cunha, "Carta de D. Luis da Cunha em resposta do aviso," p. 86.

However, this new knowledge ought not be produced on an individual basis, making the *Academia* a privileged space for intellectual sociability and the exchange of ideas. "The academy is instituted under the banner of communication, implying an ideal of collaboration, one which condemns solitary work."[50] The existence of the *Academia* meant that its members, though spatially distant, remained connected. However, the production of a collective body of knowledge did not mean that there were no differences or divergences in the academics' speeches, or that the institution was in any way monolithic.[51] On the contrary: debate was the essence of Enlightenment knowledge.

Much of D. Luis da Cunha's daily routine was concerned with producing texts demonstrating his mastery of modern history, as was to be expected of a member of the *Academia* and an ambassador in particular. Alongside the other participants in an enlightened Republic of Letters, he was part of a movement that sought to make its mark on the transformation of the world. The *philosophe* was, above all, engaged. The cult of writing, the certainty that they had a civilizing mission at the service of the state, their taste for controversy, the certainty that their opinions would be heard, all gave these Enlightenment intellectuals the conviction that they were participating in history. However, their activity "was situated in the context of historical development," and their counsel should merely serve as *mirrors for princes*.[52]

D. Luis da Cunha maintained close bonds of friendship and wove an intense sociability and intellectual correspondence with some of the members of this Portuguese Republic of Letters, chief among them the 4th Count of Ericeira, D. Francisco Xavier de Meneses (1673–1743).[53] He had participated in the *Academia dos Generosos* alongside D. Luis and his father, and then founded in 1717 a private academy that met in his library in his palace in Lisbon, where he once again demonstrated his vast erudition and his desire to build an intellectual sociability in the mold of a Republic of Letters. He was also one of the founding members of the *Academia Real da História Portuguesa*, taking part in the planning of the institution and the writing of its statutes.

The Count of Ericeira was one of the great epicenters around which the *savant* elite of Portugal revolved. D. Luis would say that he had "the most noble and elevated place, not in a disorganized Republic of Letters, but in his own

---

50 Mota, *A Academia Real da História*, p. 54.

51 Norberto F. da Cunha, *Elites e acadêmicos na cultura portuguesa setecentista* (Lisbon: Imprensa Nacional/Casa da Moeda, 2006), p. 11.

52 Masseau, *L'invention de l'intellectuel dans l'Europe du XVIIIe siècle*, p. 15.

53 Mota, *A Academia Real da História*, pp. 148–54.

well-regulated empire," in which his "numerous and selective library"[54] took pride of place. As was fitting for a *savant* of his time, his intellectual connections were not limited to Portugal; in 1738 he was admitted into the Royal Society in London. Experimentalism being the basis of this new science, the count had in his house a physics laboratory (or *cabinet*), a museum of natural history, and a vast library of over 15,000 volumes, rivaling that of D. João v. Cardinal da Cunha referred to him as "an august man versed in all languages and sciences."[55] Scientific issues, especially Newtonian physics, were increasingly important to Portuguese men of letters in scientific academies, just as in England and in France.

With D. Luis, the Count of Ericeira struck up an invigorating intellectual exchange founded on freedom of thought and carried out in a vast and continuous stream of letters in which the two discussed and traded readings—not always orthodox ones—in an invariably critical and intelligent conversation. In 1741, for example, the count confessed to the ambassador that he had been reading Voltaire (François-Marie Arouet, 1694–1778), calling him "the anti-Machiavelli of the King of Prussia."[56] As a result of this intellectual camaraderie, D. Luis was to lament on January 20, 1744, the loss of the friend with whom he had shared so many identities: "I was left pierced by the keenest pain by the death of my old friend, the Count of Ericeira, in whom Your Majesty has lost such a faithful vassal, and who in his incomparable erudition gave great credit to Portugal."[57]

Another icon of this Republic of Letters was the 2nd Count of Assumar, João de Almeida (1726–1802), another close friend and lifelong correspondent of D. Luis da Cunha's. The letters they exchanged reveal their companionship, mutual respect, and intellectual exchange. They spoke of everything, especially about issues of Portuguese politics after the War of Spanish Succession (1701–1714), which both had taken part in—the former as an officer on the front line of battle, the latter as a diplomat in the treaties that followed the war. João de Almeida became a member of the *Academia Real da História Portuguesa* in 1721 and was an adviser of the state; it was to him that D. Luis sent his work *Tradução e paráfrase dos Tratados de Paz e Comércio, celebrados em Utrecht, Baden e Anvers*, so that the count might make it public in Portugal.[58]

54 ANTT, MNE, caixa 779, ff. 205v–06, June 2, 1705.
55 AUC, CLC, 641, December 26, 1743.
56 ANTT, ML, 1944, September 1741, f. 15.
57 ANTT, MNE, caixa 561, January 20, 1744, f. 42.
58 Cluny, *D. Luis da Cunha e a ideia de diplomacia em Portugal*, p. 90.

Through their correspondence, Count of Assumar and D. Luis grew closer, shared and discussed ideas, and traded news as well: the former reporting what was happening in Portugal and in the empire, while the latter filled him in on the latest from the European theater. In 1720, Assumar's news from Brazil—where his son, Pedro de Almeida (1688–1756), the future 3rd Count of Assumar, was governor of the captaincy of Minas Gerais, in Brazil—reveals reflections on the conflicts rocking the territory. This information, seen through the lens of the Assumars,[59] helped mold D. Luis's conception (shared by other members of the group) of the centrality of Minas Gerais and Brazil to the Portuguese empire, as well as the dangers they were subjected to.

Apart from their exchange of letters, the personal friendship between D. Luis and the 2nd Count of Assumar is revealed in the task assumed by the ambassador in the 1740s: he took Assumar's two grandsons, João and Luis, sons of Pedro de Almeida, into his house in Paris in order to educate them.[60] In this way, the same friendship was extended to the next generations of the house.

The 1st Marquis of Abrantes, D. Rodrigo Anes de Sá Almeida e Meneses (1676–1733), was another figure with whom D. Luis maintained an intense friendship. The relationship between the two was characterized by intellectual camaraderie, occasionally marked by conflicting ideas. As the men almost always lived far from one another, their correspondence was the main vehicle for this dialogue. The marquis was also an important member of the *Academia Real da História Portuguesa*. He was descended from the old Portuguese nobility, and his interests concentrated on architecture and painting. The exchange between D. Luis and the Marquis of Abrantes was incessant; and, through their correspondence, the ambassador sought to influence from afar both his friend and royal politics, as the king consulted the marquis regarding many issues. Instead of convoking formal institutions, D. João v preferred to be advised rather more informally, by men he trusted. In 1729, in a letter, D. Luis makes note of his relationship with the Marquis of Abrantes, and of the influence the latter's opinions had on the king: "It seems that the Marquis of Abrantes has not yet formed his latest opinion, so that our master might make his final resolution on the subject."[61]

In this restricted circle of nobles, as an enlightened Republic of Letters, of which D. Luis da Cunha was an active part—albeit at a distance—the exchange

---

59 Laura de Mello e Souza, "Teoria e Prática do Governo Colonial: D. Pedro de Almeida, Conde de Assumar," in *O sol e a sombra: política e administração na América portuguesa do século XVIII* (São Paulo: Companhia das Letras, 2006), pp. 185–252.

60 Monteiro, ed. *Meu pai e meu senhor muito do meu coração.*

61 ANTT, MNE, caixa 789, October 4, 1731, f. 7.

of ideas, freedom of thought, and sharing of (often heterodox) readings were all prized, all with the aim of serving as mirrors for princes and influencing the politics of the Portuguese empire.

4 The Ambassador and the Art of Diplomacy

In the year 1697, D. Luis da Cunha began his long career as a Portuguese ambassador, serving until his death in a number of European courts. He worked in England until 1712, where he undertook difficult negotiations with the English, as they had promised to finance the Portuguese army in exchange for Portuguese participation in the War of Spanish Succession. The negotiations left him with a profound distrust of the British, who, as he saw it, only valued their own interests. At the end of the conflict the Congress of Utrecht was organized in order to deal with various diplomatic impasses. The preliminary talks were held in London in 1712, where D. Luis was still serving, and the congress itself was held in the Netherlands between 1712 and 1715. D. Luis da Cunha was named Second Plenipotentiary, assisting the First Plenipotentiary, the 4th Count of Tarouca, in the Portuguese delegation. Before settling down in Utrecht and The Hague, where the congress negotiations were held, he spent a short time in Paris in 1712.[62] After the congress ended in 1715 he returned to London; in 1716 he passed through Hanover, The Hague, and London again; in 1717 and 1718 he returned to The Hague; and spent 1719 in Madrid in an attempt to deal with the aftermath of a diplomatic incident at the Portuguese embassy. From 1720 to 1725 he lived in Paris as an emissary sent to the Congress of Cambrai, which he left once Portugal broke off relations with France. 1726–1727 was spent in Brussels, while from 1728 to 1736 he returned to The Hague, where he bought a house and planned to settle down; but finally, from 1736 to 1749, he stayed in Paris, diplomatic relations having been reestablished, and it was there that he died. If "the seventeenth century was described as one of wars, the eighteenth century was much like a succession of congresses."[63] D. Luis was involved in all the great negotiations and European treaties of the first half of the century.

Diplomatic life was not only a space for political negotiation; it also became an important element of the formation and socialization of the Republic of Letters on the European continent. Ambassadors were educated men, and, at congresses and foreign courts, they struck up a profound cordiality. The first European congress of the eighteenth century, in Utrecht, emerged as a turning

62 Cluny, *D. Luis da Cunha e a ideia de diplomacia em Portugal*, pp. 74–86.

63 Lucien Bély, *Espions et ambassadeurs au temps de Louis XIV* (Paris: Fayard, 1990), p. 12.

point, shaping as it did new forms of diplomatic policy and molding a new generation of diplomats, for many of whom this was their first experience of such a congress.[64] Many of the same men continued to encounter each other throughout their lives at the various courts where they served or at other peace congresses, forging an international network of men of letters who dealt with foreign affairs. The diplomat was above all a man of letters, and ample culture was essential to their training.

In addition to negotiating, ambassadors were obliged to represent the sovereigns or princes they served.[65] The eighteenth century brought a novel conception of diplomacy, now "considered as a specific activity, with its own rules, its defining spaces and its own 'disciplinary' and 'legal' statutes,"[66] which thus demanded specific knowledge and even, as D. Luis put it, a specific language. Because of this, he confessed that "when I left the court of Lisbon to come to London, I could do no more than dispatch a report (...) and so it became necessary to learn another language; and take up a new study."[67] Above all, diplomacy was an art. The art of knowledge, of prudence, of political intelligence, of mediation, of civility,[68] but also that of deceit. Among the qualities demanded for the position were "penetrating observation, perspicacity, a just appraisal of men, circumstances, and opportunities, self-control, moderation, and balance."[69] In addition, an ambassador ought to be "very familiar, open, and munificent."[70]

Diplomacy, for which the art of conversation was an important expedient, helped to build and deepen ties of identity, friendship, and business. The network of intellectual sociability emerging around D. Luis da Cunha, for example, whose first threads were woven in his youth, in his father's house under the auspices of the *Academia dos Generosos*, built itself up in his adult life principally around his diplomatic posts abroad, through which he came into contact

---

64 Bély, *Espions et ambassadeurs au temps de Louis XIV*, p. 327.

65 Pedro Cardim, "A prática diplomática em Europa do Antigo Regime," in Luis Nuno Rodrigues and Fernando Martins, eds., *História e relações internacionais: temas e debates* (Lisbon: Edições Colibri, 2004), pp. 42–46; Cluny, "Os Diplomatas de negociação e de representação em Portugal no século XVIII," in Zília Osório de Castro, ed., *Diplomatas e diplomacia: retratos, cerimônias e práticas* (Lisbon: Livros Horizontes, 2004), pp. 53–68.

66 Daniela Frigo, "Ambassadeurs et diplomatie à l'époque moderne: modèles et pratiques entre l'Italie et l'Europe," in Castro, ed., *Diplomatas e diplomacia*, p. 31.

67 Cunha, *Instruções políticas*, p. 190.

68 Frigo, "Ambassadeurs et diplomatie à l'époque moderne," p. 31.

69 José Calvet de Magalhães, "A acção diplomática no pensamento dos diplomatas portugueses dos séculos XVII e XVIII," in *A diplomacia na história de Portugal: actas do colóquio* (Lisbon: Academia Portuguesa da História, 1990), p. 14.

70 BNF, M, MP 18, 40–41.

with many figures from the Portuguese and European Republic of Letters. This ability to establish political and intellectual relationships as well as friendships was essential to penetrate lofty hierarchies, not only in one's court of origin (in order to garner posts and favors), but the court where the ambassador happened to be serving (in order to defend his country's interests). The Marquis d'Argenson, who was preparing to represent France in Portugal, in 1737, knew that "an ambassador (...) must be a man of society so that he might relate to the people in the government where he is serving; all goes well for our nation when one is pleasantly received everywhere."[71]

Diplomacy was organized in different hierarchical posts: ambassadors, plenipotentiaries, emissaries, residents, and secretaries,[72] all of whom constructed horizontal and vertical relations amongst themselves. The ambassador's house was staffed by various auxiliaries; as lower employees rose through the diplomatic ranks, they were supposed to return the favors and protection of their superiors with loyalty. In this way, the lifelong professional networks connecting these men were forged.

When he was in Paris in 1724, D. Luis da Cunha wrote a letter of recommendation for Francisco Mendes de Góis (1670–1753), who sought a royal favor.[73] The letter allows us to observe this web of sociability and service that remained between ambassadors and their old employees. He wrote as follows:

> I certify that in the year 1710 Francisco Mendes de Góis came to London, where I met him, as he was practically living in my house, with good behavior, and so I entrusted him with various duties, judging to see in him the necessary ability for them (...); in all this time and in all respects I have always observed his capability, modesty, loyalty, and selflessness.[74]

Francisco Mendes de Góis allows us to see how a junior diplomatic official, in serving different ambassadors, might act as a connection between them, in this case bringing the triad together. Diplomacy was in many cases only a first step towards more important administrative posts. It was not by chance that Diogo de Mendonça Corte Real, Marco António de Azevedo Coutinho, and Sebastião José de Carvalho, having served as ambassadors, were subsequently

71 D'Argenson, *Journal et mémoires du Marquis d'Argenson*, p. 226.

72 Cardim, "A prática diplomática em Europa do Antigo Regime," pp. 42–46; Cluny, *D. Luis da Cunha e a ideia de diplomacia em Portugal*, p. 42.

73 Cluny, *D. Luis da Cunha e a ideia de diplomacia em Portugal*, pp. 171–77. Cluny believes that this refers to his being named special agent in Paris.

74 ANTT, MNE, caixa 1, maço 2, 1716–1719, f. 2.

recruited for and took on the post of Minister of Foreign Affairs—the post that D. Luis most desired. Diplomacy was a source of strategic knowledge, accumulated with an eye to benefit the state. This new Portuguese intellectuality, gathered around D. João v and with D. Luis da Cunha as one of its representatives, made a point of Portugal's need to configure itself in a new form within the European political orchestra, seeking out a more important role, as accorded by the wealth and power enjoyed by the nation.

5 The Salons and the Dinner Table

The enlightened men of the Republic of Letters were not solitary figures. They existed in society, and through conversation. For that reason, "following the standards of the literary establishment," one had to master this skill in order to exchange knowledge and opinions.[75] The Portuguese doctor António Nunes Ribeiro Sanches (1699–1783), a great *savant* of his time who belonged to D. Luis's circle, despite his "average appearance, had a spiritual physiognomy, two small but lively eyes, and a fine smile that seemed in conversation to be a guarantee of his intelligence, or the interpreter of his thought." Louis Bénigne François Bertier de Sauvigny (1737–1789),[76] who wrote these lines in praise of his friend, emphasizes that Ribero Sanches's conversation "was always interesting, at times lively and animated," going on to observe that "he had no love for argument, but he gave himself happily to the discussion and always brought up relevant and interesting facts to support solid reasoning."[77] From these comments, we can see that mastery of rhetoric demanded consistent reasoning and that the art of conversation was based on intelligent discussion, as differentiated from mere argument.

In this way, participation in lettered society "was realized, in great part, in select company who shared the pleasures of conversation, encounters, of both the salon and the dinner table."[78] The salon was these lettered men's space of sociability par excellence; when D. Luis lived in Paris, the city was crawling with these "small societies where learned men loved to converse and debate."[79] He often appeared at "the house of Me Tencin, which is a sort

75 Roger Chartier, "O homem de letras," in Michel Vovelle, ed., *O homem do Iluminismo* (Lisbon: Editorial Presença, 1997), p. 133.
76 Louis Bénigne François Bertier de Sauvigny was Intendant in Paris.
77 Louis de Sauvigny, "Précis historique sur la vie de M. Sanchès," in *Catalogue des livres de feu M. Sanchès* (Paris: Chez Bure, 1783), pp. 25–27.
78 Chartier, "O homem de letras," p. 128.
79 Chartier, "O homem de letras," p. 129.

of museum, over which Mr. de Fontenelle has presided for many years."[80] Claudine Alexandrine Guérin de Tencin, or Me Tencin (1682–1749), Baroness of Saint-Matin and mother of Jean le Rond d'Alembert (1717–1783), gathered the cream of French Enlightenment intellectuality in her house's famous salon and Mr. de Fontenelle or Bernard Le Bouyer (or Le Bouvier) de Fontenelle (1657–1757), a famous writer and a man of science, friend of Voltaire's, and perpetual secretary of the *Académie Royale des Sciences de Paris* from 1697 to 1740, presided over the sessions. There could be seen figures such as the Abbé Prévost (Antoine François, 1697–1763), Montesquieu (Charles-Louis de Secondat, 1689–1755), Pierre Carlet de Marivaux (1688–1763), and Claude Adrien Helvétius (1715–1771), amongst others.

Another salon attended by the ambassador in Paris was that of the wealthy financier Pierre Crouzat (1665–1740) on Rue Richelieu, which featured principally artists and art collectors. Crouzat was also part of the circle of Philippe, Duke of Orleans and Regent of France, whose secretary was D'Anville. He had amassed one of the most notable art collections of his time, displayed at his palace, the most precious pieces kept in an octagonal gallery inspired by the *Tribuna* of the Uffizi in Florence.[81] Through his contact with artists, collectors, and in observing these paintings, D. Luis gained his reputation as a connoisseur, one widely recognized by contemporaries and future generations, among the intellectual elite of Portugal as well as the rest of Europe.[82]

In fact, in his letters, D. Luis demonstrates a wide-ranging knowledge of painting and a special predilection for Flemish painters, which would result in a striking number of these artists' works in a collection assembled by D. João v in Portugal, many having been acquired by the ambassador in the courts where he served. His comments on the excellence of Flemish art are many, based on his deep knowledge of it. In presenting a book he acquired for the king—of etchings reproducing Peter Paul Rubens's (1577–1640) paintings for the Luxembourg Palace, which portrayed the life of Maria de Medici (1575–1642)—he said it was "one of the best Your Majesty will have in your library, due to the excellence of the author." In reference to other etchings of the same painter bought from an individual, he declared, "a similar collection cannot be found." As for two panels painted by Rubens donated by the

80 ANTT, ML, 1944, September 23, 1743, f. 312.

81 Angela Delaforce, *Art and Patronage in Eighteenth-Century Portugal* (Cambridge, MA: Harvard University Press, 2002), p. 44.

82 Cluny, *D. Luis da Cunha e a ideia de diplomacia em Portugal*, pp. 113–18; Nuno Saldanha, *Poéticas da imagem* (Lisbon: Caminho Editorial, 1995), p. 286; Furtado, "Collectionisme en Smaak," pp. 411–25.

Infanta Isabel (1566–1633) to the Cathedral of Brussels, one of them depicting Jesus's descent from the cross, he said he had sought out their new owners and "*observed* them with *great attention.*" After studying them, he recommended that the king buy them "because it is no small matter to find two originals by this painter, and of this quality."[83]

Since his time in London (1697–1712), at the beginning of his diplomatic career, D. Luis da Cunha had given signs of his knowledge and erudition. He was one of the great book-buyers for D. João v's library. But he did not simply buy uncritically whatever the *savants* of the realm told him; his education having given him a critical spirit towards what was being printed, he was quite shrewd in his observations on readings and the books he acquired. Throughout his research in libraries and auctions, he demonstrated his ability to recognize rarities, a talent indispensable to any connoisseur and one admired by men of letters. In England in 1697, Captain John Stevens (1662–1726), translator of a book by the Spaniard Francisco Manuel de Mello (1608–1666)—*The Government of a Wife, or Wholsom and Pleasant Advice for Married Men*—dedicated the book to the ambassador, stressing his intellect and learning.[84]

As a man of letters and a frequent presence in literary salons, D. Luis da Cunha was also a master of the art of conversation. The Cavaleiro de Oliveira (Knight of Oliveira) reported that:

> his conversation is extremely pleasurable. The freedom with which he speaks of all subjects is equaled only by the learning with which he understands them and discourses on them, (…) having resided with great approval in all the principal courts of Europe.[85]

D. João v observed that, "in Paris, (…) all respect him and assemble in his house, and none hesitates in confiding one's secrets in him, either for advice or for confidence, trusting in his probity."[86] D. Luis was a master in the art of persuasion, a skill honed by long years of exercise in his diplomatic career. Here, the monarch emphasized an important aspect of the art of conversation as employed in diplomacy. This was not merely an exercise of erudition; rather, it could be part of a strategy to discover secrets or convince the interlocutor of the interests

83 ANTT, MNE, caixa 794, ff. 101v, 109, 109v (italics are the author's).

84 Francisco Manuel de Mello, *The Government of a Wife, or Wholsom and Pleasant Advice for Married Men*, translated by John Stevens (London: Printed for Jacob Tonson and R. Knaplock, 1697).

85 Oliveira, *Memórias das viagens de Francisco Xavier de Oliveira*, vol. 1, p. 137.

86 Ferreiro, ed., *Correspondência de Dom João v*, p. 195.

of the court the ambassador represented. Achieving the desired goal—that is, extracting secrets from the representative of a rival nation—demanded a delicate process that D. Luis carried out masterfully. As he wrote, "a secret is a moral being with many enemies, against which it must defend itself in order to be kept."[87]

An ambassador was obliged to master a true art in order to draw out the most closely guarded secrets; this, according to D. Luis, consisted in knowing how to make the most of men's natural inclination to reveal their secrets, or taking advantage of the human vanity that forced some to say more than they should, or in having the cunning to guide conversation to the desired end, or, finally, in knowing when bribery is the only means of opening a tightly closed mouth. In the exercise of this art one had to sense the best moment at which to bring up a particular topic, requiring a certain dissimulation on the ambassador's part. For this reason, familiar with D. Luis's verve and oratorical skill, the British ambassador in Utrecht, Bishop John Robinson (1650–1723), was ordered by the Earl of Oxford to

> provide D. Luis with an occasion for you two to speak, but I must beg of you that you not be the first to talk; but if he should be the one to offer you an occasion, let him know that I agree with what he has proposed me via an intermediary of his friend.[88]

D. Luis has left us a memorable self-portrait. In this work, he says that despite his "melancholy, scorched temperament," he was "not tiresome company" and was adept at entertaining ministers of state. Rather than frequenting the court, he preferred to invite guests over to his house; there, after all, he could freely "discourse on past and present matters." With his ability to listen patiently to all he was told, he was able to defend the interests of Portugal, "without contradicting what displeases him, thus avoiding all manner of disputes and debates" which he found abominable.[89]

In their salons and embassies, these men of letters tended to congregate around the dinner table. Intimate dinners or spectacular festivities were unparalleled opportunities to establish intellectual and political contacts, making them, at least for diplomats, indispensable in their work. When D. Luis da Cunha wrote to Marco António de Azevedo with instructions about the organization of the secretariat of state, he recommended that the latter set

---

87 Cunha, *Instruções políticas*, p. 194.
88 Bély, *Espions et ambassadeurs au temps de Louis XIV*, p. 389.
89 Cunha, *Testamento político de D. Luis da Cunha*, pp. 181–83.

aside a day—as he had seen in Paris—to receive each foreign minister, "making use of them to know what is happening; and if, from time to time, *you have them to dinner, this will be much better*."[90]

Well aware of the importance of the dinner table to foster camaraderie, bring souls closer together, and open the most tightly shut mouths, D. Luis made sure that a large part of his intellectual, diplomatic, and political sociability revolved around food.[91] "The art of the dinner table fit naturally into diplomacy," and was "an appreciation of the meal as political symbol."[92] But "not a few *philosophes*, when they sat down to eat, forgot the 'doctrine of proportion' and abandoned themselves to excess,"[93] as "digestion facilitated confidences."[94]

Over the course of his correspondence, D. Luis da Cunha recounts several dinners and gives details of the guests and the topics of conversation, especially in the last years of his life when illness kept him practically confined to his home and many came to visit him there. Around his table sat princes and nobles, ministers of state (especially those of foreign affairs), diplomats from any number of courts, many of whom he had known since the turn of the century, principally through the Congress of Utrecht, intellectuals, and friends, many of them Portuguese and passing through the court he was serving. It was at this time that he had the opportunity to take advantage of the occasion to speak of subjects that interested him or dealing with Portugal. At his own table, or those he frequented at congresses or foreign courts, D. Luis took part in unparalleled moments in which, employing the art of conversation, he did manage to extract the information he desired or convince his interlocutor of Portuguese interests. Let us examine just a few.

In London, in January 1712, he accompanied Prince Eugene, Duke of Savoy (1663–1736), to a series of dinners preceding the Congress of Utrecht. The duke had fought in the War of Spanish Succession, where his military talents shone. Both in London and then in the Netherlands he was to play an important role in the postwar negotiations, living in close contact with D. Luis. The prince was also an enlightened *savant* and boasted precious collections, principally of stamps and maps; his court was renowned for its splendor. He was a hero in his time and fought in several epic battles, including in the war

90 Cunha, "Carta de instruções a Marco António de Azevedo Coutinho," p. 195 (italics mine).

91 Piero Camporesi, *Hedonismo e exotismo: a arte de viver na época das luzes* (São Paulo: Unesp, 1996).

92 Bély, *Espions et ambassadeurs au temps de Louis XIV*, p. 395.

93 Camporesi, *Hedonismo e exotismo*, p. 14.

94 Bély, *Espions et ambassadeurs au temps de Louis XIV*, p. 397.

against the Turks. During his stay in London, "and throughout the two and a half months, the prince was visited every day" in his house.[95] D. Luis was his tour guide in London, accompanying him in the various gatherings which, as was to be expected, revolved around dinners. The dinner table eased guests' spirits; from politics the conversation moved to more pleasant topics, and so the ambassador wrote that he invariably left very late, as was the case at a dinner offered by the Duke of Shomberg (probably Meinhardt Schomberg, 3rd Duke, 1641–1719).[96] This comment illustrates the excesses indulged in around the table, creating an ambience given to intimacy and confidences. D. Luis, however, had a less than glowing evaluation of the political results of these get-togethers. He writes that:

> Prince Eugene held his first conference Sunday and I do not believe, *judging from some things I heard from him*, that he left them more satisfied than when he entered, but he does not lack for invitations, which do little good for his business and are quite bad for my health, as I am usually invited to these meetings and my years and ailments cannot stand these repeated excesses.[97]

The congress was already in progress in Utrecht when the Prince Eugene arrived in April 1712. At the time, Cardinal Domenico Silvio Passionei (1682–1761), the Papal Nuncio, invited him to dine at his house. There were 13 guests around the table—an evocative number—D. Luis da Cunha among them. Despite the different nations represented there, Passionei's dinner table "brought together a circle of loyalty centering on Eugene," and the Portuguese, represented by D. Luis and the Count of Tarouca, were among the most loyal. "An illustrious guest made of the meal an event."[98] D. Luis was aware of the importance of the art of entertaining for negotiation, and—through contact with the imperial group around Eugene of Savoy—he used "private circles to obtain information and advantages during the congress,"[99] despite the orders he had received from Portugal to limit himself to meetings with the English. A few months later, on a new visit in November, the prince stayed at the Count of Tarouca's house. He arrived in his yacht, where the count went to meet him

95 Cluny, *O Conde de Tarouca e a diplomacia na época moderna* (Lisbon: Livros Horizonte, 2006), p. 243.

96 BNP, R, cód. 11.209, January 26, 1712.

97 BNP, R, cód. 11.209, January 26, 1712 (italics are the author's).

98 Bély, *Espions et ambassadeurs au temps de Louis XIV*, p. 396.

99 Cluny, *O Conde de Tarouca e a diplomacia na época moderna*, p. 264.

with all due pomp—in three carriages, each pulled by six horses.[100] Eugene's days in Utrecht were busy ones, being invited to eat by different diplomats in their houses.[101] This geography of the dinner table reflects the "direct relationship (established) between social encounters and the establishment of political contacts."[102]

In 1716, since he was in the Netherlands, D. Luis went to dinner at the house of King George I (1660–1727), then at the court in Hanover. It is important to note that the news of the event came from the Spanish ambassador in a letter to Lisbon, in which he states that "the minister (Luis da Cunha) went out to greet His Majesty, who was sitting down to supper and had him sit at his table and then had a long conference with him, which lasted close to four hours, dealing with these matters."[103] The matter in question was the development of the text of the treaty signed with the Spanish at Utrecht. This had to do with the Spanish ships impounded in Rio de Janeiro in March 1704, which the Portuguese did not wish to include in the bill of war reparations. The king's apparently spontaneous invitation was not merely an act of deference to D. Luis as Portuguese ambassador, but fostered a certain camaraderie between them.[104] We may note that the conversation lasted four hours, certainly time enough for him to present his point of view and possibly influence his interlocutor.

Dinner-table sociability continues to appear in Luis da Cunha's diplomatic life. In 1736, when he was serving in The Hague, he took care of thorny issues regarding the Spanish blockade of Colônia do Sacramento and the resumption of diplomatic relations with France as he prepared to serve as ambassador in Paris. One Wednesday in the month of August, he went to dine at the house of the British ambassador, where he took advantage of the occasion to hear the latest from that court. In Paris in 1746, amidst maneuvering to ensure the appointment of Portugal as a mediating power at the Congress of Breda, at which he was the Portuguese representative, his dinner table was a place for encounters and political articulation.[105] Seated at it were the two main French ministers, craftsmen of the policy of the time. "Mr. Argenson had come

100 Carriages were important symbols of dignity; and, as ambassadors were representing their princes, they ought to put on the best possible appearance. ANTT, MNE, caixa 788, February 12, 1715, f. 281.

101 Bély, *Espions et ambassadeurs au temps de Louis XIV*, p. 396.

102 Cluny, *O Conde de Tarouca e a diplomacia na época moderna*, p. 271.

103 AGS, E, legajo 7091, f. 1–1v, February 23, 1717.

104 BNP, R, cód. 11.209, ff. 441–45, September 27, 1714.

105 AUC, CLC, 359, August 17, 1746.

to dine at this house, and shortly afterwards Mr. de Chavigny arrived."[106] The former was Marc-Pierre de Voyer de Paulmy, Comte d'Argenson (1696–1764), Minister of Foreign Affairs of France and the latter, Léon Bouthilier, Comte de Chavigny (1608–1652), had been French ambassador in Lisbon. D. Luis took advantage of the opportunity to inform them of the Portuguese position in the conflict with the Spanish and make them aware of D. João V's wishes.

These friendships between ministers from different courts were passports that allowed the weaving of personal networks connected to their countries of origin. This meant that ambassadors were the first hosts of those who planned to travel to the country they represented or simply get to know members of the elite of that country. D. Luis, for example, wrote letters of recommendation for those whom he felt obliged to return gestures of friendship, and the dinner table was more often than not the stage for that retribution. When the Count of Huescar, Fernando da Silva y Álvarez de Toledo (1714–1776), the Spanish ambassador in Paris, returned to Madrid in 1749, D. Luis recommended him in a letter to the Viscount of Vila Nova de Cerveira, the Portuguese ambassador in the city, asking that he give the new arrival a warm embrace, as the ambassador was a good friend. With this recommendation in hand, Cerveira responded that the count had not arrived in Madrid yet but was in Aranjuez, in audience with the Spanish monarchs, adding that "I am determined to have him, his sisters, and his sister-in-law to dinner this week, beginning a more personal communication with him, as I have with all his relatives."[107] This circle was thus closed, demonstrating once again how the dinner table was an element linking all the members of the Republic of Letters.

The dinners he hosted in Paris in the name of D. João V were memorable, aimed at showing Parisian society all the grandeur of the monarch he represented. Several reports were written up in the *Mercure de France*, an important literary gazette in the European Republic of Letters. The issue from December 1721, for example, made note of a grand dinner at his house, with five fish courses offered to the guests, all first class.[108] In January 1722 it was D. Luis's turn to be one of the illustrious guests at the baptism of the son of Pedro Nolasco Convay, a great financier of the Portuguese king's expenditures in France; while the ceremony, dripping with pomp, is described in minute detail in the periodical.[109] In early 1723, the gazette reports on Luis da Cunha's election alongside André de Melo e Castro, 4th Count of Galveas (1668–1753), as a supernumerary member

106 AUC, CLC, 788, April 1, 1746.
107 AUC, CLC, 1106, May 12, 1749.
108 *Mercure de France* (December 1721), p. 158.
109 *Mercure de France* (January 1722), pp. 193–96.

of the *Academia Real da História Portuguesa*, a testament to his intelligence and academic capacity within European lettered society.[110] That same year saw another richly detailed write-up of the festivities hosted at the Portuguese embassy in Paris in honor of the birth of the 4th Infante of Portugal, son of D. João v.[111] Once again, the dinner table is a peerless stage for festive sociability. These celebrations were meant "not so much for the pleasure of the guests, but rather to display the grandeur of their organizers"; and, in the ambassadors' case, reflect the power of the monarch they served.[112]

Finally, in the December 1749 issue, D. Luis appeared in the obituary section on the occasion of his death on October 9, in Paris. The profile notes that he belonged to the Order of Christ; that he represented Portugal as his ambassador in France; that he had served in several courts as minister; that he had been a plenipotentiary at Utrecht and "was the dean of all the ambassadors and ministers of Europe."[113] Even in death, D. Luis da Cunha managed to highlight for all of Parisian society not merely his own importance as a participant in this lettered society, but also the pomp, luxury, and wealth of D. João v, the magnanimous and erudite king of whom he was, truly, an authentic representative.[114]

---

110 *Mercure de France* (March 1723), pp. 565–66.

111 *Mercure de France* (November 23, 1723), pp. 1046–17; (December 1723), pp. 1.344–54.

112 J. Habermas, *Mudança estrutural da esfera pública* (Rio de Janeiro: Tempo Brasileiro, 1984), p. 23.

113 *Mercure de France* (December 1749), p. 207.

114 Cluny, *D. Luis da Cunha e a ideia de diplomacia em Portugal*, pp. 51–53.

CHAPTER 2

Jean Baptiste Bourguignon d'Anville: Oracle of Geography

1 An Armchair Geographer and His Fortune

In 1802, while praising Jean-Baptiste Bourguignon d'Anville (Fig. 7), Baron Bon-Joseph Dacier (1742–1833) stated that "*savants*, travelers, enlightened people of all levels and from all countries, even the princes of his time, for his desire to contribute to the advancement of geography" viewed him as an oracle of that science.[1] It is interesting to note that this same oracular quality was this time assigned to him. While D. Luis da Cunha was viewed as a political oracle, d'Anville was said to be the oracle of geography. Attesting to his enormous contribution to the development of that field, the baron attributes to him over 211 manuscripts and printed maps and 23 works on geography.[2] He also owned an impressive collection of maps, amounting to more than 10,000 items.[3]

The eulogy points out that, like D. Luis da Cunha, throughout his life and even after his death d'Anville was the subject of a number of commentaries that praised his professional competence. An important member of the *savant* community, beyond France and his own time, he played an important role in the art of geography and cartography of the eighteenth century.[4] Condorcet (1743–1794) stated "that all nations agree to consider him the first geographer of Europe."[5] Cardinal Domenico Silvio Passionei, founder and curator of the Vatican Museum Library, called him the "god of geography,"[6] justifying the title

1 Bon-Joseph Dacier, "Éloge de M. d'Anville," in de Manne, Du Bocage, and Dacier, *Notice des ouvrages de M. D'Anville précédée de son eloge*, p. 38.

2 Dacier, "Éloge de M. d'Anville," p. 38; C. Du Bus, "Les Collections d'Anville à la Bibliothèque Nationale," *Comité des travaux historiques et scientifiques: bulletin de la section de géographie*, 41 (1926), pp. 93–145.

3 Catherine Hofmann, "La 'Collection d'Anville' au Ministère des Affaires Étrangeres (1772–1828): modalités et enjeux d'une approriation," in Haguet and Hofmann, eds., *Une carrière de géographe au siècle des Lumières*, pp. 361–66.

4 About all people and aspects of the art of cartography in the eighteenth century see the recent entries in Edney and Pedley, eds., *The History of Cartography*, vol. 4, *Cartography in the European Enlightenment*.

5 Jean-Antoine-Nicolas de Caritat Condorcet, "Éloge de M. d'Anville," in Arthur Condorcet O'Connor and M.F. Arago, eds., *Oeuvres de Condorcet*, vol. 2 (Paris: Firmin Didot Frères, 1847), pp. 536–37.

6 Jean Mascart, "La correspondance de d'Anville," *Ciel et terre*, 33 (1912), p. 10.

 | DOI:10.1163/9789004710764_004

FIGURE 7 Portrait of Jean Baptiste Bourguignon d'Anville, unknown author (oil on canvas, 17??). Musée National du Château de Versailles et du Trianon, Versailles/RMN/Other Images
REPRODUCED COURTESY OF MUSEUM OF VERSAILLES

by the fact that "in his work one could find a degree of perfection that one sought in vain in others."[7] Denis Diderot (1713–1784) held him in high regard and, for this reason, d'Anville was one of those chosen to contribute entries in

7 M. Gabriel Marcel, "Lettres inédites du Cardinal Passionei à d'Anville," *Bulletin de géographie historique et descriptive*, 3 (1904), pp. 418–38 (p. 433).

the *Encyclopédie*, writing about the Etesian winds.[8] In his *Plan d'une université pour le gouvernement de Russie*, Diderot recommended that his students study the works of d'Anville on historic cartography.[9]

In 1855, Vivien de Saint-Martin (1802–1897), founder of the *Société Française de Géographie*, took stock of the current state of European cartography on the occasion of the Universal Exhibition held in Paris, placing d'Anville at a higher level, even compared to Jacques Cassini and Guillaume Delisle. It was no mean compliment. The latter two men came from a long line of *géographes de cabinet* or armchair geographers, considered icons of the advanced stage of development that cartography had reached in France in the eighteenth century, and left behind important, recognized work that had contributed to the transformation of the art of cartography worldwide. Cassini planned the project and was responsible for initiating the first topographic survey of France,[10] which resulted in a map that bears his name, "a memorable work that remains the first model for the large chorographic works carried out since then in Europe."[11] In addition to his extensive cartographic work, Delisle played an important role in developing more accurate ways of representing the world in maps according to Enlightenment methods of cartography. However, for Saint-Martin, while "Delisle only touched the edges of the whole and the outer edges; d'Anville embraced every detail in its infinite diversity." The scope of his work was due to his ability to "associate the perfect elegance of his design with the proportion of the detail and harmony of the profound examination of the sources, the exactness of the nomenclature, and accurate determination of the positions."[12]

Like D. Luis da Cunha, d'Anville cultivated an image of himself as an oracle, presenting himself as a spirit innately fit for the art of geography. This image was uncritically incorporated by his biographers, but it was part of a strategy carefully articulated by d'Anville and typical of his time, in which, as we shall

---

8 Etesian winds were the name the ancients gave to the winds which blew every year in the six or seven weeks after the summer solstice. Anville, "Ethésiens," in Denis Diderot and Jean le Rond d'Alembert, eds., *L'Encyclopédie*, vol. 6 (Paris: Briasson, 1756), pp. 50–51; Charles W.J. Withers, *Placing the Enlightenment* (Chicago: Chicago University Press, 2007), p. 175.

9 Frank Arthur Kafner and Serena Kafner, "Anville, Jean-Baptiste Bourguignon d'," in *The Encyclopedists as Individuals: A Biographical Dictionary of the Authors of the "Encyclopédie"* (Oxford: Voltaire Foundation, 1988), p. 11.

10 Monique Pelletier, *Les cartes des Cassini: la science au service de l'état et des régions* (Paris: Éditions du CTHS, 2002).

11 Vivien de Saint-Martin, *De l'état actuel de la cartographie en Europe*, Extrait du Bulletin de la Société de Géographie (Paris: Imprimerie de L. Martinet, 1855), p. 248.

12 Saint-Martin, *De l'état actuel de la cartographie en Europe*, pp. 248–49.

see, low-born intellectuals sought social advancement by building their careers around their talent. But different from D. Luis da Cunha, who was much praised after his death, d'Anville was neglected by scholars until very recently.[13]

Jean-Baptiste Bourguignon d'Anville[14] was the last of a long succession of great French *géographes de cabinet* or armchair geographers, among whom Nicolas Sanson (1600–1667) Delisle and Gilles Robert de Vaugondy (1688–1786) stand out. In eighteenth-century France, there were at least four different professional groups specialized in the production of maps, whom we refer to indiscriminately today as "cartographers." Military engineers performed topographical surveys on the ground carrying out empirical measurements and making maps of small regions with the help of instruments and direct observation (though the results were compiled in offices or *cabinets* afterwards). *Arpenteurs* were property surveyors who worked privately for landowners or landowning corporations (such as religious institutions) and worked very closely with the former group. In addition, the century also witnessed the development of geodetic surveying, the process of locating places on the curved surface of the Earth by means of geodesic and astronomical observation and trigonometric triangulation, a method specially promoted and performed by the geographers' members of *Académies des Sciences de Paris* with the purpose of resolving the great scientific questions on the format of the Earth and the extension of the continents. Finally, armchair geographers like d'Anville, heirs to the Renaissance model of geography and confined to their *cabinets* in Europe, produced maps of great swathes of land—countries, continents, planispheres—by compiling information from a variety of sources. To produce their maps, the armchair geographers began their work with the critique of the available and often contradictory sources, composing what I call a cartographic "puzzle," in which, however, the pieces were not previously defined.[15]

Once ready, for a map to be recognized by the scientific community as precise, this process of critical geography was detailed in texts, called geographical

13 Haguet, "Splendeur et décadence d'un 'grand homme': réception et postérité de d'Anville et Son Œuvre," in Haguet and Hofmann, eds., *Une carrière de géographe au siècle des Lumières*, pp. 311–60.

14 For his biography see Furtado, *Oráculos da geografia iluminista*, chapter 4; Furtado, "Jean-Baptiste d'Anville: oráculo da geografia iluminista," in *O mapa que inventou o Brasil*, pp. 65–93; Stéphane Blond and Haguet, "Anville, Jean-Baptiste Bourguignon d'," in Edney and Pedley, eds., *The History of Cartography: Cartography in the European Enlightenment*, vol. 4, pp. 111–15.

15 See Furtado, *Quebra-Cabeça africano*, pp. 149–50; Haguet and Hofmann, "Introduction," in Haguet and Hofmann, eds., *Une carrière de géographe au siècle des Lumières*, pp. 113–16.

memoirs which were then presented jointly to some of the scientific academies of the time, some even being published.[16] These memoirs also served as mechanisms for the authentication of the value of the map with the consumer public and they can be used by the historian of cartography to unveil their production process, since they describe the selected sources of geographic information, in other words, the pieces of the geographic puzzle chosen by the geographer.[17]

In the eighteenth century, geography and the art of producing maps were transformed. "Enlightenment *savant* geography, as practiced by d'Anville, was at the same time an 'ancient' and 'modern' discipline."[18] Along with the essential erudition for the compilation of geographic information from various sources, it became necessary to guarantee the quality of the knowledge produced, that was also a fruit of empiricism, of the direct observation, and in the case of maps, of the direct measurement of the territory, through increasingly precise mathematical instruments.[19] This form of knowledge was defended in the recently opened scientific academies, which collected around them the *savant* public and producers in London, Paris, Madrid, or Lisbon, where the new knowledge to be produced and accepted had to be legitimated by the new empirical and experimental methods they advocated.[20]

Their maps are a fruit of this transition process and fused the characteristic method of the armchair geographers, where they selected according to the speculative method the information obtained from various indirect sources,

---

16 In relation to D'Anville's cartographic method of production, see Furtado, "Espelho do mundo," in *Oráculos da geografia iluminista*, pp. 147–210; Georges Tolias, "Une science de pure érudition: la géographie critique et comparée," in Haguet and Hofmann, eds., *Une carrière de géographe au siècle des Lumières*, pp. 155–78.

17 On the Amazon region in d'Anville's *Amérique méridionale*, see Furtado and Cintra, "Bourguignon D'Anville's *Carte de l'Amérique méridionale*," pp. 273–316.

18 Tolias, "Une science de pure érudition ...," p. 155.

19 Noëlle Bourguet and Christian Licoppe, "Voyages, mesures et instruments: une nouvelle expérience du monde au siècle des Lumières," *Annales: economies, sociétés, civilisations*, 5 (1997), pp. 1115–51.

20 The first clash between these two forms of production of knowledge took place in the seventeenth century in relation to the *Royal Society*, see Steven Shapin and Simon Schaffer, *Leviathan and the Air-Pump* (Princeton: Princeton University Press, 1985). The second involved the discussion of the shape of the Earth and lasted until the nineteenth century, see J.B. Shank, *The Newton Wars and the Beginning of the French Enlightenment* (Chicago: Chicago University Press, 2008). In Portugal, this role was occupied by the *Academia Real da História Portuguesa*, the *Biblioteca Régia*, and similar institutions, see Isabel Ferreira da Mota, *A Academia Real da História: os intelectuais, o poder cultural e o poder monárquico no século XVIII* (Coimbra: Minerva, 2003); Furtado, "Bosque de Minerva: artefatos científicos no Colecionismo Joanino," pp. 229–73.

such as maps, travel reports, and chronicles of a historical and geographic nature, with the Enlightenment matrix which was then emerging, based on direct observation, in measuring the territory and the geometrization of space, using increasingly improved and exact mathematical instruments for the triangulation of land.[21] Knowledge was constructed in a cumulative form, the tradition in force being occasionally renewed or discarded by the most recent empirical information. Even though both methodologies could complete each other, in the eighteenth century, the conflict between them began to worsen and armchair geographers sought cartographic sources based on rational and empirical criteria, in other words, objective ones anchored on ever more precise measurement instruments, capable of validating the information considered true and which was then incorporated into their maps.[22] In the mix methodology used by armchair geographers, cartography and the texts of third parties form the puzzle to assemble their own maps, but different from the modern game, the pieces were not defined beforehand,[23] being chosen by the geographers after submitting them to criticism, discarding some and incorporating others.

2 Origins and Education

Jean Baptiste d'Anville was born in Paris on July 11, 1697 and died in the same city in 1782 at the age of 85.[24] He was not of noble origin and he did not come from a traditional mapmaking family—a liberal art—which is indicative of the upward mobility within that new Enlightenment intellectual society, of those who were intellectuals and stood out from others of low birth because of their knowledge. He was the son of a merchant, a master tailor[25] and had a brother two years younger, Hubert-François, known as Gravelot (1699–1773),[26] who devoted himself to drawing, while his youngest brother called Paul dedicated

21 Bourguet and Licoppe, "Voyages, mesures et instruments."

22 Shank, *The Newton Wars*.

23 Furtado, "Espelho do mundo," in *Oráculos da geografia iluminista*, pp. 147–210.

24 A. Archier, "Anville, Jean-Baptiste Bourguignon d'," in *Lexikon zur Geschichte der Kartographie. Von den Anfängen bis zum Ersten Weltkrieg* (Vienna: Franz Duticke, 1986), pp. 18–21.

25 Mary Sponberg Pedley, *The Commerce of Cartography* (Chicago: Chicago University Press, 2005), p. 259; Condorcet, "Éloge de M. d'Anville," p. 528.

26 Ronald Vere Tooley, "Gravelot (Bourguignon), Hubert-François," in *A Dictionary of Mapmakers*, vol. 2 (London: Map Collectors' Circle, 1965), pp. 204–05.

himself to finance.[27] D'Anville and Gravelot, who were close in age, sometimes worked together since the latter drew many decorative cartouche images for d'Anville's maps as in the case of *Amérique méridionale*. They both chose different surnames from their parents whose family name was Bourguignon. 1717 appears to be the year Jean Baptiste adopted the surname d'Anville which appears for the first time in an abbreviated form in his *Description de l'Isle de Ceylan*. This was not uncommon in the intellectual world of the Enlightenment. It was thus that François-Marie Arouet became Voltaire. D'Anville's change of name, as in the case of Voltaire, "indicates a readiness to change skin and start with a clean slate, leaving the past behind. Using a new skin would distinguish him symbolically from his father, his family, and his bourgeois origins," signing his name in a positive form among the *savant* community, "providing a simulacrum of belonging to nobility."[28] Embellishing his literary pseudonym with a particle indicative of nobility, in this case the "D'," served to "relate his future literary activity to a place in the world, that of a refined elite, which was also the social elite."[29]

D'Anville's biography is in large part known from the praise written about him just after his death. It became an aspect of Enlightenment tradition to glorify great men of intellect, recording their major accomplishments in post-mortem laudatory texts, which intended to highlight their innate talents for the art they performed, capable of elevating them and distinguishing them both in life and eternity.[30] Several members of the French *Académie Royale des Sciences de Paris* and the *Académie Royale des Inscriptions et Belles-Lettres* were the object of such laudatory writings.[31] Although this type of text contains empirically verifiable information, it should be read as a literary genre that, in the context of the Enlightenment, "allowed the Parisian public to witness a carefully staged presentation of the academy(ies)'s work, one that ideally projects the institution's dual identity as a learned body and a civilized

---

27 Anville, "Éloge de Monsieur Gravelot," in *Le nécrologe des hommes célèbres de France (1772–1775)*, vol. 9 (Paris: L'Imprimerie de Moreau, 1974), pp. 132–45; Anville, "Histoire de l'art en France. Gravelot," in *L'Artiste* (Paris, April 1853), p. 70.

28 Nicolas Verdier, "Entre publicité, débat scientifique et vulgarisation: Jean-Baptiste D'Anville dans les Journaux," in Haguet and Hofmann, eds., *Une carrière de géographe au siècle des Lumières*, pp. 241–44.

29 Pierre Lepape, *Voltaire: nascimento dos intelectuais no século das Luzes* (Rio de Janeiro: Zahar, 1995), p. 33.

30 See the collection: *Le nécrologe des hommes célèbres de France* (Paris, 1767–1782).

31 Alembert, *Éloges lus dans les séances publiques de l'Académie Françoise* (Paris: Panckoucke, 1779).

company of *honnête savants*."[32] In the case of d'Anville it exaggerates in building an image of the "self-made man," rising purely through the talent of the individual. D'Anville's trajectory served this genre well, but one should read the information written in his biographies of the time with care and suspicion, since without weighty patronage, especially royal, such a rise would not have been possible in the *ancien régime*. It was part of social customs at the time for these scholarly officers to rise socially under the patronage of the powerful, placing their intellect at the service of the state and obtaining recognition and a place to exercise their talent in the academies, where they stood out from the rest of society, finding a place near the highest social tiers. D'Anville's journey represents exactly this type of trajectory.

D'Anville, of lower birth, without direct access to the educational establishments reserved for the nobility and gentlemen, and with a father with scarce resources to afford a high-quality education, started his studies at a small boarding school whose name is unknown. Not belonging to family dynasty of geographers he did not acquire any cartography skills at home. However, the neighborhood he lived in—Les Halles—contained many artisans, but also housed members of parliament and the judiciary and advisers of the king, an aristocratic clientele which included Louis Léon Pajot, later the Count of d'Onsenbray (1678–1754),[33] who probably used his father's services as a tailor. Later in the 1720s D. Luis da Cunha would be in touch with him, in relation to the purchase of a telescope for D. João v as both were part of the Duke of Orleans's circle.[34] Further research on the guild of master tailors will reveal a more nuanced picture of the members of his father's clothing guild and the connections that d'Anville used later when started his career as geographer.

Between the ages of 12 and 15, d'Anville drew his first map, finished in 1712. It was a small *Carte de la Grèce* (Fig. 8), in which he highlighted its main regions and ancient cities. In it, d'Anville manifested his inclination towards the field to which he would devote himself later, and which would become "a sort of passion": conjugating geography and history. He thus presumably began studying the movement of armies and discovering battlegrounds, among other vestiges of Greek civilization.[35] But as d'Anville stated himself, although both he

32 Shank, *The Newton Wars*, pp. 61, 78.

33 Pascale Mormiche, "Participer à l'éducation des princes: D'Anville et son élève Louis xv (1718–1730)," in Haguet and Hofmann, eds., *Une carrière de géographe au siècle des Lumières*, pp. 25–27.

34 Furtado, "Elargir ses réseaux, diversifier ses commandes," pp. 92–93.

35 Anville, "Histoire de l'art en France. Gravelot," p. 69; Michaud, "Anville," p. 97.

FIGURE 8 *Carte de la Grèce*, J.B.B. d'Anville, in Louis-Charles-Joseph de Manne, *Œuvres de D'Anville*, 1834 (Paris, printed by F.-G. Levrault, 1712)
REPRODUCED COURTESY OF BNF

and his brother showed an aptitude for drawing and both received the proper training very early on, he was more inclined to engage in formal education.

According to biographical texts, d'Anville's skills opened the doors to the *College des Quatre-Nations*,[36] one of the colleges that made up the University of Paris. His acceptance, which is described as something natural, deserves further examination. In fact, he did not meet the normal social requirements for admittance, since it was mostly reserved for the nobility and gentlemen. There was also the issue of maintenance costs, which were much greater than the earnings a tailor could expect, even though d'Anville said that his father spent all he could on his children's education.[37] However, the college did provide free tuition for external students who were day students[38] and d'Anville was

36 Kafker and Kafker, "Anville, Jean-Baptiste Bourguignons D'," p. 11.

37 Anville, "Histoire de l'art en France. Gravelot," p. 69.

38 Mormiche, "Participer à l'éducation des princes," p. 27; Marie-Madeleine Compère, in *Les Collèges Français 16e–18e siècle, répertoire*, vol. 3 (Paris: Institut National de Recherche Pédagogique, 2002), pp. 245–60.

probably one of the beneficiaries of this system. He undoubtedly had some influential benefactor to enter the college, and everything points to this being the Abbé Louis du Four de Longuerue (1652–1733), who is known for certain to have protected d'Anville later on. It is important to also mention the connection between d'Anville and the Abbé Longuerue's secretary Pierre-Joseph Alary (1689–1770), who was also the young Louis XV's *precepteur* or tutor in history, linking d'Anville to both men.[39] His biographers insist on making this rise through talent seem natural and simple regardless of his social origin, as was the custom in the Enlightenment, but everything indicates that Longuerure either supervised him since the production of his first map of Greece or, delighted with him, helped the boy enter the college.

When he started to show an interest in the study of geography, in particular the production of maps, d'Anville was introduced to Parisian *savants*, winning the favor of Longuerue, who became his great instructor in the art of geography and was considered one of the most intelligent people of his time.[40] He mastered Latin, Greek, and the main oriental languages, and dedicated himself to the study of philosophy, sacred and profane Antiquity, and *belles-lettres*, and was the author of a number of works on those topics.[41] It was the abbot who awoke d'Anville's interest in the intersection between history and geography, which found an expression in cartography, becoming his first patron when he decided to devote himself entirely to that art.[42]

Longuerue also established a prestigious scholarly salon, of which d'Anville became a member.[43] Participation in Parisian salons was a fundamental activity for Enlightenment *savants*. They were a forum for the discussion of intellectual and especially scientific subjects, of course, but also made possible the socialization necessary for those who, like d'Anville, came from lower classes. It was in these salons, many of them headed by women, where the worlds of the court, intellectual life, and ordinary entertainment came together. D'Anville, for example, became a *habitué* of Madame Geoffrin's salon on Rue Saint-Honoré, one of the most celebrated in Paris. A painting of the period portrays one of these weekly meetings; present is the cream of the Parisian intelligentsia of the Enlightenment, many of them members of the *Académie*

---

39 Mormiche, "Participer à l'éducation des princes," pp. 25, 27, 31, 37.

40 Dacier, "Éloge de M. D'Anville," p. 19; Michaud, "Longuerue (Louis Dufour, abbé de)," in *Biographie universelle ancienne et moderne*, vol. 25, pp. 79–80.

41 Michaud, "Longuerue (Louis Dufour, abbé de)," pp. 25, 80.

42 Kafner and Kafner, "Anville, Jean-Baptiste Bourguignon d'," p. 11.

43 "He had the pleasure of being welcomed and admitted into the company of the Abbot Longuerue, in which the conversation was, for him, an unending source of knowledge." Dacier, "Éloge de M. D'Anville," p. 19.

Royale des Sciences de Paris. In addition to d'Anville, one can make out Rousseau (1712–1778), d'Alembert (1717–1783), Bernard le Bovier de Fontenelle, Charles-Marie de La Condamine (1701–1774), Claude-Adrien Helvétius (1715–1771), and René Antoine Ferchault de Reaumur (1683–1757), gathered around a bust of Voltaire, who was in exile at the time.[44]

D'Anville's education as a geographer differed from that of most armchair geographers and military engineers of his time because he did not belong to a family lineage of great French geographers, and because he did not frequent the royal hydrography courses. Nor was he a member of the Jesuit Order whose members were great masters of that science in France at the time.[45] In 1717 d'Anville published his first geographic memoir in the *Mercure de France*, a French gazette and literary magazine.[46] It was then that he printed the abbreviation of Anville as part of his surname. The text described the Ceylon islands' geography and displayed in 60 pages all his armchair geographers erudition.[47] His first six printed maps were produced for a book by his patron, Longuerue, entitled *Description historique et géographique de la France ancienne et moderne*.[48] His hand-drawn drafts were presented to the Dauphin, the future Louis XV, when he was still underage, by the abbot himself, in 1718.[49] The future monarch seems to have been impressed with d'Anville's cartographic work, since he appointed him to assist Delisle, who was instructing him in the art of cartography and historical geography, according to the teaching program defined by one of his tutors, Louis, Duke of Burgundy (1682–1712).[50] Deslile and Chevallier, who was appointed mathematics professor of the future king in 1717, must have undoubtedly approved of the choice of their young pupil. D'Anville then produced in 1719, for the Dauphin's private instruction, his first

44 Musée National des Châteaux de Malmaison et de Bois-Préau, Anicet Charles-Gabriel Lemonnier, "Lecture de la tragédie de L'Orphelin de la Chine de Voltaire dans le salon de Madame Geoffrin," 1812.

45 François de Dainville, *L'éducation des Jésuites* (Paris, 1978), pp. 427–72.

46 Verdier, "Entre publicité, débat scientifique et vulgarisation," pp. 237–59.

47 *Mercure de France* (November 1717), pp. 70–130; Verdier, "Entre publicité, débat scientifique et vulgarisation," pp. 246–47.

48 They represented, respectively, ancient Gaul, ancient France, modern France and neighboring countries, and three regional maps of France. Louis du Four Longuerue, *Description historique et géographique de la France ancienne et moderne* (Paris: J.-H. Pralard, 1719).

49 Chantal Grell, "De l'antiquité aux temps modernes: l'histoire utile aux princes de France, au siècle des Lumières," in M.A. Bernier, ed., *Parallèle des anciens et des modernes* (Québec: Presses de l'Université Laval, 2006), p. 93.

50 Mormiche, *Devenir prince: l'école du povoir en France* (Paris: CNRS Éditions, 2015); Mormiche, "Participer à l'éducation des princes," pp. 23–51.

map of ancient France—Gaul.[51] The purpose of the map was to make his study on *Caesar's Commentaries*,[52] by Turpin de Crisse, understandable, creating an association between history and geography that was becoming his great specialty.

In 1719, at the age of 22, d'Anville became geographer to the King of France. His title was "ingénieur et géographe ordinaire du Roi" (royal engineer and geographer).[53] It was an honorary position which did not always result in direct payment from the royal coffers, but which conferred great prestige.[54] The title was offered to those who had produced at least one cartographic document approved and considered good enough by the king himself. Dacier states that d'Anville's aptitude was so clear that he was granted the title without anyone seeing any of his work.[55] Once again his biographer insists on the exclusively meritorious nature of his rise, the result of his self-evident skill. However, the title could have only been granted to him in recognition of the map of Gaul he had made for the Dauphin's private use, but which was only published a year later, in 1719, as well as the six maps he produced for Longuerue's *Description historique et géographique de la France ancienne et moderne*, which was being printed at the time. Although the public was still unaware of his work, since these maps only came to light in 1719, the future king had known of it since the previous year and granted him the title.

However, Longuerue's book was plagued with controversy. The entire first edition was seized by order of the Duke of Orleans, then Regent of France, because its author "showed himself favorable to the empire's claims to some provinces dependent on the old kingdoms of Burgundy and Arles."[56] Nevertheless, D'Anville's and Longuerue's careers do not seem to have suffered from this incident. With the help of a commission, the abbot revised the book, which was rereleased in 1722. This edition contained four more maps by d'Anville.[57]

---

51 AAIBL, "Mémoire de Mr. D'Anville sur ces cartes de l'ancienne Gaule"; BNF, DCP, Ge DD-2987 (365), Anville, *La France ancienne dedié au Roi*, 1719.

52 Turpin De Crisse, *Les commentaires de César* (Paris: Vve J. Camusat et P. Le Petit, 1650); Pelletier, *Cartographie de la France et du monde de la Renaissance au siècle des Lumières* (Paris: Bibliothèque Nationale de France, 2001), pp. 95–96.

53 Dainville, *Le langage des géographes*, p. IX.

54 When granted, the pensions varied "between 100 and 400 *livres* per year." *Livre* was the French currency at the time. Pedley, *The Commerce of Cartography*, p. 30.

55 Dacier, "Éloge de M. D'Anville," p. 26.

56 Michaud, "Longuerue," pp. 25, 80.

57 It represented the Netherlands, Alsace and Lorraine, Switzerland and Savoy, and the provinces of Lyon and Burgundy. Louis Defour Longuerue, *Description historique et*

When the book was first released, d'Anville printed on the cartouche of each map the title of "royal geographer" and dedicated the maps to the future King Louis XV and next, in 1719, a map of the kingdom of Aragon was dedicated to Philippe, the Duke of Orleans.[58] His strategy of dedicating his first works to the two most important figures in France—the Dauphin and the First Prince of the Blood and regent—yielded the expected results, and from then on he enjoyed both men's patronage, which was essential for his promotion. Patronage, especially the royal kind, was part of the ascension strategies of these educated officials in the European Republic of Letters.

D'Anville began enjoying the patronage of the House of Orleans early in his career. First under the auspices of Philippe and later, after 1723, in the service of the new duke, Louis the Pious (1703–1752), who was a patron of science and the arts in general, which evidently included geography, since this was one of the essential fields of knowledge in the education of the nobility, princes, and the king himself.[59] The geographer started out as instructor of the future duke, Louis-Philippe D'Orleans, then Duke of Chartres (1725–1785).[60] Later on Louis promoted d'Anville to the position of his secretary, and his patronage was essential for the cartographer's career. In the words of d'Anville himself, this "great prince stimulates my studies with his generosity." For this reason, many of his geographical charts and printed works were dedicated to him, including *Amérique méridionale*, which he made at "the desire of this great prince."[61]

In 1725 d'Anville presented the *Académie Royale des Sciences de Paris* with "a map he made for the King of Portugal. It contained all of southern Africa from the Equator downwards. He also read an essay in which he explained this map."[62] It is worth noting that this was his first significant endeavor in the field of modern geography, financed by D. João V and mediated by D. Luis da Cunha. It was perfectly normal to seek the recognition of the academy's *savants* in order to obtain a place at the institution in the future, a strategy that he would repeat in the following years. In November 1726, he presented a new map of

géographique de la France ancienne et moderne, enrichie de plusieurs cartes géographiques (Paris, 1728).

58 BNF, DCP, Ge D-10605, Anville, *Aragon et Pays Voisins entre Campsan, Monzon, Deca et Pampelune*, c.1719.

59 Numa Broc, *La géographie des philosophes: géographes et voyageurs français au XVIIIe siècle* (Lille: University Paul Valéry de Montpellier, 1972), p. 21.

60 Anville, *Mesure conjecturale de la terre sur l'équateur, en conséquence de l'étendue de la Mer du Sud* (Paris: Chaubert, 1736), p. 559.

61 *Mercure de France* (November 17, 1749), p. 169.

62 AASP, PV (September 1, 1725); Furtado, "Evolving Ideas: J. B. D'Anville's Maps of Southern Africa, 1725/1749," pp. 202–15; Furtado, *Quebra-Cabeça africano*.

France to it, which as usual the members examined.[63] The following month, they vouched for its quality, according to the latest observations of latitude and longitude coordinates, and concluded "that it was correctly and appropriately drawn and the detail of the coasts was precise."[64] In January 1727 he presented the academy his *Carte de l'Afrique*,[65] which was again examined.[66] In it he "combined the old and the modern," adding new knowledge on the continent but also including "what was interesting to Antiquity."[67] To do so, he used both information furnished by new voyagers, many of them Portuguese, and information from ancient authorities such as Ptolemy (90–*c*.170), Pliny (23–79), and Strabo (63 BC–AD 23).[68] This was a method that he would successfully use in all of his cartography. Like d'Anville's other maps, there were several editions and successive corrections of his *Afrique*, and later he documented all of the sources he used and the way he submitted them to the rigor of his critique in the *mémoires* of the *Académie Royale des Inscriptions et Belles-Lettres*.

In 1728, the Jesuit R. P. Jean Baptiste du Halde (1674–1743) partnered with d'Anville to produce a work on China and Chinese Tartary, bringing together descriptions by 27 Jesuits who had visited the region, entitled *Description géographique, historique, chronologique, politique et physique de l'empire de la Chine et de la Tartarie chinoise*.[69] It is possible that the introduction was made by D. Luis da Cunha who had connections with the Society of Jesus Order.[70] This project put him in touch with the Jesuits, who were the greatest producers of astronomical, cartographic, and natural science knowledge of the regions in which they were active. Du Halde provided the geographer with everything related to the astronomical observations made by the Jesuits in their missions

63 AASP, PV (November 27, 1726).
64 AASP, PV (December 4, 1726).
65 BNF, DCP, Ge DD-2987 (7772), Anville, *Carte d'Afrique*, 1727.
66 AASP, PV (January 21, 1727).
67 Anville, *Considérations générales sur l'étude et les connaissances que demande la composition des ouvrages de Géographie* (Paris: Imprierie de Lambert, 1777), pp. 63–64.
68 Tolias, "Une science de pure érudition," pp. 155–78.
69 Jean-Baptiste Du Halde, *Description géographique, historique, chronologique, politique et physique de l'empire de la Chine et de la Tartarie chinoise, enrichie des cartes generales & particulières de ces pays, de la carte générale & des cartes particulières du Thibet & de la Corée & ornée d'un grand nombre de figures & de vignettes gravées en taille-douce* (Paris: P.G. Lemercier, 1735).
70 "Du Halde et D'Anville. (Cartes de La Chine)," in *Recueil de mémoires orientaux: textes et traductions publiés par les professeurs de l'Ecole Spéciale des Langues Orientales Vivantes à l'occasion du XVIe Congrès international des orientalistes réuni à Alger, Avril 1905* (Paris: E. Leroux, 1905), p. 395.

to the East,[71] which allowed him to produce the maps "with the clarity and precision for which he was known."[72]

D'Anville's collaboration with Du Halde was not limited to Asia. In 1733, his map of Paraguay appeared, part of volume 21 of *Lettres édifiantes et curieuses*.[73] Published in 34 volumes between 1702 and 1776, the organization of this gigantic work was initially entrusted to the Jesuit Charles le Gobien (1653–1708), procurator in Paris for the Jesuit missions in China. After his death in 1709 Du Halde was assigned to organize the collection and, given the successful collaboration with d'Anville on the cartography on China, he hired the geographer to produce a map and a geographic *mémoire* of Paraguay when the volume on the missions in that region was being developed.[74] D'Anville wrote to the clergyman:

> I cannot live up to the honor Your Reverence bestows on me by commissioning the map *Le Paraguay où Les RR. PP. de la Compagnie de Jésus ont Répandu Leurs Missions* [Fig. 9], to be part of the *Recueil des Lettres édifiantes*, in which I will invest all efforts that depend on me to endow this chart with every perfection.

To produce it he compiled information from various sources and, as he himself stated, "I made use of many charts brought to light by the Jesuit missionaries in that country to compose the map of *Paraguay*."[75]

Mercure de France raised some controversy about this map.[76] In addition to some stylistic faults, d'Anville was accused of extending to the Rio de la Plata the territory of Brazil, not making clear the boundaries between the Jesuit missions and the other Spanish provinces, and even reducing their area. He explained himself in a letter addressed to the gazette. On the borders of the missions' region, he argued that there was no reduction in relation to the preceding

---

71 "Du Halde et D'Anville. (Cartes de La Chine)," p. 395.

72 Du Halde, *Description géographique, historique, chronologique, politique et physique de l'empire de la Chine et de la Tartarie chinoise*, p. xlviij; "Du Halde et D'Anville. (Cartes de La Chine)," p. 397.

73 *Lettres edifiantes et curieuses, ecrites des missions etrangères*, vol. 9 (Toulouse: Noel-Etienne Sens/Auguste Gaude, 1810–1811), p. 254.

74 Anville, "Observations géographiques sur la *Carte du Paraguai*," in *Lettres edifiantes et curieuses, ecrites des missions etrangères*, pp. 254–66.

75 NL, VB, Ayer MS 1918, doc. 549, f. 1.

76 *Mercure de France* (April 1734), pp. 715–20. Verdier, "Entre publicité, débat scientifique et vulgarisation," pp. 247–49.

FIGURE 9 *Le Paraguay, où Les* RR. PP. *de la Compagnie de Jésus ont Répandu Leurs Missions*, J.B.B. d'Anville (Paris, printed, 1733). BNF, DCP, GD-10881
REPRODUCED COURTESY OF BNF

maps and that it was circumscribed by the Paraná and Paraguay rivers between the province of Buenos Aires and Asunción. D'Anville also maintained that it was not necessary to draw their borders clearly as these were marked with crosses. On the frontier of Brazil, a much more delicate matter involving two rival powers, he apologized saying that he had not been informed of the recent settlement of the Spaniards in the Montevideo area, as a result of the recent conflicts in the region. Consequently, he admitted that Portuguese territory, unlike what was shown in the map, did not extend to Colônia do Sacramento, situated on the banks of the Rio de la Plata. He attempted to extricate himself from any accusation of a political nature, stating that "I do not think geography can admit complacency that was not required of me,"[77] but the map took a clearly Portuguese position in this conflict—the same one defended by D. Luis da Cunha to the French authorities, arguing that it was the Spanish settlement in Montevideo that was illegitimate.[78] We cannot forget that since 1724 he was working close to D. Luis da Cunha for the King of Portugal, and that possession of the Colônia do Sacramento was a key point for the Portuguese.

3 Between an Oblong and an Oblate Earth

At the session held on September 1, 1734, the scholars of the *Académie Royale des Sciences de Paris* debated the topic of the shape of the Earth. Once again, the followers of the two main philosophical systems fought with each other in a new chapter of what came to be named the *Newton Wars*. Starting around 1690 these shook the European scientific community that supported or opposed Isaac Newton's (1643–1727) and René Descartes's (1596–1650) theories of celestial mechanics: "by 1732 the shape of the Earth would become the great measure of the validity of the Newtonian and Cartesian systems of natural philosophy."[79] The French *Académie* split into two groups. On one hand, there were the Newtonians Jean Richer (1630–1696), Abbot Jean Picard (1620–1682), and Pierre Louis Moreau de Maupertuis (1698–1759). The latter clearly expressed their position in his *Discours de la figure des astres*, in 1732.[80] On the other hand were the Cartesians Fontenelle, Jacques Cassini, and Delisle. Both systems were based on a mechanical theory that explained the motion of celestial bodies and defended the idea that Earth was a spheroid but the philosophical and epistemological principles on which they based their conclusions about Earth's true shape were different.[81] While the Cartesians maintained a prolate (elongated) spheroid shape, on which the length of a degree of latitude decreased towards the poles, the Newtonians advocated for an oblate (flattened) spheroid, on which the length of a degree of latitude increased towards the poles.

77 "Lettre de M. D'Anville, géographe ordinaire du Roy, sur une carte du Paraguay, du 21 mars," *Mercure de France* (April 1734), p. 719.

78 I disagree with Nicolas Verdier's interpretation that with his position he sought to escape protection from the Jesuits who had commissioned the map as he was taking the side of the Portuguese branch of the Jesuit Order. Verdier, "Entre publicité, débat scientifique et vulgarisation," p. 249.

79 Shank, *The Newton Wars*, p. 324.

80 Broc, *La geographie des philosophes*, p. 44.

81 Shank, *The Newton Wars*.

In 1735 Maupertuis presented a memoir to the *Académie Royale des Sciences de Paris* where he intended to "examine the means that astronomy and geography provide to determine the shape of the Earth." He advocated that using mathematical and geometric calculations it would be possible to determine the shape of Earth based on the theory of universal gravitation. According to Newton's propositions, the English school claimed that the Earth was spherical, flattened at poles—an oblate, or grapefruit-like shape—which was a result of the action of gravity on any planetary body such as the terrestrial globe. On the other hand, the French defenders of Cartesianism maintained that mathematics were not enough to prove a philosophical system and that the Earth's shape could be solved only by using the new techniques to measure longitude, which had been undergoing improvements, and it depended on direct observation of territories, using increasingly sophisticated and improved geodesic and astronomical instruments. Maupertuis advocated the first method also because he claimed that direct measurements at the necessary locations "would be of an extreme practical difficulty," which led him to reflect on the mathematical means to solve the problem, making use of the few observations that had been performed up to that point.[82]

The problem was that the existing measurements were inconclusive and pointed to the limitations of the methods and technology available. Between 1668 and 1670, Abbot Jean Picard made measurements of an arc of the Paris Meridian and began advocating that the Earth was a sphere. But in 1672, Jean Richer established the Cayenne Meridian and observed there that the length of a pendulum with a period of one second was shorter there than in Paris. The evidence gathered by Richer, Jean Varin, and Jean Des Hayes (16?–1706), who measured lower latitudes with pendulums between 1681–1683, also seemed to endorse this last observation.[83] Isaac Newton and Christiaan Huygens (1629–1695) quickly concluded that these observations proved their theory that the Earth was flattened at the poles.[84] But other experiments pointed in the opposite direction. This was the case of Jacques Cassini's measurement of the southern part of the Paris Meridian. In 1718, he started to measure a succession of 28 triangles between the Paris observatory and Dunkirk—nine were the same as measured by Picard and 19 were new. His results pointed to an elongated sphere along the poles. He presented his calculations at the

82 AASP, PV (September 1, 1734), pp. 585–616.

83 Nicholas Dew, "Scientific Travel in the Atlantic World: The French Expedition to Gorée and the Antilles, 1681–1683," *British Journal of the History of Science* (BJHS), 43, 1 (March 2010), pp. 1–17.

84 Broc, *La geographie des philosophes*, p. 42.

Académie and maintained that the Earth was probably spheroid, as Newton proposed, but elongated at the poles, that is an oblong shape like a lemon.[85] In 1733 he made new measurements farther north from Paris, which demonstrated that latitude degrees diminished rapidly as they neared the poles, thus further sustaining the idea of an elongated spheroid Earth.[86]

Since these observations were contradictory and did not allow a definitive conclusion, in 1734 both groups, especially Cassini, and La Condamine, and even Maupertuis began more vigorously defending at the *Académie* that the only way to resolve the dispute was by organizing two expeditions to measure the angles at an interval in latitude at two extreme different points of the meridian—one at the Equator and another one at the pole. As Maupertuis showed in the diagram of his memoir presented in the *Académie* titled *Sur la Figure de la Terre*, the angles of the meridians would be different on both points depending on whether the Earth was prolate or oblate.[87] At the beginning of 1735 they became engaged in an open campaign to convince the king of the importance and feasibility of the expeditions, while simultaneously discussing how to resolve the technical problems that hindered their implementation. Several memoirs were presented at the *Académie* discussing methods, techniques, and instruments that would be used in the two expeditions that started to be organized when Louis XV gave orders for them to go ahead.[88]

In 1735, when the first Franco-Spanish expedition left for Peru,[89] commanded by Louis Godin (1704–1760) with the participation of Pierre Bouguer (1698–1758) and La Condamine—all members of the French *Académie*—d'Anville published a book he had written the previous year on the issue of the Earth's shape. Entitled *Proposition d'une mesure de la terre dont il résulte une diminution considérable dans sa circonférence sur les parallèles*,[90] it was dedicated to Louis, the new Duke of Chartres, son of the Duke of Orleans, his protector and patron. The book had the approval of Fontenelle, secretary of the *Académie*, and a convinced Cartesian. At its beginning, the geographer explains that up to that point he had maintained a "profound silence" on that great issue that stirred up the *savants* of the time, but that he had resolved to break it with that publication, which was produced at the urging of Abbot Jean-Paul Bignon

85 Shank, *The Newton Wars*, p. 325.

86 Lloyd A. Brown, *The Story of Maps* (Boston: Little, Brown and Company, 1949), pp. 250–15.

87 AASP, PV (January 26, 1735), (February 4, 1736).

88 AASP, PV (January 8, 1735), (January 26, 1735), (August 13 and December 10, 1735).

89 Neil Safier, *Measuring the New World: Enlightenment Science and South America* (Chicago: Chicago University Press, 2008).

90 Anville, *Proposition d'une mesure de la terre dont il résulte une diminution considérable dans sa circonférence sur les parallèles* (Paris: Chaubert, 1735).

(1662–1743), librarian to the king. As can be seen, people representative of the world of knowledge and related to the *Académie* were behind the publication. He then states that his "particular opinion and the foundations on which it is based had nothing in common with those (two) different opinions." After explaining the two great systems—the Newtonian and the Cartesian—he, in accordance with the Newtonians, disagrees that it would be necessary to use astronomical measurements to solve the problem of the shape of the Earth, contrary to what had become hegemonic among the members of the *Académie*. As Maupertuis defended before, d'Anville himself pointed out that the instruments available at the time were neither sufficient nor reliable enough to guarantee that the results obtained with them on the territory would be irrefutable.[91] As an armchair geographer, he maintained that the traditional geographic methods using measurements of geographical distance taken from different sources, mostly from maps and travel itineraries, were enough. After making his calculus, he concluded with the Newtonians that there was "an irregularity in the Earth's shape," but like the French Cartesians maintained the Earth had its elongation in the direction of the poles.[92]

In fact the issue of the shape of the Earth was still far from being resolved, and d'Anville's position was in tune with the ongoing debates in the *Académie*.[93] In 1736, at the same time that Maupertuis was to leave for the North Pole, he published the *Carte réduite de la Mer du Sud selon la nouvelle hypothése de la longitude* (Fig. 10) accompanied by a small text.[94] The map was made taking into consideration the best measurements taken by Spanish, English, and French expeditions, and he concluded that the South Sea or Pacific Ocean occupied half of the size of the Earth and because of that he partly changed his mind and accentuated also the elongation of the region near the Equator, stating that this could be observed, as represented in the attached map, by an elongation of the South Sea. He was not in favor of an oblate (flattened) spheroid at the poles but maintained a spheroid elongated both at the poles and at the Equator. At the same time he wrote a work for D. Luis da Cunha about the Tordesillas Line, engaged in the belief of an oblong spherical Earth, also wider at the Equator and extended in the direction of the poles, which could be turned in favor of Portugal's interests in its territorial disputes in America and in the South Sea: "I have a particular reason to express myself thus and

91 Anville, *Proposition d'une mesure de la terre*, p. xv.

92 Anville, *Proposition d'une mesure de la terre*, pp. xi, 12, 147.

93 AASP, PV (June 15, June 18, August 6, August 13, 1735), (February 4, 1736).

94 BNF, DCP, GE DD-2987 (9682), Anville, *Carte réduite de la mer du sud selon la nouvelle hypothése de la longitude*, 1736.

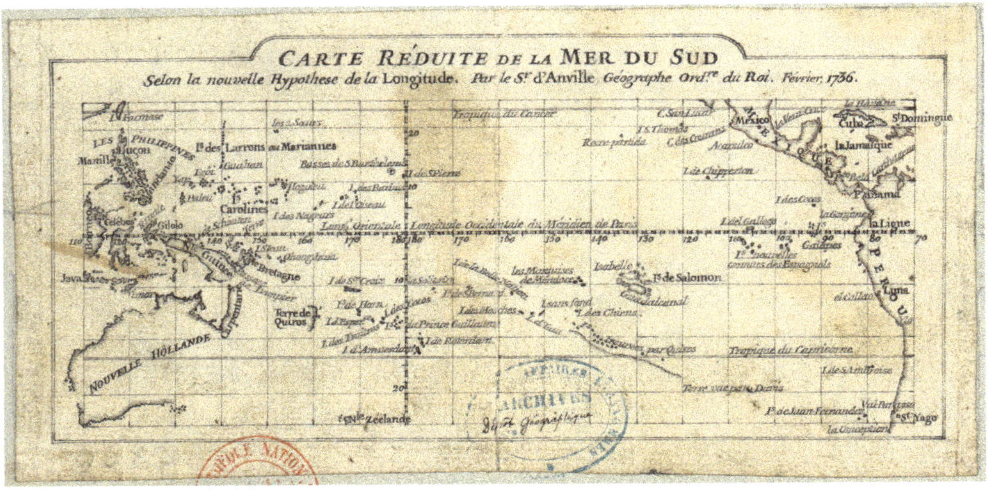

FIGURE 10 *Carte réduite de la Mer du Sud selon la nouvelle hypothèse de la longitude*, J.B.B. d'Anville (Paris, printed, 1736). BNF, DCP, GE D-10890
REPRODUCED COURTESY OF BNF

that allows me to recognize the belief, or rather, the conviction that the Earth is elongated in the direction of the Equator, rather than narrower."[95]

In 1738 d'Anville published another book on the subject of the shape of the Earth,[96] even though on November 13, 1737, Maupertuis had read at the *Académie* his conclusions on the geodesic measurements made at the North Pole and, as Newton maintained, claimed that the Earth was flattened at the poles.[97] D'Anville's book was an answer to a memoir sent to the *Académie* by M. Simonin, a professor of hydrography in Bayonne, who argued against d'Anville's *Mesure du Mer du Sud* and all theories that were contrary to a perfectly spherical Globe as his own ocean travels around the world were undisputable proof that the Earth was a perfect sphere. Little is known about Simonin, other than that over the following years he inundated the *Académie* with six memoirs, where he developed arguments in favor of the Earth being perfectly spherical in shape. In addition to his criticism of d'Anville's work, he argued against anyone who supported Newton's proposition on the shape of the Earth, and of Cassini and Picard's measurements of the Paris Meridian.[98]

95 NL, VB, Ayer MS 1918, doc. 549, f. D3.
96 D'Anville, *Réponse de M. D'Anville au mémoire envoyé à l'Académie Royale des Sciences* (Paris, 1738).
97 AASP, PV (Nov. 13, 1737), p. 202.
98 AASP, PV (February 26, 1738, February 25, 1739, Oct 4, 1740); BNF, M, MF. 22.233, pp. 203–04.

On February 26, 1738, the *Académie Royale des Sciences de Paris* took a stance in this debate. Now the members were no longer divided and maintained that the *Académie* "never believed that proofs derived from measurements of itineraries and the speed of vessels made by travelers were sufficient to deduce the oblong or flattened shape of the Earth." For them the Earth's shape defended by both Simonin and d'Anville, on the opposite side, "are based on the same principle." Because of that, they decided to "leave it to M. Simonin and M. d'Anville to discuss the validity of their conjectures."[99] The verdict was painful to d'Anville as it was clear that it was not in favor of geography based on erudition in which he was a master to solve the shape of the Earth. For its members it "is in the sky that one should search for knowledge of the Earth,"[100] in other words, in astronomical and mathematical geography based on measurements taken by scientific instruments—geodetic surveying.

In his 1738 *Réponse de M. D'Anville au mémoire envoyé à l'Académie Royale des Sciences, contra la mesure conjecturale des degrez de l'équateur, en conséquence de l'étendue de la Mer du Sud*, d'Anville defended the cartographic method to discover Earth's true shape based on erudition and argued that it was based "on more positive and less equivocal land measurements," as well as maritime ones. He was responding to both Simonin and the *Académie* and compiled information gathered by Dutch, Spanish, and English ships, but especially data obtained recently by the French vessel Saint Antoine, commanded by M. Frondat, and other nautical charts extracted from *Dépôt des Cartes et Mémoires de la Marine*. Based on that he maintained that the Earth's oblong spherical shape was wider at the Equator.[101] Later on, Dacier, when praising d'Anville, regretted that error, even though he claimed that it was "more useful than harmful to his reputation," arguing that "his conjectures were not improbable and had many followers."[102] This error must have undoubtedly partially contributed to d'Anville's difficulties in entering the *Académie Royale des Sciences de Paris* as he continued to stand for the armchair geographer's method and the *Académie* was in favor of geodetic surveying to solve the problem of the shape of the Earth. This dispute with the members of the French *Académie* impacted his plans to enter the institution, an important *locus* for the recognition of his career based on individual talent.

Early in 1721, when he was only 24, d'Anville wrote a long manuscript in which he summarized his geographic method. But it would only be published

99 AASP, PV (February 26, 1738).
100 Dacier, "Éloge de M. d'Anville," pp. 28–29.
101 Anville, *Réponse de M. D'Anville au mémoire envoyé*, pp. 4–5, 45–46.
102 Dacier, "Éloge de M. d'Anville," pp. 28–29.

many years later, in 1777, under the auspices of *Académie Royale des Inscriptions et Belles-Lettres*.[103] The book was entitled *Considérations générales sur l'étude et les connaissances que demande la composition des ouvrages de géographie*.[104] Georges Tolias shows that in it d'Anville makes "a critique of mathematical geography and a panegyric of erudition, praise for critical meticulosity, an almost religious appreciation for the exactness of facts" and "contrasts the limits of the still incipient astronomical geography."[105] Throughout his life, d'Anville remained fond of his cartographic method.

As the tradition of the armchair geographers declined, another one emerged, of which the Cassinis, with their herculean venture of making a map of France based on a measurement of the Paris Meridian,[106] were the greatest examples. This new methodology was grounded on the primacy of the direct observation of the terrain, using mathematics and geometry to measure land, and of the celestial bodies to define latitudes and longitudes, and employing increasingly precise mathematical astronomical instruments for that purpose. Although they were not directly opposed methods, since the armchair geographers depended on maps produced based on geometric measurements made on the land, but also based on measurements of itineraries, the two methodologies were presented as opposites to the enlightened *savant* public, as Jacques Cassini illustrates when he stated that he was proud "to have undertaken a great reform in geography, to have even created a brand new science, associating geometry and astronomy in a very intimate manner."[107] Throughout the eighteenth century, these two traditions coexisted side by side, complementing and sometimes confronting each other, until the former was finally eclipsed by the latter.

4 Paper Earth[108]

In his *Description abrégée des principaux monuments de l'Égypte*, Louis-Madeleine Ripauld (1775–1823) observes that, among all the maps that

103 In the session of June 17, 1755, the academy "conceded to m. D'Anville permission to inscribe the title of academician on all his maps and make use of the privilege of this Company as sr. m. Delisle made use of the Academy of Sciences." *Journal des délibérations et des assemblées de l'Académie des Inscriptions et Belles-Lettres*, 1755, June 17, f. 85.

104 Anville, *Considérations générales sur l'étude*.

105 Tolias, "Une science de pure érudition," p. 175.

106 Brown, *The Story of Maps*, p. 254.

107 Jean Domenique Cassini, *Mémoires pour servir a l'histoire des sciences et a celle de l'Observatoire Royal de Paris* (Paris: Bleuet, 1810).

108 José de Aguilar, *Historia de la cartografía: la tierra de papel* (Buenos Aires: Codex, 1968).

Napoleon's troops took to Egypt, d'Anville's was the most essential for movements in the region because of its accuracy. The geographer's skill elicited constant admiration: "He marked with a perfectly astounding degree of precision the position of old cities and towns, as well as the course of canals, in a country he had never visited."[109] Jean-Denis Barbié du Bocage (1760–1825) stressed that such a degree of accuracy had been attained "only through the force of his critique,"[110] that is his armchair geographers' method.

In 1754, because of his dedication to the study of Antiquity, associating history and geography, d'Anville was elected member of the *Académie Royale des Inscriptions et Belles-Lettres*. His appointment was part of the typical career of a scholar in the European Republic of Letters. It provided prestige and opened new doors. To obtain it, one needed the patronage of the greatest men in the kingdom, which he enjoyed from the king and the dukes of Orleans. Not without pride did d'Anville start boasting the title he had received from that institution[111] in his maps' cartouches after 1755. The *Académie des Inscriptions et Belles-Lettres* embodied devotion to and study of *belles-lettres*, but also of history associated with the geography of Antiquity and more rarely the modern era. Medals, coins, inscriptions, monuments, and maps were considered important objects as repositories of memory.

The cartographic method used by d'Anville, who was an heir to the tradition of armchair geographers who preceded him, was based on his ability to critique the extensive documentation—maps, travel narratives, works by authors from Antiquity, and so forth—he collected to produce each of his maps and in his access to the most modern and up-to-date sources from the region to be mapped.[112] However, although he exercised this ability with great mastery, he found no heirs with comparable skill. At the end of his life, he regretted, not without a certain resentment, not having been able to "train a student, to give me (as people like to say) a successor."[113] Nevertheless, in 1728, when d'Anville produced his maps of China for the Jesuits, he contracted the young

109 Louis-Madeleine Ripauld, *Report of the Commission of Arts to the First Consul Bonaparte on the Antiquities of Upper Egypt and the Present State of all the Temples …* (London: J. Debrett, 1800), p. B2.

110 Bocage, "Avant-Propos," p. 13.

111 In the session of June 17, 1755, the academy "conceded to m. D'Anville permission to inscribe the title of academician on all his maps and make use of the privilege of this Company as sr. m. Delisle made use of the Academy of Sciences." *Journal des délibérations et des assemblées de l'Académie des Inscriptions et Belles-Lettres*, 1755, June 17, f. 85.

112 Haguet, "La bibliothèque cartographique, outil de travail du géographe de cabinet," in Haguet and Hofmann, eds., *Une carrière de géographe au siècle des Lumières*, pp. 117–54; Tolias, "Une science de pure érudition," pp. 155–78.

113 Anville, *Considérations générales sur l'étude*, p. 110.

Pierre Helle (17?–1767).[114] He was the first apprentice who would work with him throughout his career. However, Barbié du Bocage had no hesitation about claiming, nonetheless, that he was "the only *protegé* that monsieur d'Anville ever had."[115] Helle was nephew of Robert-Alexandre d'Hermand (1670?–1739) an important link with D. Luis da Cunha, who had strengthened his connections with Hermand when, in 1720, D. João v ordered the ambassador to purchase books and mathematical instruments in Paris, which will be discussed in the next chapter. As well as the maps of China, Helle worked on the maps of Africa which d'Anville had begun to produce to illustrate a project sketched out by the ambassador for the Portuguese crossing of the south of the continent.[116] Between 1728 and 1732, d'Anville's African maps illustrated the books of the Dominican Jean-Baptiste Labat (1663–1738), and the Oratorian Joaquim Le Grand (1653–1733), whose publications were orchestrated by D. Luis da Cunha, who made the Portuguese sources available, in what I call a "*savant* publishing project."[117] Although d'Anville had not named a successor, some geographers who came after him did not hesitate to state they followed him and applied his method, such as Edme-François Jomard (1777–1862), an engineer who accompanied Napoleon to Egypt, and the Englishman James Rennell (1742–1830).[118]

Armchair geographers made their maps by comparing old and new maps with textual geographic descriptions, taken from both old books (from Antiquity and the Arabic world) and more recent travel reports. Not by chance was d'Anville portrayed not handling a map or any mathematical instrument, not even a quill pen (Fig. 7); in the painting the author placed a terrestrial Globe at his side on the table, the only direct reference to the world of geography. With his gaze lost in the distance, he seems to reflect on the content of the book he was reading with interest a few moments before, before the painter captured his image. On the shelf behind him there are other books whose titles cannot be distinguished on their spines. It is neither a map printer's workshop nor a royal cartography classroom, since the work of a geographer was quite different from that of engravers or engineers. Geographers such as d'Anville

114 Haguet, "Splendeur et décadence d'un 'grand homme,'" pp. 322–23.

115 Bocage, "Avant Propos," p. 15. Du Bocage started working for d'Anville at the age of ten, and after d'Anville's death continued working on the catalog and organizing the collection of maps that the geographer had sold to the king. See C. Du Bus, "Les Collections d'Anville à la Bibliothèque Nationale," pp. 98–99; Hofmann, "La 'Collection d'Anville' au Ministère des Affaires Étrangeres (1772–1828): modalités et enjeux d'une approriation," in Haguet and Hofmann, eds., *Une carrière de géographe au siècle des Lumières*, pp. 361–96.

116 Furtado, *Quebra cabeça africano.*

117 Furtado, "O projeto editorial *savant*," in *Quebra Cabeça-Africano*, pp. 305–30.

118 Haguet, "Splendeur et décadence d'un 'grand homme,'" pp. 324–25.

neither printed the maps they drew nor performed measurements of the land areas they represented in them. The many books on the shelves around him are references to treatises, reports, memoirs, and others, which he collected in large numbers to draw his maps. It involved an abstract, erudite knowledge rather than a practical one, typical of a scholarly official rather than a manual one. It reveals the value given to this type of knowledge and to the people who practiced it during the Enlightenment. The painting gives a good sense of this by representing the geographer's studio as a place of reflection and knowledge.

To update their cartography of vast expanses of land—countries, continents, the world—they consolidated information obtained from the most varied sources. Because of that,

> they depended essentially on their information networks for the progress of their work. Their contacts in the court and government offices were to them an advantageous means for the fast and systematic collection of novelties on the new worlds. These contacts also allowed them to solicit collaborations from informants spread out in the provinces and even outside the kingdom.[119]

Armchair geographers constantly searched for the latest information, especially on lands about which little or almost nothing was known. Firsthand access to previously unknown geographical information could give a map unprecedented quality and perfection, which was important for establishing one's reputation, making the map a unique work and a masterpiece. The connections that d'Anville established with the Portuguese (through D. Luis da Cunha)[120] and Spanish crowns, the Jesuits through Du Halde, and the Dominicans through Jean-Baptiste Labat, and the Oratorians through Joaquim Le Grand, and through French diplomats to other European *savants* were essential, providing him with reports of travelers, sailors, soldiers, and missionaries.[121] The *Palais Royal*, which belonged to the Duke of Orleans, was a point of encounter for travelers returning from their voyages abroad, a fact that geographers connected to the house of that prince took advantage of, as was the case with

119 Nelson Martin Dawson, *L'atelier Delisle: l'Amérique du Nord sur la table à dessin* (Québec: Editions du Septentrion, 2000), p. 17.

120 Furtado, "Elargir ses réseaux, diversifier ses commandes," pp. 53–83; Furtado, "A Nascente do Nilo," in *Quebra-Cabeça africano*, pp. 283–304.

121 Mascart, "La correspondance de d'Anville"; Marcel, "Lettres inédites du Cardinal Passionei à d'Anville," p. 433.

d'Anville and Delisle, who thus gained firsthand access to geographical information about distant lands.[122]

In addition to contacts within the country and with travelers who ventured to the most inhospitable regions of the planet, these geographers established a wide and diversified information network. An example of this was the correspondence initiated by d'Anville with Pierre-Michel Hennin (1728–1807) when the latter was made an official in the French embassy in Poland. Since d'Anville was developing a map of that country at the time, he expressed interest in receiving information on everything that had to do with its geography. Despite some interruptions, the correspondence between the two provided d'Anville with information on places that Hennin visited, as well as the maps to which the traveler had access, providing the geographer with recent and precise information on the territories he was mapping.[123] Connections with members of the European Republic of Letters, some of them ambassadors, were useful in providing him with reports and the maps existing in their valuable libraries. Examples of these include Longuerue, Rollin, and D. Luis da Cunha, in France; the 4th Count of Ericeira, in Portugal; and Ribeiro Sanches, in Russia. Lucile Haguet explores the role of his global network to form his collection of maps.[124]

Enlightenment maps were thus permanently open works, which explains why so many editions of the same map were printed, each of them correcting the previous one. Regarding the map of India,[125] d'Anville stated that

> in any case, the advancement of geography is much dearer to me than the Map of India; [for this reason] I hope it will only set the stage for a more exact and complete one, which will destroy it in all ways and leave it with no other merit than to have given way to an improvement. I will be more impassioned than anyone in researching everything that may allow that advancement.[126]

However, having access to those sources was not enough. The next step was figuring out how to conduct a critical analysis of that documentation. It was necessary to "rigorously and critically handle the information gathered." To do so it was necessary to have "a meticulous, methodical, and critical attitude that

---

122 Pedley, *The Commerce of Cartography*, p. 39.

123 M. Gabriel Marcel, "Correspondance de Michel Hennin et de D'Anville," *Bulletin de géographie historique et descriptive*, 3 (1904), pp. 441–83 (pp. 446–48).

124 Haguet, "La bibliothèque cartographique, outil de travail du géographe de cabinet," pp. 117–54.

125 Wulf Bodenstein, "D'Anville's 1752 Map of India, His Commentary and English Translation," *Maps in History*, 47 (September 2013), pp. 19–25.

126 Anville, *Éclaircissemens géographiques sur la carte de l'Inde*, p. vi.

allowed the cartographer to reduce his margin of error."[127] While this approach was essential, each geographer developed his own methodology, which allowed him to decide which of two reliable pieces of information should be discarded and which should be used. To a great extent this methodology depended on a rare skill, based on each person's individual ability and inclination, which made it difficult to train a successor, to d'Anville's chagrin. It was a symbiosis between art and nature, or a gift which is why Jacques-Nicolas Bellin (1703–1772), hydrographer to the King of France, regrets that "a good geographer is somewhat rare, for it is necessary for nature and art to come together to make one."[128]

Bellin sums up the vast and complex qualities necessary to be a good geographer, which in many ways also applies to his working method, as follows:

> The first is memory, love for the work, patience, and a spirit of order and arrangement; another one is the necessary knowledge of geometry and astronomy, after which comes the long and sterile study of travelers, a critical discussion of their reports and travel diaries, which are continuous sources of uncertainty and errors, where often the most assiduous work will fail: add to this some knowledge of foreign languages.[129]

Dacier lists 211 maps made by d'Anville during his lifetime.[130] To produce them, the geographer gathered an impressive collection of more than 10,000 cartographic documents,[131] which were essential as sources of information for his own production. In 1772, in the context of his election as *adjoint-géographe* (royal associate geographer) to the *Académie Royale des Sciences de Paris*, he first proposed that his collection be acquired by the king to enrich his library.[132] It is quite probable that this was in fact the strategy which made his appointment possible. The proposal was mediated by Emmanuel-Armand de

127 Dawson, *L'atelier Delisle*, p. 18.

128 Jacques-Nicolas Bellin, "Remarques de M. Bellin… sur les cartes et plans qu'il a eté chargé de dresser. Dissertation préliminaire, sur l'origine des Amériquains," in Pierre-François-Xavier de Charlevoix, *Histoire et description générale de la Nouvelle France avec le journal historique d'un voyage fait par ordre du roi dans l'Amérique septentrionnale*, vol. 3 (Paris: Chez Pierre-François Giffart, 1744), p. xii.

129 Bellin, "Remarques de m. Bellin," p. xii.

130 Manne, Bocage, and Dacier, *Notice des ouvrages de M. D'Anville*, pp. 45–102.

131 Paul Poindron, "Les cartes géographiques du Ministère des Affaires Étrangères (1780–1789): Jean-Denis Barbié du Bocage et la Collection D'Anville," *Sources, études, recherches, informations des Bibliothèques Nationales de France*, 1, 1 (1943), pp. 46–72.

132 ADO, SHAO, 252J224/87; Catherine Hofmann, "D'Anville et la Bibliothèque Royale/Nationale: forces et ambiguïtés d'un héritage," in Haguet and Hofmann, eds., *Une carrière de géographe au siècle des Lumières*, p. 398.

Richelieu, Duke of Aiguillon (1720–1788), then Minister of Foreign Affairs. The offer was accepted, giving rise to a long negotiation for the transfer of the collection to the crown, which only took place after his death in 1780. Initially the collection was housed at the *Ministère des Affaires Étrangères*, though nowadays it is hosted at the BNF.[133]

D'Anville himself began producing a first inventory and a catalog of the collection, which was only completed in 1827 long after his death. More recently the collection was reorganized in the BNF, including the scanning of images and online access. The collection was divided as follows:

TABLE 1 D'Anville's map Collection in the *Bibliothèque Nationale de France*

Larger divisions	**Number of items**	
	1780[a]	2018[b]
Uranographie	–	57
Cosmographiques	57	–
Mappemondes	78	85
Cartes de voyages	8	–
Générales d'Europe	44	
Divers États d'Europe	4872	6310
Cartes de France	1395	
Asie	1300	1261
Afrique	740	731
Nouveau continent	24	–
Amérique du Nord	633	
Amérique du Sud	462	1112
Grandes mers	74	–
Hydrographie	–	73
Géographie ancienne	534	
Géographie historique	153	812
Cartes de géographie sacrée	128	
Total	10,502	10,441

a Du Bus, "Les Collections D'Anville à la Bibliothèque Nationale," pp. 41, 93–145.
b Hofmann, "La 'Collection d'Anville' au Ministère des Affaires Étrangeres," p. 394.

133 Du Bus, "Les Collections D'Anville à la Bibliothèque Nationale," pp. 41, 93–145; Hofmann, "La 'Collection d'Anville' au Ministère des Affaires Étrangeres (1772–1828)," pp. 361–96.

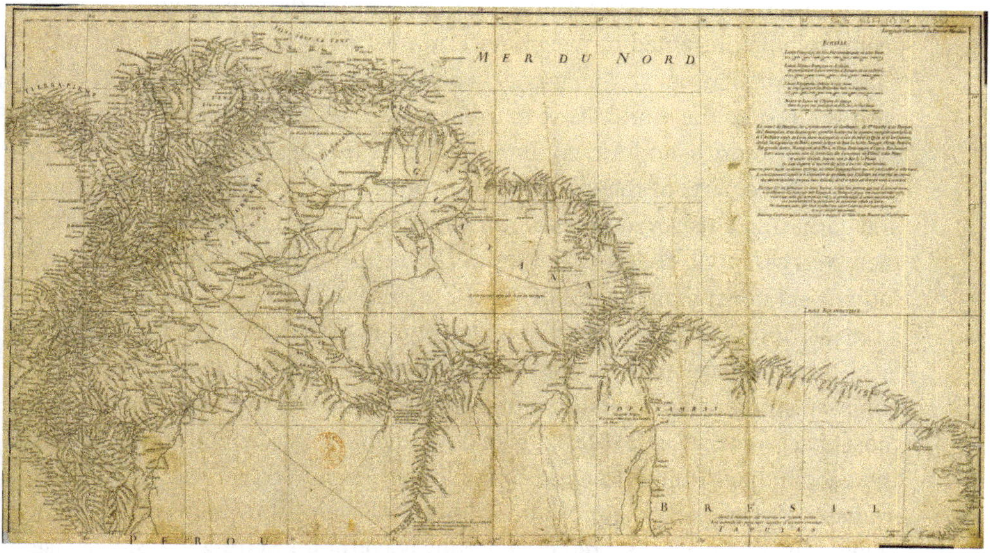

FIGURE 11 Page 1 of *Amérique méridionale*, J.B.B. d'Anville (Paris, manuscript, ink on paper, by the author at the Louvre galleries, 1748). BNF, DCP, GE C-10657 (1)
REPRODUCED COURTESY OF BNF

The total of around 10,000 maps included ones of various natures and typologies. First, there were printed or manuscript maps, used and consulted by d'Anville to produce his own cartography. These could be originals or copies made by himself or by others. These are the most numerous in the collection. Second, there were fragments and drafts of manuscripts made by him, which are studies to create parts of maps in development. They are generally drawn at a smaller reduced scale than the final map, which allowed the geographer to work on a region with an enlarged view, thus providing greater detail. In the third place there are the originals of his maps. In these, such as in the *Amérique méridionale*'s manuscript (Fig. 11), one can sometimes see corrections, additions, or observations noted at a later time. Finally, there are his own printed maps in their various versions, according to the changes he made throughout the years.

By gathering such a significant collection, "of which more than 500 are hand-drawn," with "a very particular devotion in the course of 60 years of diligent study," d'Anville intended to "achieve perfection" in the study of geography.[134] His collection allows some trends to be noted. Based on the initial

134 Anville, *Considérations générales sur l'étude*, p. Aiij.

classification, we can see that in modern geography the European continent seems to be represented in the greatest number of maps. This predominance clearly reflects Europeans' greater acquaintance with the Old Continent than with the rest of the world. Next in order were Asia, portrayed in maps since Antiquity, America, and Africa, since advancing the knowledge of the interior of these three parts of the world was one of the great challenges of the time, and gathering a reasonable set of documents on which to base this development was essential. The so-called New Continent refers to Oceania, which was much less known than the rest of the continents and of which there were only 24 maps. Louis-Antoine de Bougainville (1729–1811) who travelled to the South Sea (Pacific Ocean) in the second half of the eighteenth century took with him copies of several maps from the royal collection, and the one he considered the most exact was the *Carte d'Asie* drawn by d'Anville in 1752, which demonstrates the geographer's ability to consolidate a map even with very scarce sources available.[135] Despite the importance of historical geography in his production, ancient, historical, and sacred cartography account for 815/812 maps. Because of their complexity to produce, cosmographic charts and world maps are present in smaller numbers.

D'Anville's methodology was not limited to comparing maps. An immense number of handwritten documents were also consulted and examined, which required mastering several languages. The enormous collection of this type of source gathered by him was not as successful surviving the passage of time, but some items still remain. In his "geographical paper" collection,[136] we can see that each document was carefully copied, classified, and later numbered by Barbié du Bocage when he made the inventory to sell the collection,[137] and commented on the margins when appropriate. The nature and origin of these texts were quite varied. For example, they could be a compilation of observations taken from various sources, including cartographic ones. One of them, numbered 364, classified as relating to Greece, referred to information "according to a letter written from Constantinople to M. the Duke of Orleans by M. Belles, physician, from the king's ship," which was used for comparison with the observations contained in "the *Map of Ancient Greece* by Delisle."[138]

135 Louis Antoine de Bougainville, *Voyage autour du monde* (Geneva: Éditions de Crémille, 1973), p. 280.

136 BNF, M, NAF 6502–6503; NAF 17381–17383.

137 BNF, M, NAF 22186, 11–68; Hofmann, "D'Anville et la Bibliothèque Royale/Nationale," pp. 424–25.

138 BNF, M, NAF 6502–6503. "D'Anville—papiers géographiques," pp. 364, 127.

More recently some of his papers were discovered in the *Fonds de la Société Historique et Archéologique de l'Orne*, by Lucille Haguet.[139]

The maps' cartouches manifested the importance of these documents for their creation. In them the geographers always mentioned their main sources, highlighting the fact that they incorporated the most recent observations made by famous travelers, many of them sponsored by the *Académie Royale des Sciences de Paris* or the Royal Society of London. But none of these observations would make sense if the process of criticism to which these sources were subjected was not made explicit. For this reason, maps were generally accompanied by geographic memoirs, where the geographer explained how they were constructed, in order to submit them to judgment by their peers.[140] D'Anville made clear the importance of this type of document to confer authority on his own maps. According to him the memoir "allows one to discern what is represented in a more or less solid manner" and make known what was formerly unknown, thus contributing to the construction of a new geography.[141] It not only reveals the process of the map's production, but also guarantees the principle of authority, since at a time when authors' rights were not clearly protected and copying was common, the "real author of a map was he who knew *how* to make the map,"[142] although the author could manipulate the information he gives as d'Anville did with the one written about his *Amérique méridionale* map.[143] For that reason, when the English geographer John Green (*c*.1688–1757) published his commentaries on his map of North America, he complained that "d'Anville, in his elaborate analysis of his map of Italy, promised to publish explanations of all his maps; yet since then none has appeared together with his maps, such as those of Africa and America." The importance of memoirs as bases for maps was such that Green justified himself stating that "if however I disagreed with him in some aspects, or did not follow him more often than I did, it is probably due to those oversights."[144]

Incorporating such diverse documentation meant having to impose some uniformity, such as, for example, establishing a correspondence between measurements by standardizing them. For this reason, d'Anville devoted himself

---

139 ADO, SHAO, 252J224.

140 Pedley, *The Commerce of Cartography*, p. 109.

141 Anville, *Eclaircissemens géographiques sur la carte de l'Inde*, p. iiij.

142 Pedley, *The Commerce of Cartography*, p. 110 (original emphasis).

143 Furtado and Cintra, "Bourguignon D'Anville's *Carte de l'Amérique méridionale*," pp. 273–316.

144 John Green, *Remarks in Support of the New Chart of North and South America* (London: J. Jefferys, 1753), p. 6.

painstakingly to this task, a lesson he learned from Delisle.[145] This dedication is expressed in several works, such as where he delved, for example, into Roman and Gallic measurements to represent France,[146] and Hebraic measurements to represent Jerusalem and its surroundings.[147] It also appears in the legends of several maps, where he presents the conversion between the ancient and modern measurements used. All the knowledge he accumulated was brought together in his *Traité des mesures itinéraires anciennes et modernes*, dedicated specifically to converting the measurements in itineraries.[148] The immensity of their variations, both past and present, explains one of the most significant French Enlightenment endeavors, which was the establishment of a universal measurement standard, the meter.

Once the work of gathering and critiquing the map's sources was completed, there were still some operations that preceded the work of drawing it. D'Anville listed them and explained the rules and criteria that oriented his practice in his 1777 book *Considération générales sur l'étude et les connoissances que demande la composition des ouvrages de géographie*.[149] Compiling a map was to him "a destiny," whose meticulous and difficult work was "constrictive and tyrannical, because of the necessity of making choices, at the risk of erring, regarding minute details," in which no mistakes were allowed. The lengthiness of the work to produce a map often compromised profitability;[150] because of that, it was essential to have powerful patrons, such as the Dukes of Orleans, or D. João v. Regarding his map of Gaul, he wrote that it had cost him many years of work and that profit was not what motivated him. He made it "solely out of vocation."[151] Although d'Anville tried to make it seem like cartography was a mission, it was a business, and payment was an inherent aspect of it and the object of innumerable negotiations and renegotiations, depending on the complexity of the work. We can get an idea of how much he received in payment for his work from the correspondence with Du Halde about the maps of China. It was initially established that d'Anville would receive 600 *livres* to make 15 small maps, where half of the payment would be in silver and the

145 Anville, *Considérations générales sur l'étude*, pp. 11–12.
146 Anville, *Eclaircissemens géographiques sur l'ancienne Gaule* (Paris: L'Imprimerie Royale, 1741).
147 Anville, *Dissertation sur l'étendue de l'ancienne Jérusalem et de son Temple, et sur les mesures hébraïques de longueur* (Paris: Prault Fils, 1747).
148 Anville, *Traité des mesures itinéraires anciennes et modernes* (Paris: L'Imprimerie Royale, 1769).
149 Anville, *Considérations générales sur l'étude*.
150 Pedley, *The Commerce of Cartography*, pp. 37–41, 53.
151 Pedley, *The Commerce of Cartography*, p. 42.

other half would be 30 copies of the book and of some of the maps (1728). As the project grew and the work increased, the amount was renegotiated; an additional 120 *livres* in silver were agreed upon in 1729, and in 1734 a new agreement was signed, for a total of 1000 *livres*.[152] When D. João v hired him to help with the stamp collection that the king was acquiring and with the restoration of an atlas that had suffered water damage on its way to Portugal, D. Luis da Cunha mediated his request for "20 *pistoles* [guns] as payment for his work," regarding which the ambassador suggested: "I hope that His Majesty will agree to send them."[153]

The first step of a series of tasks following the critique of the documents and prior to the map's design was the choice of projection, "which consists of researching the form or manner of tracing the parallels and meridians adequate to the object one intends to represent." Like all cartographers, d'Anville was conscious of the fact that, since the surface of the Globe is convex, flattening it on "one or more sheets of paper" resulted "in greater or lesser corruption, as the object to be represented is stretched."[154] The best-known projection is the Mercator one, developed by the renowned cartographer Gerard Mercator (1512–1594) in 1569. It is accomplished by projecting the Earth's sphere on a cylinder, rolled out as it is projected on a sheet of paper, and is therefore a cylindrical projection. In this type of projection, the distance between parallels increases as one moves away from the Equator. The resulting map presents many east–west deformities. The Cassini projection, invented by him to draw a map of France in 1746, derives from the Mercator projection, but takes the pole as a starting point rather than the Equator, diminishing east–west deformities of cartographic representation, but increasing them in the north–south direction.[155]

D'Anville was aware of the fact that any projection distorted the object, and throughout his production he did not stick to a single one but varied his choice depending on the location of the area represented on the terrestrial Globe. For the map of Paraguay, for example, he wrote to Du Halde that "after having chosen for this map the most favorable projection, with which the intersection of meridians and parallels happens almost as regularly as it does on the convex surface of the Earth, I regarded numerous points astronomically fixed on the South Sea."[156] This projection, called "square plane" and employed since

152 Du Halde, D'Anville, *Cartes de la Chine*, pp. 394–96.
153 ANTT, MNE, caixa 789, f. 23v, January 25, 1726.
154 Anville, *Considérations générales sur l'étude*, pp. 20–21 (original emphasis).
155 Dainville, *Le langage des géographes*, pp. 28–37.
156 NL, VB, Ayer MS 1918, doc. 549, f. 2.

the fifteenth century, was frequently used by d'Anville when he represented regions close to the Equator, and is characterized by the uniformity of sections between parallels and horizontals, configuring a quite simple projection that had advantages and disadvantages relative to the Mercator projection. On the other hand, to draw a hemisphere he used a conical or concave representation of the Earth, using as the starting point only one hemisphere, like an open cone.[157]

The next step, still prior to drawing the map, was determining the scale used and the map's orientation, since according to him "the meticulousness of refinement depended on the accuracy of precision." An "infinitely beautiful map for its background of work and its execution" was one in which the cartographer had managed to find a perfect balance between the object and its representation.[158] Therefore, there was no *a priori* model to follow, but whatever the choice was, it required the geographer to have a deep knowledge of mathematics and geometry. Maps of large areas, especially continents, generally had a north–south orientation after Mercator. In other words, the north is on the upper side of the map and the south on the lower one. Maps of smaller extensions of land did not follow a consistent pattern, and the compass rose was used to point to the cardinal point that orients the map.

Once the projection, scale, and orientation were chosen, the process of drawing the map began—a complex activity involving several stages. D'Anville usually started by representing specific parts, separately designed at a smaller reduced scale. This way, the region did not end up too small, making it possible to add greater detail and to better consolidate the information extracted from the sources. Then, each part was incorporated into an early draft of the map, in pencil, and then patiently corrected and perfected. It was then necessary to reduce the size of each of the parts, using a larger reduced scale—the same one used for the final map. The map paper's base grid on which geographical elements were drawn facilitated unified scales and measurements.

An important part of the process of "giving birth to a map" was the "placement of the writing it should contain." Cartographic perfection could not be achieved if it was not "pleasant at first sight."[159] For d'Anville, this was not merely a question of aesthetics, but it was also a matter of conferring order and rationality on the cartographic space, adapting it to the new demands of

157 Anville, *Considérations générales sur l'étude*, pp. 25–26.
158 Anville, *Considérations générales sur l'étude*, p. 54.
159 Anville, *Considérations générales sur l'étude*, p. 66.

Enlightenment inclinations, according to its rational new taste.[160] Letter size had to obey a standardization that reflected the hierarchy and nature of geographic features. Moreover, letters should maintain a regular and symmetric distance within each word, and italics should be used for specific place names. The task of "lettering" a map was so complex and specialized that there were specific engravers for Roman and italic letters.[161] There was also the issue of the language itself, and d'Anville argued that the names of geographical features should respect the native tongue.[162] However, this required mastering a huge number of languages and a careful study of the documentation, to ensure that the spelling of words did not contain errors. One of the characteristics of the copies of d'Anville's maps made in England is the "Frenchification" of toponyms, despite he himself advocating that toponyms should follow the local language of the space represented. His obsession with the form of names resulted in his maps being recognized for their elegance and clarity.

Maps could contain illustrations that complemented the cartographic information presented. This was a remnant from sixteenth- and seventeenth-century cartography, which, having little or no information about a given region, used to cover the inner spaces of continents with drawings that reflected the images that Europeans had of that region. Eighteenth-century cartography tended to move this type of representation to the cartouche or the edges of the map. D'Anville's cartography was characterized by almost entirely abolishing figurative images. He was a strong advocate of a sober and rational cartography, and the few and rare image representations that appear are limited to his maps' cartouches. The maps that he drew to accompany the *Description de la Chine* were embellished, according to Du Halde's wishes, by M. Humblot, in a style that was "perfectly in accordance with the taste of the paintings made by the Chinese,"[163] in direct reference to the regions represented revealing the visions that Europeans in general, and the cartographer in particular, had of the Far East. Many of the cartouches of d'Anville's maps were produced by his brother Gravelot. In them he advocated a clean and simple style and restraint in the use of decorations according to the eighteenth-century taste in and style of maps.

160 For the eighteenth-century taste in maps, see Nicolas Verdier and Jean-Marc Besse, "Art and Design of Maps," in Edney and Pedley, eds., *The History of Cartography*, pp. 117–25; Furtado, "Um novo gosto," in *Quebra-Cabeça africano*, pp. 394–97.

161 Pedley, *The Commerce of Cartography*, p. 45.

162 Anville, *Considérations générales sur l'étude*, pp. 68–81.

163 Du Halde, "Preface," in *Description géographique, historique*, pp. xlviii–xlix.

Cartouches were important for making complementary information clear, such as authorship (of the map, the printing, the cartouche itself), date, dedication, sources, production location, scale, the geographer's titles and address, among many other things.[164] All of these are important clues for reconstructing the production process, the context in which the map was made, and understanding the map itself. Cartouches, like other parts of the production and printing process, required specialized work by skilled designers. They were drawn separately and incorporated into the map during the printing process, as can be seen in the manuscript of the *Amérique méridionale*, by Gravelot, which does not have the cartouche present on the printed version.

Once all of those stages were completed, the map was ready to be sent for printing. The printers transferred the image to one or more copper plates (depending on the final size), from which the map was printed. Proofs would then be made and, when necessary, corrections could still be incorporated by making incisions on the plate, removing or adding the desired object. Finally, after so many investments, uncertainties, errors, choices and decisions, the geographer could show off his creation, which brought together the work of many men devoted to the making of maps. Nonetheless, despite all these efforts, the work was destined to be only a fleeting representation of that land, replaced as soon as a more "exact" one came to light.[165]

5 Royal Geographer

D'Anville's background and education led him along quite a different path to that of traditional French families of geographers. His first office was situated at "rue Saint-Honoré, vis-à-vis the rue de l'Arbre sec à la Coupe d'or." This address is printed for the first time in his *Carte de la partie Occidentale de l'Afrique*, in 1727.[166] It was not by chance that this location was very close to the Portuguese embassy located at the same *rue Saint-Honoré* as at this point d'Anville was commissioned and being paid to work on D. João V's projects.[167]

In 1746, he moved to the Louvre galleries adjoining the royal palace, which belonged to the king, in the former office occupied by Delisle, and many of his

164 Chet Van Duzer, *Frames that Speak* (Leiden: Brill, 2023).

165 David Woodward, *Five Centuries of Map Printing* (Chicago: Chicago University Press, 1975); Pedley, *The Commerce of Cartography*, pp. 35–70.

166 Haguet, "La bibliothèque cartographique, outil de travail du géographe de cabinet," p. 145.

167 Furtado, "Elargir ses réseaux, diversifier ses commandes."

maps from then on contained this information.[168] It was his right as royal geographer to have an office in the palace galleries, and it constituted a privilege. The palace was also the royal residence, of part of the court and its main administrators, whose proximity was desirable for geographers, at a time when royal patronage was the greatest source of prestige and upward mobility for *savants* committed to the Enlightenment. Many artists who enjoyed royal patronage established themselves in the galleries, along with several royal institutions. One of them was the *Académie Royale des Inscriptions et Belles-Lettres* itself, and another was the *Académie Royale des Sciences de Paris*, which brought together the elite of that Parisian Republic of Letters along with the royal neighborhood. In the same building there was also a gallery that exhibited the crown's main military maps and the Royal Library. Robert-Alexandre d'Hermand, the king's mathematician, had a highly popular *cabinet* of curiosities there. When Cardinal da Cunha, João Cosme da Cunha (1715–1783), was in Paris, D. Luis da Cunha took him to visit that *cabinet* of curiosities, and Hermand demonstrated the workings of his interesting invention.[169] The immense globes (a terrestrial and a celestial one) that the Italian geographer Vincenzo Coronelli (1650–1718) built by command of Louis XIV, which bedazzled Paris and, by extension, Europe, were housed there from 1715 to 1722.[170] As we can see, d'Anville positioned himself geographically at the heart of the main French Enlightenment institutions, under royal patronage. This way he not only had contact with the members of the Republic of Letters who frequented the building every day, but also demonstrated publicly that he was a part of it and that he enjoyed a privileged patronage.

In fact, it took a long time for d'Anville to enter the *Académie Royale des Sciences de Paris* and when he did, he was granted the position of *adjoint-géographe*, which was the lowest, non-voting rung of the *Académie*'s membership. The position of *adjoint-géographe* had been created for Philippe Buache (1700–1773) in 1730.[171] d'Anville's appointment as a corresponding member of the *Académie des Sciences de Saint-Pétersbourg* on September 17, 1748 was an important step in this regard. This last nomination was also aided by D. Luis da Cunha; the geographer having been recommended for it by D. Luis's doctor and friend António Nunes Ribeiro Sanches, who had lived in that city and was

168 Haguet, "La bibliothèque cartographique, outil de travail du géographe de cabinet," p. 145.
169 ANTT, MNE, livro 792, f. 333, August 23, 1722.
170 Hélène Richard, *Les globes de Coronelli* (Paris: Bibliothèque Nationale de France, 2006), p. 27.
171 Monique Pelletier, "Philippe Buache," in Edney and Pedley, eds., *The History of Cartography*, pp. 228–30.

a member of the Russian society. As pointed out, since 1725 d'Anville had systematically presented his maps of modern geography to the *Académie Royale des Sciences de Paris* hoping for recognition of his career, but the delay in getting the title was revealing of the disputes that had occurred in the *Académie*, between a geography based on erudition—the compilation of sources—which d'Anville defended, and mathematical geography, which had become hegemonic in the institution. As noted, the first round in this dispute took place during the debates on the shape of the Earth, when d'Anville argued in his publication that the erudite method was sufficient to resolve the question and the *Académie*, in turn, argued that geodesic expeditions were the only solution. Not by chance, Condorcet questioned his mathematical and astronomical competence, stating publicly that his authority in these areas was limited.[172] This accusation was not true, and what lay behind it was the conflict between erudite geography, whose greatest living symbol was d'Anville, and the mathematical and astronomical geography defended by the *Académie*.

D'Anville had coveted the position as *adjoint-géographe* of the *Académie Royale des Sciences de Paris* for years and achieved it with great difficulty, one of the reasons probably being the fact that family groups had long monopolized it, the other possible reason being his position on the Earth's shape debate. There were elections in 1751, 1753, 1754, and 1772, and he was not even named as a candidate in any of them.[173] But finally in the session that took place on Wednesday, March 17, 1773 the vacancy of that position was announced after the death of Philippe Buache. "The candidates chosen for the position of associate geographer were Jean-Baptiste Bourguignon d'Anville, Rigobert Bonne (1727–1794), Jean-Nicolas Buache de la Neuville (1741–1825), Didier Robert de Vaugondy (1723–1783), and an unknown man called Hugin."[174] All but Hugin and d'Anville were heirs of important families of French geographers and all were much younger than him. Despite his not very favorable family origins, d'Anville beat the Vaugondy and Buache heirs, with Rigobert Bonne in second place. It is quite likely that this dispute gave rise to the rivalry which would later oppose d'Anville to Bonne and, finally, to Vaugondy. In d'Anville's victory it appears that talent and merit had finally overcome family traditions,[175] but one should not forget that mobilizing powerful patrons was essential to gaining access to the *Académie*, and he enjoyed the protection of powerful figures, such as the

---

172 Haguet, "Splendeur et décadence d'un 'grand homme,'" pp. 318–21.

173 Pedley, *The Commerce of Cartography*, pp. 281, 285.

174 Pedley, *Bel et utile: The Work of the Robert de Vaugondy Family of Mapmakers* (Great Britain: Map Collector Publications, 1992), pp. 114–15.

175 Pedley, *Bel et utile*, p. 115.

kings of France and Portugal, the Dukes of Orleans, and the *abbés* Longuerue and Bignon, among others. Finally, in 1773, d'Anville was elected as first geographer to the *Académie* which gave him the title of "Première Géographe du Roi" (royal first geographer), a title that he began to have proudly printed in the cartouches of his maps after that year. The appointment provided him with an income of about 1000 *livres* a year.

D'Anville had to earn at great cost a place in a community of geographers marked by family traditions, and at that point in time, from the heights of his experience and the titles attained, he felt he had the right to put himself and his critics in the place in which he believed each belonged. That is to say, the small world of Parisian cartography reflected, as one might expect, the great issues convulsing European society of the time—between the hierarchies established on the basis of family privilege and the opening to individual talents, of which d'Anville grew to be one of the principal representatives, labeling himself the oracle of Enlightenment geography.

PART 2

Encounters

CHAPTER 3

Minerva's Forest: the Ambassador, the Cartographer, and D. João v's Royal Library

1 An Enlightened King and His Library

In the early 1720s, D. João v began the process of forming a voluminous library in his palace, at the Paço da Ribeira. The Royal Library ought to gather books, prints, engravings, maps, scientific instruments, as well as everything else necessary for the development of human knowledge. But it should also be a public demonstration of the importance that the king, great patron as he was, gave to knowledge and to the culture of the Enlightenment, as did so many enlightened monarchs of the period. The library at *Paço* was not the only initiative on this front; D. João v also created other libraries and expanded existing ones, all with an eye to modernization and the progress of the kingdom (two structuring concepts of the Enlightenment worldview). These book collections ought to boast collections of classic works, but also display the best and newest publications.[1]

Starting in 1724, D. João v devoted himself to expanding the library of the Oratorians, at the Monastery of Nossa Senhora das Necessidades that played an important role in his scientific initiatives. There, the Oratorians taught the new mechanic philosophies of Newton and Descartes; and the books amassed in their library were essential for the renewal of scientific education. The facilities were completed with the construction of a study for natural history, an astronomical observatory—the site of important observations with regard to a renewed study of the cartography of Portugal—and a collection of scientific instruments.[2]

At first the Royal Library, partially destroyed in the 1755 Lisbon earthquake, was installed on the third floor of the royal palace, in the pavilion of a building jutting out over the Tagus River.[3] As the collection grew, its installations

1 Furtado, "Elargir ses réseaux, diversifier ses commandes," pp. 53–83; Furtado, "El embajador portugués, el cartógrafo francés y la Biblioteca Real de Don João v na construcción de la cartografía de la América," in Mauricio Nieto Olarte and Sebastián Díaz Ángel, eds., *Dibujar y pintar el mundo: arte, cartografía y política* (Bogotá: Uniandes, 2016), pp. 131–36.

2 Delaforce, *Art and Patronage in Eighteenth-Century Portugal*, p. 113.

3 ANTT, ML, maço 729, f. 511v.

 | DOI:10.1163/9789004710764_005

became more accommodating, taking up spaces in an adjoining building.[4] By the 1730s it already stretched all the way to the Hall of the Ambassadors on the fourth floor. There, the king ordered the construction of long walnut shelves "which meet such as roads of four directions," estimated to contain some 20,000 volumes.[5] The library ought to encompass all fields of knowledge and was put under the protection of Minerva, the goddess of sciences and knowledge. It would be "a warehouse of the sciences, treasury of the most precious gems, and the most useful in the world."[6] This was Minerva's palace, or Minerva's forest, and "each book of that numerous library (…) is a tree eternally bearing mature fruits of wisdom." In this sense, the Royal Library was the intellectual garden of the royal palace.[7]

The Royal Library, however, was not merely a "place with many books on shelves." It constituted an epicenter of D. João V's scientific program, open to the Enlightenment, characterized by the renewal and opening of knowledge and thought.[8] To achieve these ends, the institution planned to bring together "the rarest editions" on the one hand; but, on the other, it also ought to gather "a great number of manuscripts, mathematical instruments, admirable timepieces and many other rare things."[9] It combined the luxury "of precious things, admirable and rare," with those useful to the spirit, such as paintings by the Masters, manuscripts, timepieces, and mathematical instruments necessary for observing nature and the universe. The library was organized such that scientific books and instruments were made available to be consulted simultaneously.[10] From this point of view, one might say that the Royal Library was the consolidation of a larger scientific program of an encyclopedic and Enlightenment bent.

---

4 Delaforce, *Art and Patronage in Eighteenth-Century Portugal*, p. 67.

5 Joaquim Veríssimo Serrão, *História de Portugal: a restauração e a monarquia absoluta*, vol. 5 (Lisbon: Verbo, 1982), p. 407.

6 Rafael Bluteau, *Suplemento ao vocabulário português e latino que acabou de sair à luz, Ano 1721*, vol. 2 (Lisbon: Patriarcal Officina da Musica, 1727–1728), p. 225; Delaforce, *Art and Patronage in Eighteenth-Century Portugal*, p. 68.

7 Manuel Teles da Silva Alegrete, *Colecçam dos documentos estatutos e memorias da Academia Real da Historia* (Lisbon: Academia Real da Historia, 1727), p. 10.

8 Bluteau, "Livraria," in *vocabulário português e latino*, vol. 5 (Coimbra: No Collegio das Artes da Companhia de Jesus, 1728), p. 163.

9 Antônio Caetano de Sousa, *História genealógica da casa real portugueza: desde a origem até o presente, com as familias illustres, que procedem dos reys, e dos serenissimos duques de Bragança*, vol. 8 (Lisbon Occidental: Officina de Joseph Antonio da Sylva, 1741), pp. 272–73.

10 Delaforce, *Art and Patronage in Eighteenth-Century Portugal*, p. 8; Cândido Marciano da Silva, "D. João V patrono do astrónomo Bianchini," in *Estrelas de papel: livros de astronomia dos séculos XIV a XVIII* (Lisbon: Biblioteca Nacional de Portugal, 2009), p. 53.

This was not merely an effort to buy books, but also to produce them. As was the case in other nations, "these imposing 'libraries', along with the encyclopedias and the dictionaries, constituted a major part of the grand publishing ventures of the 18th century."[11] That being the case, a vigorous assortment of books was published inside and outside Portugal in this period, an editorial push concurrent with the development of the book trade in the Portuguese empire. The effort to build the Royal Library included not solely the acquisition of books but also the purchase of a print shop, donated to the *Academia Real da História Portuguesa*, and the hiring of printers and engravers for the production of books.[12] As a reflection of this vitality, "the number of new titles published annually tripled, going from 300 to 900, as did the number of printers, from 13 to 46 in Lisbon alone."[13] At the same time, many of these books were translated and published in foreign languages, promoting Portuguese culture and the intellectuals producing it outside the borders of the realm. In these senses, the Portuguese intelligentsia found themselves aligned with the model of rationalist thinking which characterized the European Republic of Letters. Under the auspices of the library, the most varied cultural artifacts arrived and were produced in Portugal, promoting a new model of taste in the country.[14]

But what constituted this taste? For Enlightenment thinkers, as the eighteenth century wore on, taste was a matter of a new sentiment slowly taking over all the fields of art, literature, science, and, by extension, cartography. d'Alembert emphasizes the development of taste as one of the principal aims of the *Académie Royale des Inscriptions et Belles-Lettres*. In reference to the term, he says that it signifies "the faculty of distinguishing (…) the various qualities which are adapted to excite pleasure or disgust," completely opposed to the old conventions of "common and trivial ideas expressed with affectation."[15] This new aesthetic, born of a triumphal rationalism, set itself against the excessive

---

11 Roger Chartier, *A ordem dos livros: leitores, autores e bibliotecas na europa entre os séculos XIV e XVIII* (Brasília: UNB, 1998), p. 71.

12 Luis Ferrand de Almeida, "Dom D. João V e a Biblioteca Real," in *Páginas dispersas: estudos de história moderna de Portugal* (Coimbra: Instituto de História Econômica e Social/Faculdade de Letras da Universidade de Coimbra, 1995), pp. 213–14; Isabel Cluny and Paulo J.S. Barata, "A propósito de um documento da política cultural joanina," *Leituras*, 3, 3 (April/October 1998), p. 133.

13 Cluny and Barata, "A propósito de um documento da política cultural joanina," p. 133.

14 Ana Hontanilla, *El gusto de la razón: debates de arte y moral en el siglo XVIII español* (Madrid: Iberoamericana, 2010), p. 11.

15 Alembert, "Preface," in *Éloges lus dans les séances publiques de l'Académie Françoise* (Paris: Panckoucke, 1779), p. XXVI.

theatricality that preceded it, and which this new artistic and cultural program sought to bestow with a more naturalist, less artificial tone.

Maps were also changed by the new demands of a more naturalistic taste, without exaggeration or rococo. Taste "appears as a criterion for assessing the quality of the map, but also as a general rule of production, applicable to the entire cartographic method."[16] This meant not only the pursuit of geographical accuracy, but also harmony of form, expressed in restraint, proportion, and the hierarchy of both graphic and textual elements. In cartographic style, this meant banishing all decorative excesses from the map. As d'Anville sought a more accurate representation of the continent's geography, he developed his methodology and improved his techniques for composing cartographic elements, seeking increasingly universal and rational symbols of representation. His maps of the continents produced from the 1740s onwards—such as *Amérique méridionale*—show that, by this time, he was able to master not only the critical methodology used to produce them, but also the visual refinement required to represent his territories.[17]

Indicating literary and artistic works, as well as scientific instruments, came to mobilize esteemed members of Lusitanian Republic of Letters. In Portugal, members of the privileged circle around the king and the *Academia Real da História Portuguesa*—such as the 1st Marquis of Abrantes and the 3rd Marquis of Alegrete—Manuel Teles da Silva (1641–1709), the 4th Count of Ericeira, D. Francisco Xavier de Meneses, Father Bartolomeu de Gusmão (1685–1724) and his brother Alexandre (1695–1753), Cardinal da Mota and Martinho de Mendonça Pina e Proença (1693–1743, charged with overseeing the formation of the collection)—drew up lengthy lists of works. At another point in this network, it was up to the ambassadors serving in foreign courts to collect books, paintings, prints, tapestries, maps, mathematical instruments, and whatever else might serve the intellectual curiosity of the Portuguese *savants*. All diplomats received orders for royal purchases, chief among them D. Luis da Cunha due to his vast erudition. France, the southern Netherlands (Belgium), and the Netherlands or Seven Republics of the Low Countries (mostly modern-day Holland), places where D. Luis da Cunha served from 1717 on, would top the list of the largest suppliers to the Royal Library.

D. Luis da Cunha was constantly pressed to recommend as well as buy works, paintings, and instruments, and even to suggest changes in the country's educational curriculum. His status as expert or connoisseur was widely recognized, even by generations to come, among Portuguese and European

---

16 Verdier and Besse, "Art and Design of Maps," p. 118.

17 Furtado, "Um novo gosto," p. 394.

intellectuals more broadly.[18] It was during the Congress of Utrecht that he, alongside the Count of Tarouca, began setting himself to royal purchases with greater energy. Within the congress, diplomats sought a pretext to strengthen relations between a variety of European *savants*, who exchanged tastes and shared in the prevailing enlightened culture. "Gathered in Utrecht, ambassadors from a variety of countries contributed to an international dialogue founded on the circulation of goods, people, and ideas."[19] A "diplomacy of spirit" was formed around a European Republic of Letters.[20] The congress configured a network of cultural sociability among the lettered men gathered there,[21] including the two Portuguese ambassadors, whose homes became spaces for intense, erudite bohemia. Not coincidentally, the Count of Tarouca's house, in all its magnificence, was compared by Nicolas Chevalier (1661–1720) to Jupiter's palace, where Hercules was born, recalling the Brazilian gold rush.[22] This was not idle praise; Chevalier, a renowned collector who dubbed himself a *marchand* of books and medals, owned a famous *cabinet* of antiquities, which he had brought with him when he moved from Amsterdam to Utrecht.[23] He reflected the collectionism prevalent at the turn of the eighteenth century which, in addition to its interest in curious and rare objects (serving as a theater of the world), began to reflect the scientific panorama of the European intelligentsia, inspiring other collections, such as those being constructed by enlightened monarchs like D. João v.[24]

It was also in Utrecht that D. Luis da Cunha was to meet an important specialist, one later recruited to the ranks of the Portuguese's monarch's collecting initiatives. This was Pierre Jean Mariette (1694–1774), young heir to an important dynasty of print- and booksellers in Paris. Around 1717, with the aim of expanding his tastes and his contacts in the field, Pierre arrived in Utrecht. Around then, his father, Jean Mariette (1660–1742), had sent him on a Grand

18 Cluny, *D. Luis da Cunha e a ideia de diplomacia em Portugal*, pp. 113–18; Nuno Saldanha, *Poéticas da imagem* (Lisbon: Caminho Editorial, 1995), p. 286.

19 Cluny, *O Conde de Tarouca e a diplomacia na época moderna*, p. 301.

20 Marc Fumaroli, *La diplomatie de l'esprit: De Montaigne à La Fontaine* (Paris: Hermann, 1994); Fumaroli, *Diplomatie de l'esprit: colloque l'Europe des Traites de Westphalie* (Paris, 1968).

21 Fumaroli, *La diplomatie de l'esprit*.

22 Nicolas Chevalier, *Relation des fêtes que Son Excellence Monseigneur le Comte de Tarouca a donné au sujet des naisssances des deux princes de Portugal, et de Plusiers autres fêtes aissi donnés par son excellence en differentes occasions* (Utrecht: N. Chevalier, 1714), p. 80.

23 Chevalier, *Recherche curieuse d'antiquités Venuës d'Italie, de la Grece, d'Egypte, & Trouvées à Nimegue, a Santen, au Château de Wiltenburg Proche D'Utrecht, dans le Château de Britten Proche de Leyde, & a Tongres* (Utrecht: N. Chevalier, 1709).

24 Chevalier, *Recherche curieuse d'antiquités*, p. IV.

Tour of Europe, which began in the Netherlands and Germany.[25] It was only natural, then, that the ambassador and young Mariette should be drawn to one another, given their common tastes and interest in the arts, as well as the first commissions sent through by D. João V.[26] In fact, a few years later in 1724, when D. Luis da Cunha was already in Paris and the king ordered him to acquire a collection of engravings,[27] the ambassador found in Pierre Jean Mariette[28]—now heading up the family business—the ideal supplier to meet royal demand.[29]

One of the peaks in royal purchases to stock the library came in the 1720s, its epicenter being Paris—where D. Luis was then serving. In this the ambassador had the help of his assistant Francisco Mendes de Góis, who threw himself into the task. Some of these purchases were meant to manifest the luxury of D. João V's court, but many were related to culture and erudition, in keeping with the scientific function of the Royal Library that should be understood here in a broader sense, referring to all aspects of the development of human knowledge.[30] The 1722 arrival in Lisbon of the two so-called Priest Mathematicians, the Jesuits Giovanni Battista/João Baptista Carbone (1694–1750) and Domenico/Domingos Capacci (1694–1736), had a significant impact on the organization of the Royal Library when it came to astronomy and geography, as well as the teaching of those subjects,[31] bearers of the necessary knowledge for the cartographical surveying of the empire currently being undertaken in order to guide the diplomatic negotiations around the territorial limits of Portuguese America, as D. Luis da Cunha insistently pressed. The structuring of physical and mathematical sciences under D. João V also meant the "organization of museums of the three realms of nature, astronomical

25 Cluny, *D. Luis da Cunha e a ideia de diplomacia em Portugal*, pp. 113–15.

26 Friendship was a structural concept of the *ancien régime* and took on unequal, hierarchical dimensions. António Manuel Hespanha and Ângela Xavier, "As redes clientelares," in Hespanha, ed., *História de Portugal: o antigo regime*, dir. José Matoso, vol. 4 (Lisbon: Editorial Estampa, 1993), pp. 381–93.

27 Royal instructions arrived on June 12, 1724. Fernando de Castro Brandão, *História diplomática de Portugal: uma cronologia* (Lisbon: Livros Horizonte, 2002), p. 132; ANTT, MNE, LL, livro 14 (July 12, 1724). See Mandroux-França and Préaud, eds., *Catalogues de la collection d'estampes de Jean V*; Furtado, "Colecionismo e Gosto," in *Oráculos da geografia iluminista*, pp. 208–38.

28 Roseline Bacou, "Mariette: la collection de Dessins," in *Le cabinet d'un grand amateur, P.J. Mariette, 1694–1774* (Paris: Musée du Louvre, 1967), pp. 17–23.

29 FBN, M, I-14,04,017, March 23, 1725.

30 Paula Findley, *Possessing Nature: Museums, Collecting, and Scientific Culture in Early Modern Italy* (Berkeley: University of California Press, 1996).

31 André Ferrand de Almeida, *A formação do espaço brasileiro e o projecto do novo atlas da América portuguesa* (Lisbon: Comissão Nacional para a Comemoração dos Descobrimentos Portugueses, 1991), pp. 85–100.

observations, and laboratories of physics, where observations and experiments were undertaken with the use of mathematical instruments," all structured by "Cartesianism, empiricism, and Newtonian physics."[32] From then on, "the Royal Library at the Paço da Ribeira became a center of scientific experimentation, and a symbol of D. João v's program of educational reform."[33]

Related to that scientific effort, D. João v started the cartographical surveying of the kingdom, especially Brazil (in the service of diplomatic negotiations with Spain), and stepped up the acquisition of any and all materials that might serve astronomy and cartography, from books and maps to mathematical, astronomical, and geodesic instruments.[34] Hereafter D. Luis da Cunha was to dedicate himself more directly to the task of producing the geographical maps key to a better understanding of the Portuguese empire. Since the Treaty of Utrecht, the ambassador had insisted on the need to build a significant cartographical base, allowing diplomats to shed greater light on the subject of the negotiations and put an end to questions or contentions.

2 Delisle and the New Measuring of the Tordesillas Meridian

The royal demand for books, scientific instruments, and maps intensified when, on November 27, 1720, Guillaume Delisle[35]—then the first geographer to the King of France—read a geographic memoir in the *Académie Royale des Sciences de Paris*, composed according to the latest longitude measures and using new techniques, referencing the eclipses of the satellites of Jupiter.[36] Along with this memoir, he presented a new map of the Globe, drawn up for Louis xv.[37] The impact of the map and the full text, entitled *Détermination géographique de la situation et de l'éntendue des diferentes parties de la terre*,[38] was felt far beyond the walls of the *Académie Royale des Sciences de Paris*; this rearrangement of lands around the Earth would transform the entire art of cartography.

---

32 Almeida, "Dom D. João v e a Biblioteca Real," p. 209.
33 Silva, "D. João v patrono do astrónomo Bianchini," p. 53.
34 Silva, "D. João v patrono do astrónomo Bianchini," pp. 51–54.
35 Dawson, *L'atelier Delisle*.
36 Dava Sobel, *Longitude: The True Story of a Lone Genius Who Solved the Greatest Scientific Problem of His Time* (New York: Penguin Books, 1995).
37 AASP, PV (November 27, 1720).
38 Guillaume Delisle, "Determination géographique de la situation et de l'etendue des diferentes parties de la terre," in *Des sciences* (Paris: Académie des Sciences de Paris, 1722), pp. 365–84.

Among innumerable changes in the geographic configuration of the continents through the repositioning of meridians, especially the Meridian of the Île de Fer and the Paris Meridian, Delisle proposed also shifting the Tordesillas Line—with evident implications for claims to the regions of the Colônia do Sacramento in America, disputed by the Spanish and Portuguese, and the Moluccas Islands in the South Sea (the Pacific). He also adjusted the Meridian at Cabo do Norte (now Amapá state, in northeastern Brazil), then being disputed with the French (Fig. 26).[39]

The issue of measuring longitudes was a ferociously contested subject at the time. Effectively, since the early eighteenth century, with accurate celestial measurements being used to pinpoint new positions for the meridians, this cartographical fever in France led Louis XV to complain that these geographers' precise measurements had made the country lose more territory in such a short time than all her enemies over many years.[40] D. Luis da Cunha immediately understood the effect that this news would have on the Luso-Spanish tussles in South America. He related that he came into possession of "a geographical dissertation, (in) which Mr. DeLisle, geographer to the king, determines the location and extension of different parts of the Earth."[41]

As a matter of fact, Delisle's *Détermination géographique* precisely inverted what was happening in terms of the occupation of these territories. The region of *Cabo do Norte* (Fig. 12) had been under Portuguese control since the Treaty of Utrecht, but the exact position of the line of demarcation was still a subject of debate with the French. As for the dispute with Spain, Portugal had bought the Moluccas from Spain in the Treaty of Zaragoza in 1529, a purchase which now seemed useless; and in South America, the Portuguese occupation of the Colônia do Sacramento (Fig. 13) had spread up to the northern bank of the Rio de la Plata—territory which, according to the new measurements, belonged to the Spanish. In the 1720s, in the context of the return of the Colônia do Sacramento to the Portuguese after the agreements at Utrecht, Portuguese diplomats were still wrestling with Spain as to the legitimacy and the extension of Portuguese possessions in this region of southern Brazil.

Worried by the implications of Delisle's report on the dispute between the two crowns, D. Luis da Cunha sent a manuscript copy of it to D. João V. He

39 AASP, PV (December 18, 1720). In the text, Delisle mixes up the Treaty of Tordesillas, 1494, and the pontifical bull of Alexander VI which preceded it, in 1493, reducing the position of Tordesillas to 100 leagues from the Cape Verde Archipelago. Cortesão, *Alexandre de Gusmão e o Tratado de Madrid*, vol. 2 pp. 330–31.

40 Sobel, *Longitude*, p. 27.

41 ANTT, MNE, livro 790, f. 73, March 9, 1721.

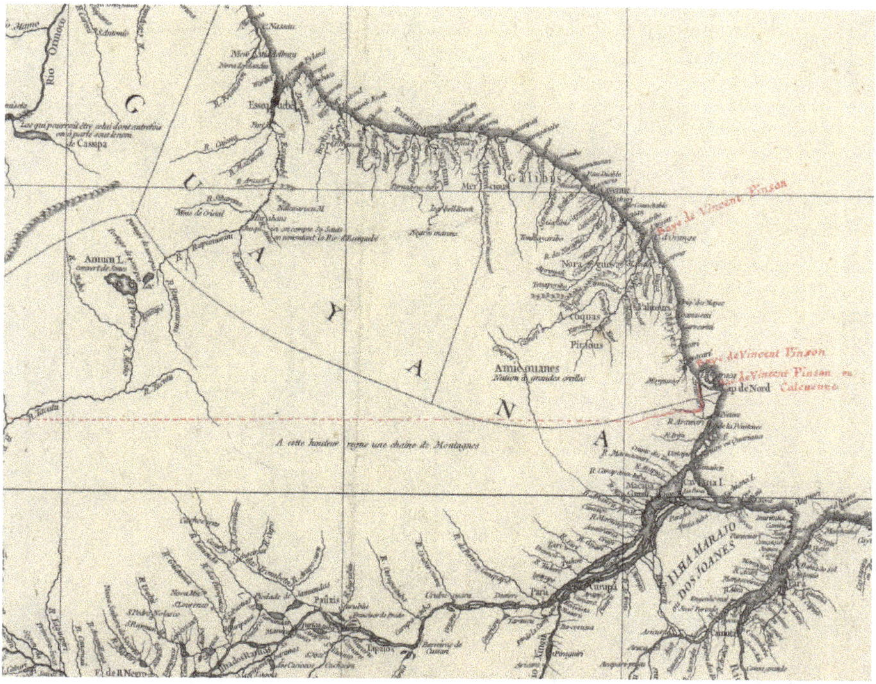

FIGURE 12 Detail of the *Cabo do Norte* in *Amérique méridionale*, J.B.B. d'Anville (Paris, printed by Gravelot, 1749). BNF, DCP, GE DD 2987 (9168B)
REPRODUCED COURTESY OF BNF

related that the regent himself, the Duke of Orleans, had read the opuscule and approved it; this being the case, D. Luis believed that the regent—always favorable to Portuguese interests—could have had no knowledge of the inclusion of these changes. His opinion was that Delisle had included the discussion on Tordesillas "ex proprio marte"—that is, on his own initiative[42]—as he believed that the regent who supported it would never approve a dissertation that would go against Portuguese interests in the region.

To understand this belief and D. Luis's first actions in Paris subsequently, we must analyze the French political context. At this time, the Duke of Orleans was regent during the minority of the future Louis XV. He was the First Prince of the Blood, which gave him hereditary rights to the throne; in the absence of a legitimate heir from Louis XIV, he could be crowned king. The duke "controlled France from 1715 until his death in 1723, in a precarious position, (which) encouraged him to look abroad for allies who could counteract the

42 ANTT, MNE, livro 790, ff. 73–74, March 9, 1721 (italics are the author's).

FIGURE 13 *Demonstracion convincente de la extension del terriorio, en que está situada la Colonia del Sacramento* (manuscript, ink and color on paper, *c.*1736). BNL, R, Papeis Origin[ais] Pertencentes à Entrega da Collonia do Sacramento, Raros e Particullares
REPRODUCED COURTESY OF BNL

Spanish threat to intervene if Louis died."[43] Louis XIV's succession by a direct descendant was not yet secure, and the regent had a chance of ascending to the throne. The only legitimate heir alive, the Dauphin (the future Louis XV), Louis XIV's great-great-grandson, was still a child and his health was still delicate, sparking trepidation. It was Filipe V's (1683–1746) accession to the Spanish throne in the midst of the War of Spanish Succession which pushed him towards a new, closer relationship with England—and, by extension, with Portugal, its traditional ally. This was because Filipe V, as great-great-grandson of Louis XIV, also had hereditary claims on the French throne; and, despite his abdication of them through the Treaty of Utrecht, the absence of a direct descendant of the deceased king might lead him to make fresh attempts to recover them, complicating the regent's regal pretensions.

And so, during the regency and under the influence of Cardinal Guillaume Dubois (1656–1723), the Duke of Orleans now made overtures to England. The cardinal was a member of the House of Orleans and had served in the previous century as ambassador to England; there he came to admire English political culture and became an advocate for a closer relationship between the Orleans and the Hanovers, now occupying the English throne. During his tenure as Secretary of Foreign Affairs, in 1717, the Triple Alliance was signed between England, France, and the Netherlands, marking their commitment to peace in Europe[44] and threatening traditional Franco-Spanish ties. In this period of the regency, Dubois underwent a meteoric rise and wound up at the highest rank of the French administration. He was made first secretary, then Councilor of State, and finally, in 1722, named Prime Minister.

D. Luis da Cunha arrived in Paris precisely around the time of this favorable push toward England and, by extension, Portugal, being driven by the Duke of Orleans and his entourage. He knew Dubois—they had been ambassadors in England at the same time—was close to the House of Hanover (in 1716 he had dined at the home of the future King of England, George I), and was part of the intellectual circles around Madame Tencin, another supporter of the House of Orleans open to better relations with England. During his time in Paris in the 1720s, the ambassador grew close to this group. Alongside a rapprochement with England, both sides argued for a similar move between France and Portugal, as "the Duke of Orleans, while he was Prime Minister, sent word

43 Jeremy Black, *European International Relations, 1648–1815* (Basingstoke: Palgrave, 2002), p. 146.

44 Pelletier and Henriette Ozane, *Portraits de la France: les cartes, témoins de l'histoire* (Paris: Hachette/BNF, 1995), pp. 197–98.

many times to D. Luis da Cunha so that he might make the King of Portugal more disposed to sign a treaty of commerce with this crown."[45] After the regent's death, Cardinal André-Hercule Fleury (1653–1743)—who became the highest-ranking minister at the time—carried on his predecessor's policies.[46] "Cardinal Fleury sought above all to maintain peace, that being to maintain an alliance with England and remain on good terms with Spain and Austria."[47] As for Portugal, D. Luis da Cunha and Francisco Mendes de Góis[48] continued to press negotiations with the aim of drawing up a Franco-Portuguese commercial treaty as well as seeking French mediation of the territorial disputes with Spain.[49] As the Portuguese court had not yet made a decision on this change of course in its foreign policy, a potential threat to its traditional alliance with England, these negotiations in France—especially regarding a possible commercial treaty—were undertaken by the ambassador and his aide with the utmost caution and discretion.[50]

With all this in mind, it made perfect sense that, in the aftermath of Delisle's *Détermination géographique*, D. Luis da Cunha should seek help from the regent—the Duke of Orleans—and his ministers; after all, the geographer himself was with the House of Orleans. The Portuguese crown felt mortally threatened, especially when it came to the still-disputed Colônia do Sacramento, and charged the ambassador to do whatever possible to ensure that the dissertation was not published or circulated. The text might give powerful ammunition to Spain, which was still dragging its heels and refusing to meet the terms of Utrecht. A repositioning of the Tordesillas Meridian could easily bolster the Spanish argument that the Portuguese did not, in fact, have the right to extend their territory as far as the northern bank of the Rio de la Plata.

The instructions from the crown, in March 1722, were that D. Luis da Cunha seek out the geographer and also speak directly with the Duke of Orleans and Cardinal Dubois, negotiating the suppression of the *Détermination géographique* and a retraction from Delisle.[51] Following these orders, D. Luis da Cunha sought out Delisle and made him tempting offers in exchange for withholding the opuscule from publication, even hiring him to execute

45 ANTT, MNE, caixa 560, ME.III-A-Fr.C1.nº.174, 1723.
46 FBN, M, I-14,04,017, March 13, 1725, f. 50.
47 Pelletier and Ozane, *Portraits de la France*, p. 199.
48 ANTT, MNE, caixa 1, maço 2, f. 71.
49 In one passage, D. Luis da Cunha appears skeptical as to the durability of the French–English alliance, despite having argued insistently for France's growing closer to Portugal. ANTT, MNE, caixa 1, maço 1, doc. 28, April 13, 1727.
50 ANTT, MNE, caixa 1, maço 1, f. 56, May 22, 1736; f. 135, September 1727.
51 Cortesão, *Alexandre de Gusmão e o Tratado de Madrid*, vol. 2, p. 333.

a seven-volume atlas to be sent to the king.[52] Though he accepted the task, the *savant* was not swayed; and the ambassador failed in his mission to stop publication of the text. Once he realized this, D. Luis switched targets; in keeping with the "second part of the orders" from Portugal, he sought out the French authorities. The first measure he took was to speak with Cardinal Guillaume Dubois (1656–1723)—who by this point was also Archbishop of Cambray—and ask him that the memoir not be printed. The latter, sensitive to Portugal's desires, promised that he would do so; but asked that the ambassador first send him a written copy of the sections that harmed the Portuguese position.[53] D. Luis showed him the relevant passages and asked him to order "Mr. Delisle to remove from his *Dissertation* [*Détermination géographique*], if he does indeed have it printed, the passages in question."[54]

D. Luis da Cunha's strategy was to separate science from diplomacy, keeping the former from interfering in what had been previously established in treaties (especially Utrecht).[55] The ambassador did not insist, as the king had ordered, that the text not be published—he knew it would be impossible to keep a *savant* from publishing discoveries already announced before the *Académie Royale des Sciences de Paris*—only that the sections with diplomatic consequences for Portugal be removed. He believed that the cardinal's alignment with England and his clearly anti-Spanish policy would make him take Portugal's side in such a dispute. But this would also be an attempt, moreover, to show that France did not have the right to claim the territories in question in Cabo do Norte, in keeping with the 1713 agreements at Utrecht. In a calming tone, D. Luis informed D. João V that the cardinal "then gave me his word that he would carry out what he had promised," preventing the publication of the opuscule.[56]

A copy of the Delisle's *Détermination géographique* was sent to Portugal on March 9, 1721, but never reached its destination. The Spanish intercepted the missive[57] shortly after it crossed the border.[58] In September, Diogo de Mendonça Corte Real, Secretary of State—still oblivious to what had happened—took D. Luis da Cunha to task for his delay in sending Delisle's *Atlas*. He also wrote to say, "His Majesty (…) had ordered me to find out if you

---

52 ANTT, MNE, livro 790, f. 388, October 27, 1721.
53 ANTT, MNE, livro 790, f. 126, October 27, 1721.
54 ANTT, MNE, livro 790, f. 125, October 27, 1721.
55 Cortesão, *Alexandre de Gusmão e o Tratado de Madrid*, vol. 2, f. 333.
56 ANTT, MNE, livro 790, ff. 126–27, May 12, 1721.
57 ANTT, MNE, livro 790, f. 389, May 12, 1721; f. 411, October 27, 1721; November 3, 1721.
58 Cortesão, *Alexandre de Gusmão e o Tratado de Madrid*, vol. 2, p. 334; Almeida, *A formação do espaço brasileiro*, pp. 67, 51.

would be able to discover there what the King of France is accustomed to giving to foreigners who make him such offers."[59] It seems that the secretary either did not understand or pretended not to understand the ambassador's strategy in ordering an atlas from Delisle and thus putting him at the service of the Portuguese king; in the same missive, Corte Real seemed to think it strange that the geographer, who had written a memoir harming Portuguese interests, could make the king such an offer. D. Luis was quick to respond that he had paid for all of Delisle's services with "a free hand (*mão aberta*)."[60] To the 1st Marquis of Abrantes, who also asked about the matter, he would insist that "M. Delisle is very well paid for his work." And he added that, "his science aside, he is a man of such avarice that he will not take a step if it is not in his interest."[61] He made it clear that to put a man of this stature in the court's service, palms had to be greased. As he himself said, "the golden key, or the key of any equivalent metal, (…) opens the most tightly locked lips"[62] and overcomes the strongest wills.

On November 10, 1721, D. Luis da Cunha sent along to Portugal a new "copy of the *Dissertation* by Monsieur Delisle." As the document had been "founded on the new astronomical observations which have been made," the ambassador believed "that, in order to convince him, it will be necessary to have other observations made, more modern than the alleged ones, so as this would not appear to impugn the opinion of a geographer." He called D. João V's attention once again to the need to map the Portuguese territories in America using actual measurements, not pure speculation.[63] Understanding that science and diplomacy could not always be compatible, D. Luis da Cunha justified Delisle's attitude to the king—it would hardly be "practical for a man who had given that speech at the *Académie Royale des Sciences de Paris*, as a member of the same *Académie*, to contradict himself without being torn to shreds."[64] However, despite all his efforts and Cardinal Dubois's promises, and as he had predicted, the *Détermination géographique* was published as was customary by the *Académie* just the next year, in 1722.[65]

Given the impossibility of the geographer's rejecting his own theory, the ambassador made a point of arguing that the crown ought to stick to what had been established at Utrecht and seek to diminish the impact a new position of

59 Cortesão, *Alexandre de Gusmão e o Tratado de Madrid*, vol. 3, pp. 222–23.
60 ANTT, MNE, livro 790, ff. 388–89, October 27, 1721.
61 ANTT, MNE, livro 790, ff. 395–96, October 27, 1721.
62 Cunha, *Instruções políticas*, p. 194.
63 ANTT, MNE, livro 790, f. 421, November 10, 1721.
64 ANTT, MNE, livro 790, f. 389, October 27, 1721.
65 Delisle, "Determination géographique de la situation."

the Tordesillas Line might have on Portuguese territory in America. After all, he had always defended the terms established in previous treaties as an important diplomatic weapon in future negotiations. In this vein, as he insisted in a letter to the 1st Marquis of Abrantes, "I do not understand how the *Dissertation* or opinion of a geographer can hinder what is stipulated by a treaty."[66] He also called the marquis's attention to the transitivity between cartography and diplomacy, an aspect to which Portugal would do well to attend in future negotiations.

3 Mathematical Instruments and Their Experts[67]

Under the impact of Delisle's *Détermination géographique*, Portuguese ambassadors at European courts began purchasing mathematical instruments with an eye to surveying Portugal and Brazil in 1721, and D. Luis da Cunha was charged with a large part of this task. First, as soon as he arrived in Paris in 1720, he received orders to acquire an assortment of scientific, mathematical, and nautical books from Mariette's printing house and complement this with the purchase of scientific artifacts.[68] Some years later, in 1729, one notes an increasing equivalence between scientific books and instruments—a balance sheet drawn up by D. Luis includes works on navigation, military science, nautical and land maps, as well as mathematical instruments, all sent to the kingdom in previous years.[69]

This transitivity also encompassed the hiring of scholars, some recruited abroad, to help in several functions: recommending and selecting objects to be purchased abroad; organizing the artifacts once acquired; producing mathematical instruments; giving instructions as to their assembly and use; and informing on the latest scientific methodologies, etc. Some were invited to move to Portugal and work in the service of the crown; others were asked to help with D. João V's purchases in their countries of origin. In Paris, D. Luis made use of the intellectual circle around the Duke of Orleans, friendly to Portugal and hostile to Filipe V of Spain, to wrangle *savants* and fulfill the orders he received, thus equipping the crown with the instruments it needed to advance scientific knowledge in the Portuguese empire.

66 ANTT, MNE, livro 790, f. 395, October 27, 1721.

67 See Furtado, "Bosque de Minerva", pp. 229–73; Furtado, "Elargir ses réseaux, diversifier ses commandes," pp. 53–83.

68 Delaforce, *Art and Patronage in Eighteenth-Century Portugal*, p. 86.

69 ANTT, MNE, caixa 1, maço 1, doc. 1.

One of the first scholars D. Luis da Cunha was to hire in Paris was Robert-Alexandre d'Hermand, mathematician and engineer to the king.[70] In 1723 D. Luis grew closer to him,[71] whom he held in high esteem. He would call Hermand "a man as esteemed as he is proud of his military and mechanical inventions," suggesting that D. João v bring him into his service.[72] Military engineering encompassed a wide range of knowledge and demanded a mastery of mathematics, mechanics, geometry, hydraulics, etc.[73] At the time, a military engineer was primarily "who produces machines and works for offensive and defensive war," and only secondarily "who makes any sort of machine or device,"[74] demonstrating the strong military connotation of the profession, which ranged from knowledge about moving troops over land, the planning and construction of military installations, developing armaments, surveying territories, and drawing up maps.

All these skills, which Hermand combined as a military engineer, would be of great value to D. Luis da Cunha in the execution of his tasks; the ambassador had been closely following the transformations in military engineering since D. João v's naming Manuel/Manoel de Azevedo Fortes (1660–1749) royal chief engineer. In 1719 Azevedo Fortes took charge of the *Aula Régia de Fortificação* (Regal Class of Military Fortification) in Lisbon, and began to prepare the changes he judged necessary for the education of Portuguese military engineers according to the most modern knowledge.[75] In the *Academia Real da História Portuguesa*, Azevedo Fortes was named by the king to deal with the area of geography and direct the organization of that part of the Royal Library. With the king's protection, he published technical manuals designed to standardize Portuguese military engineers' practice according to modern methodologies, entitled *Tratado do modo o mais fácil de fazer as cartas geográficas assim de terra como de mar, e tirar as plantas das praças* (1722) and *O engenheiro português* (1729),[76] as well as other initiatives. These manuals were not simple adaptations of the literature being referenced, especially the French and Dutch, as "various texts were

70 ANTT, MNE, livro 792, f. 333, August 23, 1723.

71 ANTT, MNE, livro 792, f. 333, August 23, 1723.

72 ANTT, MNE, livro 793, f. 357, July 10, 1724.

73 Furtado, "História da engenharia," in Heloísa M. Murguel Starling and Lígia Beatriz de P. Germano, eds., *Engenharia: história em construção* (Belo Horizonte: Ed. UFMG, 2012), pp. 21–69.

74 Bluteau, *Vocabulário português e latino*, vol. 3, p. 117.

75 See Beatriz Piccolotto Siqueira Bueno, *Desenho e desígnio: o Brasil dos engenheiros militares (1500–1822)* (São Paulo: EDUSP/FAPESP, 2011), pp. 203–06.

76 Mário Gonçalves Fernandes, ed., *Manoel Azevedo Fortes: cartografia, cultura e urbanismo (1660–1749)* (Porto: GEDES, 2006).

consulted and compiled, and the most relevant issues were then discussed, emphasizing the debate between several authors and the writer's own opinion on the matter."[77]

In early 1724, D. Luis da Cunha sent off to Portugal a spyglass or telescope made by Hermand,[78] "which allows one to see the *cortaduras* (draining ditches)[79] and other constructions that the besieged may undertake behind their walls."[80] The spyglass was an offer; the king was expected to test it and be convinced of the ingenuity of its inventor, who would then enter into his service and receive his favors. The trouble this instrument encountered after arriving in Lisbon illustrates the paradoxes generated by the torrent of scientific instruments and news pouring into the country. It seems that the device did not initially provoke much attention at the court; by the time D. Luis wrote to ask what impression it had made in Portugal, no one could say where it had wound up. In August 1724, the French expert in mining named Blumenstein,[81] owner of a lead mine on the outskirts of Lyon, who was already working for the king in Lisbon, wrote to Hermand that he had seen the telescope "lying in a corner" in the royal palace. Desolate, he lamented "that it should have gone astray, as Mr. Hermand was very proud that His Majesty should wish to see this, his curiosity."[82] Finally, in 1725, he got word that "Mr. Hermand's books seemed fine, as well as the spyglass which was found a few days ago among other orders, and which Blumenstein assembled." Grateful for the offering, D. João V ordered that "the man be recompensed as he sees fit."[83] The incident gives some sense of the volume of objects coming into the court, as well as D. João V's importance as a patron among enlightened circles in Europe and his role as a judge of these inventions, an indication of his erudition.

The Portuguese king was not, as a matter of fact, a passive observer of the intellectual renewal taking place in his country. The education he had received as a child, "as the magnificent paragon of a perfect prince,"[84] allowed him to take a critical stance on scientific matters. In astronomy, mathematics, geography, and cartography—all related sciences—he had been taught by Manuel

---

77 Bueno, *Desenho e desígnio*, p. 140 (italics are the author's).

78 ANTT, MNE, livro 793, f. 35, January 24, 1724.

79 Bluteau, *Vocabulário português e latino*, vol. 2, p. 573.

80 ANTT, MNE, livro 793, f. 357, July 10, 1724.

81 Blumenstein operated a lead mine in Lyon and was hired by D. Luis da Cunha in 1724 to work as an expert in Lisbon. ANTT, MNE, livro 792, f. 481, November 1723; ff. 485–86, December, 6, 1723.

82 ANTT, MNE, livro 793, f. 453, August 28, 1724.

83 ANTT, MNE, LL, livro 14, f. 186, July 26, 1725.

84 Delaforce, *Art and Patronage in Eighteenth-Century Portugal*, p. 35.

Pimentel (1650–1719), then royal cosmographer, and come to master these fields. Another hired scholar, the Swiss botanist Charles Frédéric Merveilleux (?–1749), left a record of his observations on the reign of D. João v and praised him as a great *savant* and collector. He suggested that experts in search of support join his service—but only those of great talent, as the monarch was a connoisseur, of a more refined taste than all his subjects.[85] "Francisco Xavier da Silva (1793–1852) described D. João v as a true bibliophile who was only satisfied when the title and edition that he sought was safely stored in his palace."[86]

But let us return to 1723, when D. Luis da Cunha got to know Hermand. That same year he got orders from the king to gather all the information he could about the organization of the French troops and their uniforms, under the supervision of the 1st Marquis of Abrantes—and there was nobody better than Hermand to advise him in this.[87] In October, he managed to get the Duke of Orleans's authorization for Hermand to work for Portugal, sending over designs to be copied.[88] In order to tend himself to D. João v's projects, and at the regent's orders, Hermand broke off compiling documents on military issues for the Dauphin's education.[89] In this project he had d'Anville's help;[90] maps would be indispensable tools to prepare the king for military maneuvers, giving him a better vision of battles, troop movements, and fortifications. War, geography, and cartography were constantly interlinked.[91]

1724 was a turning point in the royal purchases related to astronomy and geography. That year, guided by the Priest Mathematicians, the king established an observatory in the royal palace at the Paço da Ribeira, which even served for his private use, and started construction of another large observatory at the *Colégio das Necessidades*.[92] Besides these two, there was already an operating observatory in Lisbon, at the Jesuit college of Santo Antão, where the *Aulas de Esfera* (Courses on the Sphere) were held, a gathering point for

85 Charles-Frédèric Merveilleux, *Memoires instructifs pour un voyager dans les divers etats de l'Europe* (Amsterdam, 1738), vol. 1, p. 177.

86 Delaforce, *Art and Patronage in Eighteenth-Century Portugal*, p. 69.

87 Mandroux-França, "La Collection Royale Portugaise," in Mandroux-França and Préaud, eds., *Catalogues de la collection d'estampes de Jean v*, p. 59; ANTT, MNE, livro 792, f. 333, August 23, 1723.

88 Almeida, "O naturalista merveilleux em Portugal," p. 288.

89 Mandroux-França, "La Collection Royale Portugaise," p. 59; ANTT, MNE, livro 792, ff. 451–52, November 8, 1723.

90 Mandroux-França, "La Collection Royale Portugaise," p. 59.

91 Furtado, "Guerra, diplomacia e mapas: a Guerra da Sucessão Espanhola e a América portuguesa na cartografia de D'Anville," *Topoi*, 12, 23 (July–December 2011), pp. 66–83.

92 Almeida, *A formação do espaço brasileiro*, p. 92.

all the Ignatian masters in the field of astronomy.[93] In tune with the ambassador's argument that, in order to refute Delisle and survey the empire, more modern measurements were necessary, the new observatory at the Paço da Ribeira was first and foremost designed to establish the Meridian of Lisbon. In the effort to bolster its construction and arm the facility with the latest mathematical equipment, D. Luis da Cunha received the following orders in the month of August: "charge the most capable mathematician of the *Académie de Paris* (to) make the instruments indicated on the paper accompanying the same letter, and make them with the greatest speed and perfection possible."[94] This *savant* would be tasked with "making the instruments that His Majesty desires work perfectly," taking their specifications from an "included memoir in Latin." The instruments ought to be sent as soon as they were finished, there being "no need to wait until they have all been finished to send them together." D. Luis would have to choose someone to aid him in this carefully. These officials would be assisted by the "mathematician to direct them and avoid any imperfection in (the instruments), because one desires that they be very exact, and the best that can be made."[95] The Portuguese scholars, meanwhile, were tasked with determining which objects ought to be acquired and how they ought to be made, demonstrating that they were keeping up with the most recent advances in scientific knowledge.

To guide the construction of the Lisbon observatories, the ambassador was also ordered to seek out the plans for the royal observatory of the *Académie Royale des Sciences de Paris*, directed by Jacques Cassini—"the most capable mathematician of the *Académie de Paris*."[96] Quickly obeying all the orders he received, D. Luis da Cunha reported back that he was "attending to the task mentioned on the paper in Latin, which will be the exterior and interior plan of all the parts of the observatory and the designs of all the instruments found there."[97]

Along with these plans, several mathematical instruments were then ordered from Paris. The first seems to have been pendulums made by Henry Sully (1680–1729), with an office in the galleries of the Louvre.[98] Jacques Cassini

93 Throughout the second half of the sixteenth century, "the *Colégio de Santo Antão* had offered a specific course (…) in the teaching of Mathematics as applied to the Natural Sciences, Astronomy, Cosmography (i.e., the *Sphere*), Practical Geometry, Geography, and the Art of Fortifications." Bueno, *Desenho e desígnio*, p. 181.

94 ANTT, MNE, livro 793, f. 415, August 7, 1724.

95 ANTT, MNE, livro 14, f. 115, July 19, 1724.

96 ANTT, MNE, livro 793, f. 415, August 7, 1724.

97 ANTT, MNE, livro 793, f. 448, August 28, 1724.

98 ANTT, MNE, LL, livro 14, f. 199v, October 13, 1725.

made another pendulum to measure longitude, meanwhile; and Jean Le Fèvre (1650–1706) turned out a micrometer, his own invention,[99] and an azimuth half-circle. Before being dispatched to Lisbon, all of these instruments had to be carefully tested and evaluated. The orders were clear: "After any instrument is finished you will have it examined in its operations by people trained in the matter, and only with their approval will effect" payment for it.[100] Cassini was hired to supervise production and test several of the instruments,[101] chief among them a large telescope.[102]

A large telescope was ordered according to the instructions sent from Portugal to which Cassini[103] and the Count of d'Onsenbray, Louis-Léon Pajot (1678–1754), specified additional changes[104] that were discussed and partially altered in Lisbon. D'Onsenbray was a physicist and member of the *Académie Royale des Sciences de Paris*, having gained renown for the invention of an instrument to measure liquids and another to measure wind.[105] One sees that this was not merely a matter of copying existing instruments but innovating in their production—here, Portuguese scholars were apt to chime in alongside the French in their efforts to outfit the Lisbon observatory with the most modern tools possible. After considerable negotiation, the ambassador informed that, with the "same workman who accompanied Mr. D'Onsenbray to this house," he had settled "the price and the time to produce the instrument."[106] Despite all these precautions taken by the Portuguese *savants*, when the large telescope arrived in Lisbon and was finally tested, it was found to have several unfortunate defects.[107]

D. Luis was also open to hearing of new inventions which might serve the Portuguese crown, and many artisans sought him out to show him their creations.[108] On one occasion, a "maker of clocks in the English style" brought him one to test. The ambassador related that this same clockmaker

---

99 "He will deliver a micrometer, invented by Monsieur LeFevre, and as regards the most interesting mathematical instruments you may refer to the attached memoir." FBN, M, I-14,04,017, f. 88, March 13, 1725.

100 ANTT, MNE, livro 14, f. 127, September 17, 1724.

101 FBN, M, I-14,04,017, f. 88, March 13, 1725.

102 "I have been apprised that the books and telescope are coming, but the ship has not yet arrived." ANTT, MNE, livro 14, f. 181, June 6, 1725.

103 ANTT, MNE, livro 793, f. 603, December 10, 1724.

104 ANTT, MNE, livro 793, f. 625, December 25, 1724.

105 Jean-Henri Hassenfratz, *Encyclopédie méthodique: physique*, vol. 2 (Paris: Panckoucke, 1793), p. 231.

106 ANTT, MNE, livro 793, f. 625, December 25, 1724.

107 ANTT, MNE, LL, livro 14, f. 187v, July 25, 1725.

108 AASP, PV (January 27, 1725).

> also showed me another [clock] which apparently does not vary at sea, thus allowing one to find the longitude, if not precisely, then at least with a very small difference (...) and in the experiment which was held, taking it in a coach, the device corresponded to this claim, because despite the great movement the clock did not vary a second, and so another experiment ought to be held at sea.[109]

The search for an efficient way of measuring longitude during maritime voyages was a greater challenge, given that pendulums were thrown off by the constant movement of ships at sea.[110]

From Paris, D. Luis da Cunha also ordered the construction of a fourth mural circle,[111] a sextant,[112] a three-legged quadrant and—although the designs for these and the other instruments had come from Portugal—he was advised that "if any of the mathematicians of the French *Académie* should like to add or change any thing for the improved suitability of the instrument, Your Excellency should make providences."[113] Nicolas Bion (1652–1733), "an important French instrument maker,"[114] built a sextant and an armillary sphere, "composed of three circles, which serves to consult the hours, quarter-hours, and minutes in fives."[115] Each instrument was to be examined "in its operations by people trained in the matter, and with their approval will effect" payment.[116] Another armillary sphere was made by Domenico Maraldi (1709–1788), who completed it in 1728 using the specifications drawn up by Carbone, one of the Jesuit Priest Mathematicians, in Portugal.[117] All these instruments were necessary and used to clarify the position of the Meridian of Lisbon, thus beginning the surveying of Brazil, as encouraged by D. Luis.

4 Maps and Prints

One important element of the Royal Library was its collection of prints, encompassing a great variety of items and themes: artists' engravings, portraits

109 FBN, M, I-14,04,017, f. 89, March 23, 1725.
110 Sobel, *Longitude*.
111 ANTT, MNE, LL, livro 14, f. 197, October 13, 1725.
112 ANTT, MNE, LL, livro 14, f. 199, October 23, 1725.
113 ANTT, MNE, LL, livro 14, f. 127, September 17, 1724, attached page.
114 Almeida, *A formação do espaço brasileiro*, p. 86.
115 ANTT, MNE, LL, livro 14, f. 127, September 17, 1724, attached page.
116 ANTT, MNE, LL, livro 14, f. 127, September 17, 1724, attached page.
117 Delaforce, *Art and Patronage in Eighteenth-Century Portugal*, p. 86.

of distinguished figures, maps, views of cities, plans of fortifications, etc. A collection of prints constituted a broad visual catalogue, illustrating the current state of human knowledge.[118] The royal drive to collect in this area picked up after July 12, 1724, when D. João V sent an order to all his ambassadors to collect all the available engravings produced in the last 30 years in their countries of residence.[119] Six months later, on January 30, 1725, the order was expanded to take in "all the engravings that can be found, since the time in which they were invented."[120] These two instructions gave new life to the royal collecting enterprise; but as always, the purchases were not limited to prints alone, and in fact took in everything related to intellect, luxury, and taste.

Besides the prints per se, at D. João V's orders, the Portuguese diplomats commissioned or purchased a few atlases of purely cartographic content, as well as other publications which included both prints and maps. Among the commissioned publications, highlights include the seven-volume *Atlas* produced by Guillaume Delisle in 1721,[121] and another which was up for auction in Amsterdam and which, on January 13, 1722, the Count of Tarouca received orders to bid on.[122] This was the *Atlas Boendermaker*, a masterful demonstration of the ties between cartography and collections of prints.[123] It was based on "classification by country and continent and the referencing between the maps, which serve as a point of departure, and the representations of topography, architecture, historical events and the portraits of those important figures, according to Theodorus Boendermaker (1682–1722)."[124] Tarouca related that "the *Atlas* was as rare as it was splendid," because it took up 103 volumes.[125] Unfortunately, the better part of it was consumed in a fire at Tarouca's house in The Hague.[126] "Only 60 volumes were saved, (and) these incomplete at that." The ambassador bemoaned "that the collection had been left imperfect and of little value," but, to remedy the harm done, he suggested that "those maps and

118 Lygia da Fonseca Fernandes Cunha, *A coleção de estampas le Grand Théâtre de l'Univers*, vol. 3 (Rio de Janeiro: FBN, 2004), pp. CII–CXXV; Joaquim Oliveira Caetano, coord., *Gravura e conhecimento do mundo: o livro impresso ilustrado nas coleções da BN* (Lisbon: BNP, 1998).

119 Mandroux-França, "La Collection Royale Portugaise," p. 55; ANTT, MNE, LL, livro 14, ff. 122v–23, July 12, 1724.

120 Mandroux-França, "La Collection Royale Portugaise," p. 55; BNP, R, AT 26[12], March 8, 1725.

121 ANTT, MNE, livro 790, ff. 388–89, October 27, 1721.

122 BNP, R, AT 26[11], February 19, 1722.

123 BNP, R, AT 26[11], March 18, 1723.

124 Mandroux-França, "La Collection Royale Portugaise," p. 80.

125 BNP, R, AT 26[11], March 18, 1723.

126 BNP, R, AT 26[11], October 2, 1722.

prints as can be found should be collected to restore its magnificence."[127] In part following this advice and attempting to lend some unity to what remained of the precious *Atlas*, orders came from Portugal to the effect that the ambassador "buy only those prints which have disrupted the order required by the collection (…) by citing the page which is found in the *Atlas*."[128]

In early 1724, D. Luis da Cunha wrote from Paris to tell D. João v that the abbot Jean de Vayrac (1664–1734) had offered to sell an atlas.[129] In March, the ambassador received the orders to buy it for the king. Cautiously, he wrote that "I shall not buy them without first having them examined by an intelligent person, as I do not have a high opinion of the abbot's honesty,[130] who played the role of a poorly rewarded spy,"[131] referring to his connections with Spain. The examination and approval of the *Atlas* were made with all haste because by May it had already been sent off. They traveled on a ship bought in France purely for transporting royal purchases.[132] However, the ship sank; and despite first reports that the *Atlas* had not gotten wet,[133] having been rescued from the shipwreck, it was later found to have water damage.

It was thus in the context of the recovery of the wetted *Atlas*, in the first quarter of the eighteenth century and with scientific orders piling up, that D. Luis da Cunha came to know Jean-Baptiste Bourguignon d'Anville. This was hardly surprising, given the ambassador's contact with the Duke of Orleans and with Delisle, Cassini, and Hermand—*savants*, like d'Anville, associated with the House of Orleans. And, as was the case with all the others, D. Luis certainly had to obtain the duke's permission to recruit the cartographer to D. João v's service.

We do not know if the duke himself or which of these scholars introduced d'Anville to the ambassador. It could easily have been Delisle, who worked with him in educating the Dauphin; Hermand, d'Anville's neighbor at the Louvre and who had collaborated with him in producing tomes on military history for the young prince; or even Cassini, given that he had produced the plans for the royal observatory at D. Luis's request,[134] and in fact one of d'Anville's first tasks

---

127 BNP, R, AT 26[11], March 18, 1723.
128 ANTT, MNE, LL, livro 14, f. 123, August 30, 1724.
129 Cortesão, *Alexandre de Gusmão e o Tratado de Madrid*, vol. 3, pp. 252–54.
130 ANTT, MNE, livro 1, maço 1, f. 54, October 27, 1727.
131 ANTT, MNE, livro 793, f. 142, March 27, 1724.
132 ANTT, MNE, livro 793, f. 142, May 11, 1724.
133 ANTT, MNE, livro 793, f. 319, June 26, 1724.
134 Mandroux-França, "La Collection Royale Portugaise," pp. 64–65.

was to copy them out, including the designs for the observatory's mathematical instruments.[135]

While the person who introduced them and the date they met remain unknown, we do know that, at some point in 1724, D. Luis da Cunha recruited d'Anville into D. João v's service, first in charge of repairing the damage the *Atlas* suffered in the shipwreck[80] and of having him aid the other *savants* in the tasks they were already carrying out for the king, but gradually giving him a more important function. We can picture the damaged *Atlas* from the *Atlas, ou recueil de cartes géographiques, Collection de cartes géographiques en trois volumes que Son Excellence l'Ambassadeur a fait choisir pour son usage par le Sieur d'Anville* (Fig. 14),[136] an "*Atlas* which [the geographer] made in three volumes for the ambassador."[137] The first volume, entitled *World*, includes general maps, Europe and its principal divisions, details of Asia, Africa, and America, and winds up with cartographical maps of ancient history. The second encompasses France, including a general map, historical maps, plans of the provinces, rivers, and an extensive cartography of each of the country's dioceses. The third volume, meanwhile, includes Germany, Italy, Spain, and the British Isles, with general and regional maps of each country. None of these maps, however, was d'Anville's; at the time his cartographical production was incipient, practically limited to representations of the geography of Antiquity, Gaul and Greece in particular, without further forays into modern geography. But he was a young mapmaker and could be employed as a specialist or a map compiler. The maps in the *Atlas* were, with rare exceptions,[138] mainly by French authors—almost all by Nicolas Sanson and Guillaume Delisle, some by Jean-Baptiste Nolin (*c.*1657–1708), the royal hydrographer, Bernard Jean Jaillot (1673–1739),[139] and Nicolas de Fer (1646–1720).[140] Portugal appears in just two maps, as a part of Spain, as was common in mapmaking until the early eighteenth century; this explained the nationalist Portuguese interest in an independent cartography. One map of the country is by Placide,[141] from 1700, and one of the source of the Tagus is by De Fer, with a few views of Lisbon.

---

135 FBN, M, I-14,04,017, October 29, 1725.

136 FBN, At, 015, 02, 001–003.

137 ANTT, MNE, caixa 1, maço 3, f. 42, August 14, 1729.

138 There are a few maps by Wischer (among others, of Vienna, Utrecht, the Rhine, and Scotland), one of Bavaria by Homann, one of Italy by Rossi, and one of Auvergne by Vanlochan. There are also prints by Mariette of Japan, French parishes, and one of French dioceses.

139 The Jaillots were a family with several generations of geographers. In this case, it was Bernard Jean Hyacinthe Jaillot.

140 Nicolas de Fer, son of a map dealer, made over 600 maps, his most famous work being *l'Atlas curieux où le monde représenté dans les cartes générales et particulières du ciel et de la terre.*

141 Placide de Sainte Hélène le père.

FIGURE 14 Front page of d'Anville, *Atlas, ou recueil de cartes géographiques, collection de cartes géographiques en trois volumes que son excellence l'ambassadeur a fait choisir pour son usage par le Sieur d'Anville, géographe du roy* (Paris, printed, 17??). FBN, AT 015,02,001/003
REPRODUCED COURTESY OF FBN

What d'Anville did was compile the best remaining maps and have them bound by Padeloupe *le jeune* at his workshop on the Place de la Sorbonne. "Antoine Michel Padeloupe (1685–1758), commonly called Padeloupe *le jeune*, who succeeded Luc Antoine Boyet (1684–1733) as 'the king's bookbinder', was of a family with five successive generations of papermakers and bookbinders."[142] It is not unlikely that d'Anville himself attested to the quality of Vayrac's *Atlas* for D. Luis da Cunha. As a matter of fact, some years later, Marco António de Azevedo Coutinho was to make reference to his competence as an expert. When the latter wrote to Francisco Mendes de Góis in Paris, ordering an atlas for Cardinal da Mota, he advised him to seek out d'Anville, "who, as he understands the matter, knows how to choose the most accurate maps."[143] His *Atlas, ou Recueil de cartes géographiques*, given the quality of the maps he had selected, was a valuable addition to the Royal Library and gave his name greater renown among Portuguese scholars.

It was not possible to pinpoint the month in which he was hired for his next task that was to copy the plans of the royal observatory provided by Cassini, but in February of 1725 the ambassador reported back to Portugal that "d'Anville is working on the project of the mathematical instruments,"[144] from which one can infer that he had been hired some time ago. A few months earlier, in July of 1724, D. Luis informed the king that, as for the prints, the "two books containing the state of *gens d'armerie*, carabiniers, and light cavalry are finished and Mr. Hermand has showed us them."[145] The ambassador also began purchasing a collection of prints at Chez Mariette, with cartography being a related subject. And so, at the time, "monsieur d'Anville and monsieur d'Hermand were hired to continue the collection (of prints), with good taste and order in its binding, because the two have great knowledge of both."[146] For his service at this phase of print collecting and in the restoration of the *Atlas*, d'Anville "requested 20 pistols," and D. Luis—though he had hired him without an express order from the king—"hop(ed) that Your Majesty approves that they be given to him"[147] in order to ensure the quality of his work.

This partnership developed in the circles of the Duke of Orleans, on which England seems to have had a direct impact and was a mixture of science and

142 Herbert P. Horne, *The Binding of Books: An Essay in the History of Gold-Tooled Bindings* (New York: The Grolier Club, 1884), p. 44.

143 ANTT, MNE, caixa 1, maço 3, f. 42, August 14, 1729.

144 FBN, M, I-14,04,017, f. 44, February 3, 1725.

145 ANTT, MNE, livro 793, f. 357, July 10, 1724.

146 FBN, M, I-14,04,017, f. 44, February 3, 1725.

147 "Mr. D'Anville, who at my orders continues with the collection of prints and is remaking the atlas which got wet." ANTT, MNE, caixa 789, f. 23v, January 25, 1726.

politics. The most suggestive reference to d'Anville seems to have been written almost haphazardly in the margins of one of Dom Luis's papers some years later. In 1736, in The Hague, D. Luis was preparing to return as ambassador to the Portuguese embassy in Paris. At the same time, he struggled with England's hesitation to support Portugal in its military efforts to combat the Spanish, who had laid siege to the Colônia do Sacramento. The English wavered and secretly negotiated with France, which in turn supported Spain's interests, but were at the same time willing to mediate in favor of Portugal's interests. D. Luis worked toward a Franco-Portuguese alliance to replace the traditional union with England, a dream he had held since the 1720s. Over time, his experience as ambassador had led him to nurture a pathological mistrust of the English,[88] who he believed looked only after their own interests, since "this court is another species of chameleon, which not only adopts different colors, but different shapes as well."[89] In an intriguing passage of his correspondence, he refers to the fact that d'Anville's friends in France, in other words, Orleans's circle, were also his friends, and he also mentions the connection between that circle and England, which he tried to use in favor of Portugal's interests, and even claimed that d'Anville was directly paid by the English crown for services rendered. He wrote that:

> M. d'Anville, who also calls himself a minister of England, for he is paid for that, tells his friends, who are also mine, that his court [France] cannot do otherwise, because its interests in Spain make it neither willing nor able to turn its back on it.[90]

With their common interests and moving in the same social circles, the ambassador and d'Anville would strike up a long and profitable partnership. In him, D. Luis da Cunha found a new geographer keen for recognition; as he had no connections to the traditional families of French geographers, D. João v's patronage would suit his ambitions to upward mobility. The geographer, meanwhile, found in the Portuguese king an illustrious patron, magnanimous in his offers, and a gateway to the latest geographic information out of the vast Portuguese territories, essential for the construction of his armchair geography. References to this partnership appear intermittently in the papers of both men, although much documentation has been lost. When the documents in the Paris embassy were catalogued after the ambassador's death, by his nephew, a note was made of a bundle of 20 letters from d'Anville, unfortunately lost to us.[148]

---

148 ANTT, MNE, caixa 564, doc. 356, 1756.

As time passed, D. Luis drew d'Anville into all of the geographical, mathematical, and artistic projects he'd been tasked with in Paris. For example, when in November 1725 D. Luis received "copies of the observations made of the eclipse" of the moon in Lisbon in November of the past year, he sent the geographer to "make them seen at the *Académie Royale des Sciences de Paris*."[149] This was the first astronomical observation from the Italian Priest Mathematicians Carbone and Capassi, with an eye to establishing the Meridian of Lisbon and apparently already using new instruments from France, given that the report coincides with the installation of the observatory at the Paço da Ribeira.[150] As d'Anville was not a member of the *Académie Royale des Sciences de Paris*, he could not himself read the memoir. And so he reported to D. Luis, "the observation of the eclipse of the moon that you sent was read in the *Académie des Sciences* by the director and all praised it, but praised even more highly His Majesty's effort in protecting these sciences."[151] The discovery of the most precise methods for measuring longitude was, at the time, a question "attended to by all the greatest sovereigns of Europe,"[152] not least by D. João V, who demonstrated his enthusiastic patronage to the Parisian academicians and even brought some of them into his service. In the December 9 session, Cassini, the director of the academy, read the observations on the lunar eclipse from Lisbon, alongside those from Maraldi and Delisle, who had seen it from Paris. The records of the *Académie*, however, when regarding the events of that day, do not highlight the observations from Lisbon. Besides the reports from Paris, the acts only mention observations made in Lima.[153]

D. Luis da Cunha was also ordered to seek out the memoirs of the observations "made from the [Paris] observatory, as well as those made of the innermost satellite of Jupiter from the same observatory, from the past year as well as this one,"[154] to compare them to those being carried out in Portugal.[155] Via d'Anville, the ambassador once again got the requested memoirs and sent

149 ANTT, MNE, livro 793, f. 603, December 10, 1724.

150 Almeida, *A formação do espaço brasileiro*, p. 94; Rômulo Carvalho, *A Astronomia em Portugal no Século XVIII* (Lisbon: ICLP, 1985), pp. 37–53.

151 ANTT, MNE, livro 793, f. 603, December 10, 1724.

152 AASP, PV (April 15, 1722).

153 AASP, PV (December 9, 1724).

154 ANTT, MNE, livro 793, f. 583, December 4, 1724; f. 603, Paris, December 10, 1724.

155 "Capassi's work, entitled *Lusitania astronomice ilustrata*, was never printed, and collects the astronomical observations made in Coimbra, Porto, and Braga between 1726 and 1727." Almeida, *A formação do espaço brasileiro*, pp. 97–98.

them to Lisbon. The reports provided by Cassini[156] reached Portugal in 1725.[157] From there the ambassador was informed that:

> the observation of the eclipse of the moon which Your Excellency sent, made in that observatory by Mr. Maraldi, and compared by the same to that which was sent from here, and was received with great applause, both for being made with the accustomed exactitude and accuracy and for coinciding in many points with the report sent to Your Excellency.[158]

The establishment of the Meridian in Lisbon was crucial for beginning to map the Portuguese empire, thus, comparing the measurements made in Paris. Establishing the Meridian of Lisbon and all the maps to be made from then on was a state matter; this being the case, the Portuguese Secretary of State recommended to D. Luis that "from now on, [observations] will be communicated with all possible diligence and caution and in accordance with His Majesty's desire that the news of the difference might be confirmed and perfected."[159]

On February 3, 1725, the ambassador sent along the plans for the observatory at the *Académie Royale des Sciences de Paris*,[160] with the designs of its main instruments, and reported on the production of the rest of the necessary instruments currently being constructed by the principal local craftsmen to equip the Portuguese facility.[161] D'Anville aided him in all of this. He copied the plans for the observatory, supervised the design and fabrication of the instruments, and even helped in organizing the prints.[162] Alongside Hermand, he helped to organize the 20 volumes of the collection, all acquired from Chez Mariette.

In 1725, D. Luis da Cunha would see diplomatic relations between France and Portugal deteriorate after a diplomatic incident around the presentation of the French ambassador at the court in Lisbon. Worried and sleepless, he set about distracting himself by sketching a plan for an overland link between Angola and Mozambique, two Portuguese colonies, situated on the east and west coasts of

156 ANTT, MNE, livro 793, f. 603, December 10, 1724.
157 ANTT, MNE, livro 793, f. 624, December 25, 1724.
158 ANTT, MNE, LL, livro 14, f. 158, January 17, 1725.
159 ANTT, MNE, LL, livro 14, f. 158, January 17, 1725.
160 FBN, M, I-14,04,017, pp. 43–44, February 3, 1725; Delaforce, *Art and Patronage in Eighteenth-Century Portugal*, p. 86.
161 ANTT, MNE, LL, livro 14, f. 115, July 19, 1724.
162 FBN, M, I-14,04,017, f. 88, March 13, 1725.

Africa, respectively. D'Anville was his main collaborator in this enterprise.[163] Entitled *Projeto de um caminho para ligar estabelecimentos Portugueses da África* (Project to Establish a Way between the Portuguese Establishments in Africa), it contained a map and memoir, entitled *Description géographique de la partie de l'Afrique, qui est au sud de la ligne equinoxiale*, both by the geographer.[164] This was the map which on September 1, 1725, d'Anville presented to the *Académie Royale des Sciences de Paris* with D. João V's patronage.[165] The map has disappeared, but there are four manuscript fragments used to produce the original.[166]

Upon leaving Paris with precipitate haste before the end of 1725 due to a break in diplomatic relations between the two countries, D. Luis entrusted d'Anville with the task of carrying on his role in the supervision of purchases and services related to the Royal Library,[167] such as the production of indexes for the royal collection, made by the Mariettes, father and son; the geographer was meant to guarantee the project's "perfection and promptness."[168] From Brussels the ambassador followed d'Anville's work, though giving him considerable freedom in the job and supporting his decisions.[169] In late 1725, certainly having seen that the artisans hired to make the instruments were not up to the task, d'Anville decided to "hire Bion for the commission of some astronomical instruments," and D. Luis asked Portugal to approve his decision. In January 1726, D. Luis sent "the letter which d'Anville wrote me, in which you will see the reason behind" hiring Bion.[170] We do not know which justifications the geographer gave, but the confidence D. Luis puts in him is clear. In this sense, he wrote to Portugal that, with d'Anville supervising the royal purchases in Paris, "they will not lack for my presence."[171]

---

163 Furtado, "Evolving Ideas: J. B. D'Anville's Maps of Southern Africa, 1725/1749," pp. 202–15; Furtado, *Quebra-Cabeça africano*.

164 SGL, R, 3-C-17; Avelino Teixeira Mota, "Dom Luis da Cunha e a Carta da África Meridional de Bourguignon d'Anville (1725)," *Separata da revista portuguesa de história*, 10 (1962). The map is lost.

165 AARSP, PV (September 1, 1725).

166 Furtado, *Quebra-Cabeça africano*, pp. 149–78.

167 Silva, "D. João V patrono do astrónomo Bianchini," p. 54.

168 FBN. M, I-14,04,017. Paris Correspondence from 1725, Brussels, October 29, 1725.

169 ANTT, MNE, caixa 789, f. 41, March 8, 1726.

170 ANTT, MNE, caixa 789, f. 9, January 11, 1726.

171 FBN. M, I-14,04,017, f. 44, February 3, 1725.

5 The Tordesillas Meridian[172]

On May 26, 1724, D. Luis da Cunha received orders from Portugal to give his opinion "in view of the letters from the ambassador from Castille, the response which was given, and the role you played in the affair" of the dispute over Colônia do Sacramento.[173] Diogo de Mendonça Corte Real, Secretary of State, sent him a memoir that attempted to establish a historical record of all negotiations carried out between both parties, taking the dispute over that territory to the field of diplomacy and away from geography, where Portugal's position was weakened because of Delisle's *Dissertation*.[174] He maintained "that the Treaty of Utrecht had resolved the problem, this agreement substituting the difficult and controversial determination of the Line of Tordesillas. It was not, therefore, for the geographers to decide the matter."[175] The text was accompanied by a map of the region of the Rio de la Plata with the Portuguese and Spanish proposal for the Tordesillas Meridian. The latter situated Spain farther east, diminishing Lusitanian possessions (Fig. 13).

After reading it, the ambassador made a long and extensive analysis of the situation in about ten pages and tended to agree with the secretary that the geographers oughtn't to be worn out, but rather excused, and that the issue should be resolved exclusively via diplomatic agreements.[176] In fact, although he still considered cartography an important resource to guide negotiations, this defense of abandoning geography as a diplomatic tool represented a temporary and brief strategic retreat, motivated mainly by the lack of good Portuguese maps to contradict Spanish claims on the regions in question. As a matter of fact, D. Luis da Cunha would continue using cartography as a fundamental expedient in negotiations; and d'Anville, who continued to gain the ambassador's trust throughout 1724–1725, was the one who provided him with suggestions as to how to geographically guide diplomatic negotiations with Spain including and outside the context of dispute of the Colônia do Sacramento. He tried to inform himself as to the geographical positioning of the Tordesillas Meridian and its implications for the conquered Portuguese territories and demanded that d'Anville draw up not one but two specific studies on the subject of Brazil's frontier lines,[177] both being a direct byproduct of

172 Furtado, "Embates na terra de papel."
173 ANTT, MNE, livro 793, f. 284, June 18, 1724.
174 Cortesão, *Alexandre de Gusmão e o Tratado de Madrid*, vol. 3, pp. 230–43.
175 Almeida, *A formação do espaço brasileiro*, p. 69.
176 ANTT, MNE, livro 793, ff. 290–91, June 18, 1724.
177 NL, VB, Ayer MS 1918.

Delisle's *Dissertation*, of which he undoubtedly had his own copy.[178] There is no doubt that these studies were requested by D. Luis da Cunha as in 1747 Marco António de Azevedo Coutinho confirmed that, at the ambassador's request, d'Anville made "two long discourses on the dividing line."[179] Later on d'Anville reaffirmed in a 1776 letter to Charles Gravier, Count of Vergennes (1719–1787) and then the French Minister of Foreign Affairs, that "it is true, [that] some years ago I had completed a project at the request of D. Luis da Cunha, Portuguese ambassador, on the frontiers of Brazil, contested between the two crowns of Spain and Portugal."[180] Also in his *Second mémoire concernant l'Amérique méridionale*, dated 1779, he refers to being called upon to "deal with the matter."[181]

The *Discussion géographique sur la ligne de démarcation etablier pour fixer des limites entre les découvertes du Portugal et celles de la Castille*[182] is a manuscript in d'Anville's own hand, in ten folios—and, to all appearances, is a draft for the work commissioned by the ambassador. It is written on small sheets of paper, at times reused, with many cuts and corrections. Here he discusses the existing opinions regarding the actual position of the Tordesillas Line, indicating the best tack for Portugal to take in the negotiations with Spain. Another related document is the *Ligne de demarcation*,[183] a much shorter one, with only three sheets, also in his own handwriting, and although there is no addressee it is signed by d'Anville. It has a more finished aspect and seems to be the draft of a letter, with fewer corrections and addendums and much more summarized than the previous one. Both of them are from the point of view of the Portuguese. The first one discusses the arguments, suggesting the best strategy to be taken by Portugal, from which we can infer that it was directly addressed to Dom Luis da Cunha to send to Portugal.

On both documents he not only reveals that he gave his opinion on the subject, but he also reveals that he had access to official documents, both Spanish and Portuguese, among them Pope Alexander VI's bull (1493), the Treaty of Tordesillas (1494), and the Zaragoza Treaty (1529) on which he based his report. Having access to that documentation, he was able to make a correct distinction between the papal bull and the Treaty of Tordesillas, unlike Delisle, who confused both in his *Dissertation*.

178 ADO, SHAO, Delisle, *Détermination géographique de la situation*.
179 ANTT, MNE, livro 825, ff. 135v–36.
180 FBN, M, 30, 1.005 n.014, August 28, 1776.
181 RBC, 229, n. 539 (2).
182 NL, VB, Ayer MS 1918, doc. 552.
183 NL, VB, Ayer MS 1918, doc. 552.

His goal was to "lay out those geographic means in favor and those against, in the differences between" the two courts.[184] He insisted on geography's primacy as the compass for these considerations, highlighting in his conclusion "that geography guides me regarding the demarcation. This is not a matter of human will. Intention and skill can bear no weight on the circumstances being discussed."[185] This reflection, besides situating itself in the wake of Delisle's *Dissertation* on longitude, was based on the recent meridians' measurements taken by geographers from various places across the Globe, which constituted a new ingredient in the territorial disputes between the Portuguese and the Spanish. Jaime Cortesão labeled this wave a "culture of longitude."[186] For the Portuguese, the geographic fixing of the Tordesillas Line would allow them to know if the positioning of Portuguese lands would support their claim on the Colônia do Sacramento and adjacent territories; and it would allow them to debate whether or not the division was still valid, coinciding or not as it did with the dominion that both crowns had actually colonized over the past centuries. The document opens with a history of the discoveries and the first negotiations between Portugal and Spain, singling out and describing the terms of Alexander VI's (1431–1503) papal bull and the Treaty of Tordesillas, both issued in the context of the maritime expansion of the fifteenth century.

D'Anville underlines the fact that the treaty had been vague; it did not define which specific island in the Cape Verde Archipelago should be used to establish the measurement, complicating the debate over the placement of the Meridian even further. The treaty simply made reference to "370 leagues from Portugal's islands."[187] It would be in the country's interest, according to the geographer, to take the measurement from the most westerly island in the Cape Verde Archipelago, Santo Antônio; the Spanish, meanwhile, insisted on using the eastern island, São Nicolau, to their advantage. He also argues that a crucial point would be Portugal's ability to claim the lands around the Colônia do Sacramento as contiguous to Brazil (that defense of Portuguese interests could explain this same configuration on his 1733 *Paraguay* map and not a lack of documents as he excuses himself). He then embarks on a long geographic discussion, using the works of several authors and navigators, in order to fix the location of the islands of Cape Verde, Santo Antônio in particular. The next step, he suggests, would be fixing the 370 leagues. This said, d'Anville—wedded to his method—reflects on measurements; as leagues, like all other measures,

184 NL, VB, Ayer MS 1918, doc. 552, f. A1.
185 NL, VB, Ayer MS 1918, doc. 552, f. 3.
186 Cortesão, *Alexandre de Gusmão e o Tratado de Madrid*, vol. 2, p. 349.
187 NL, VB, Ayer MS 1918, doc. 552, f. B2.

"varied incessantly throughout centuries and across nations."[188] He concludes that the most convenient measure would be the Spanish league used in the treaty, equivalent to 19 French *lieues*. Taking the circumference of the Earth, this would correspond to 17.5 *lieues* per meridian degree.[189] From a clearly Portuguese-friendly point of view, he indicates this estimate as the most favorable one for the crown.

He then establishes the position of the Tordesillas Line over American territory—which, he argues, first requires a discussion fixing the position of the Brazilian coast. After using the maps of Portugal's chief cartographer, Manuel Pimentel, the astronomical observations from George Marggraf (1610–1644) in the Brazilian northeast during the Dutch invasion, the map from Portuguese brigadier José da Silva Paes, and Cassini's maps, he concludes that the coast would be located at 36° and 20′ and the line positioned at 29.5° of longitude.[190] That is to say, the Tordesillas Meridian would extend from the mouth of the Amazonas River near the Cayté River to the Cape of Santa Maria, near the mouth of the Rio de la Plata.[191]

D'Anville then moves to examine the Line from the other side of the Globe, near the South Sea (Pacific Ocean). One of his major points, which D. Luis would repeat incessantly, was to extend the discussion beyond the Americas to the Pacific, where the Moluccas were also in dispute. D'Anville warns that, given the Spanish advantage "in Meridional America, the Portuguese had equal reason to claim the Philippines."[192] In this regard he guarantees that "all the Philippines, including a part of the sea separating them from the Marianas, are also in the Portuguese part."[193] This was a great novelty, since the Portuguese had purchased the Moluccas in the Treaty of Zaragoza. The geographer was certain: not only did the Moluccas belong geographically to Portugal, making their purchase by the King D. João III (1521–1557) from Carlos V (1500–1558) unnecessary, but all the Philippines did as well, though by the terms of Zaragoza they had remained under Spanish rule.[194]

From then on, D. Luis da Cunha would generally lean on these points raised by d'Anville as the negotiations went forward: the Treaty of Tordesillas no longer applied to the region around Colônia do Sacramento; the Portuguese claim on its territory on the north banks of Rio de la Plata was guaranteed by the

188 Dacier, "Éloge de M. D'Anville," p. 25; NL, VB, Ayer MS 1918, doc. 552, f. C.2.
189 NL, VB, Ayer MS 1918, doc. 552, f. C2.
190 NL, VB, Ayer MS 1918, doc. 552, f. D3.
191 NL, VB, Ayer MS 1918, doc. 552.
192 NL, VB, Ayer MS 1918, doc. 552, f. H7.
193 NL, VB, Ayer MS 1918, doc. 552, f. J8.
194 NL, VB, Ayer MS 1918, doc. 552, f. J8.

Treaty of Utrecht, even though the area was outside the Portuguese territory established at Tordesillas; and the Moluccas were clearly Portuguese possessions. In that case, the negotiations ought to deal with all these questions, not just the territory around the Colônia do Sacramento; but the Moluccas would be a bargaining chip, warning the Spanish that they could not claim gains on both sides of the Globe. Examining the pros and cons laid out by d'Anville, D. Luis said, Portugal could use both geography and diplomacy to guarantee their possession of the Colônia do Sacramento, demonstrating the illegality of the nation's payment in exchange for the Moluccas.[195]

One last point must be considered. In his discussion of the positioning of the Tordesillas Line based on the Cape Verde Archipelago, d'Anville makes note of something that, as he put it, could be of interest to Portugal. He recommends that the distance of 370 leagues be measured not from the Equator but "on the parallel of 17 degrees of latitude," where the archipelago lay. Why was this? According to the geographer, if one assumed that the Earth was not a perfect sphere, longitude intervals would be narrower at this point, shifting the Tordesillas Meridian a few miles farther to the west. So, "this debate cannot ignore scholars' new system for considering the shape of the Earth," he warned.[196] Here we see that he was agreeing with the Cartesian mechanical theory of Earth being not a perfect sphere, although as discussed, he did not agree with the Newtonian theory of an oblate (flattened) spheroid, but a prolate spheroid along the Equator and took advantage of it in favor of the Portuguese.[197]

Despite what D. Luis da Cunha had argued up until this point, then, d'Anville's reflections suggested that border negotiations in the Americas could not be based purely on previous treaties, but also on modern geography. The partnership of the two starting in the 1720s, culminating in the production of the *Amérique méridionale*, bears witness to the ambassador's recognition of the intrinsic, crucial relationship between diplomacy and cartography. He would increasingly devote his attention to the latter, contributing directly to the production of maps of Brazil meant to intervene in future Spanish and Portuguese negotiations over territorial frontiers in South America.

195 NL, VB, Ayer MS 1918, doc. 552, f. G6.

196 NL, VB, Ayer MS 1918, doc. 552, ff. C2–D3.

197 Anville, *Proposition d'une mesure de la terre.*

CHAPTER 4

The Portuguese Imperium in D. Luis da Cunha's Geopolitical View and the Formulation of *Uti possidetis*

1 Minas Gerais Captaincy or the Old Lady's Goods

On the morning of November 29, 1807, with favorable winds, an unusual fleet left Lisbon and sailed down the Tagus toward the sea. On board, D. João, Prince Regent of Portugal (future D. João VI, 1767–1826), left the country and transferred the crown, the royal family, and a good part of the Portuguese court to Brazil.[1] It seemed to be last-minute decision, but it was not quite so. Throughout the two previous centuries, some Portuguese had suggested, for various reasons, that the seat of the Portuguese empire should be transferred to Brazil as a solution to the increasing importance that the latter was acquiring within the empire, and one of them was D. Luis da Cunha, who with greatest pragmatism, clear-sightedness, and radicalism on the matter, advocated that "Brazil (being a precious domain is worthy of greater care)" should host the head of the empire.[2]

D. Luis da Cunha played an especially important role in reorienting Portuguese geopolitics during the eighteenth century, with the aim of rearticulating the role of the various parts of the empire, privileging Portuguese America over other areas and even Portugal itself. The renowned ambassador was one of the enlightened Portuguese administrative elite who most clearly identified the process of displacement of the empire's economic axis from Asia to Brazil with the recent discoveries of the gold and diamond mines,[3] as well as the central role these acquired for Portugal's future, in the context of Portugal's increasing dependence on other European powers, especially England. This view was in accordance with his opinion on the displacement of Portugal's political center to Brazil, but to do so it was necessary to consolidate its territory in opposition to Spain's interests. To this end, according to him geographical knowledge should be improved and expressed in a more perfected

1 Patrick Wilken, *O império a deriva: a corte portuguesa no Rio de Janeiro* (Rio de Janeiro: Objetiva, 2005), p. 43.
2 ANTT, MNE, Livro 787, f. 75, April 15, 1713.
3 Furtado, "Dom João V e a década de 1720," pp. 61–110.

 | DOI:10.1163/9789004710764_006

cartography that represented especially the regions considered more sensitive and strategic to the crown.

This chapter will examine the development of D. Luis da Cunha's geopolitical thought system on the spatial configuration of Brazil and how this view is expressed in d'Anville's *Amérique méridionale*, focusing on four geographical spaces: Minas Gerais and Maranhão and captaincies, Colônia do Sacramento, and Fernando de Noronha Island. Minas Gerais became central in his reflections as he increasingly attributed the sustenance of Portugal's economy to the gold and diamond wealth originating there.[4] Colônia do Sacramento and Maranhão and Grão-Pará were the object of a number of diplomatic negotiations with Spain and France, in which D. Luis participated since the Congress of Utrecht, and Fernando de Noronha Island became a matter of concern for him following a French invasion in 1736. While regarding Maranhão and Grão-Pará,[5] the ambassador considered it important to ensure Portugal's monopoly of navigation on the Amazonas River, making sure to displace the border with French Guyana increasingly to the north, and Colônia do Sacramento was seen by him as the great bargaining chip to be negotiated with Spain in exchange for other disputed territories. The French invasion of Fernando de Noronha was a turning point in his thoughts about diplomatic arguments of territorial possession. All these issues would find themselves represented in d'Anville's map, which especially in these spaces mirrors the geopolitical project formulated by D. Luis da Cunha throughout his life for Portuguese America. Finally, it will examine how the ambassador defended the way in which negotiations of these territorial limits or possessions should be carried out, which led him to develop the notion of *Uti possidetis*, that is, the territory should be possessed by whoever colonized it and the one who discovered or first conquered it, which effectively became the main line of diplomatic negotiation used by the Portuguese with the Spanish after the Treaty of Madrid in 1750—a great diplomatic innovation in the negotiation of frontiers.[6]

Early in the 1700s D. Luis da Cunha called attention to a reorientation not only of the Portuguese empire's economic, but also political position. Throughout his correspondence, and in particular in his renowned *Instruções políticas*, the ambassador quite clearly stressed the importance of the discovery of gold and

4 Furtado, "O oráculo que Sua Majestade foi buscar"; Furtado, "Dom João V e a década de 1720."

5 This administrative unit was created in 1621 during the Iberian Union (União Ibérica, 1580–1640) by Filipe II, of Portugal (III of Spain, 1578–1621), initially under the name of the State of *Maranhão*, being separated from the State of Brazil, which included the captaincies of the east coast of the Atlantic Ocean. In 1654, it was renamed the State of Maranhão and Grão-Pará, and later, in 1751, the State of Grão-Pará and Maranhão.

6 "Portuguese America under Foreign Threat," pp. 45–74.

diamonds in the captaincy of Minas Gerais, highlighting Portugal's increasing economic dependence on Brazil.[7] As a corollary of this dependence, he advocated that "the prince, in order to maintain Portugal, is completely dependent on Brazil's wealth, and not at all on Portugal's (...) from which one might conclude that it is more comfortable, and safer, to be where one has what abounds, [than] where one desires what is lacking." That is why he defended the need to transfer the Portuguese court to the Americas, establishing it in Rio de Janeiro city from where the king would govern, assuming the title of "emperor of the Occident."[8]

In the context of an extreme rivalry with Spain for American territories, D. Luis advised D. João v and argued that if "some King of Portugal wished to be emperor of the Occident and have his court in Rio de Janeiro, it is certain that the King of Castile would fear that he could conquer the kingdom of Peru up to the isthmus of Panama" and therefore took care to warn his reader: "I know well that when your lordship reads this idea of mine, you will cross yourself, thinking it a mere and pure vision." Like no other writings of his, the *Instruções políticas* synthesized his thought system, and D. Luis hoped that they would guide Portuguese policy of the time according to the principles that were dear to him: diminishing England's economic and political influence; restructuring Portugal's position in Europe and the Americas; combating the Jesuits' power and asserting religious tolerance; and finally that which is the subject of this part of the book: recognizing the importance of Brazil and its gold production for enriching the Portuguese metropolis, foreseeing the eventual transfer of the court to the Americas.[9]

This idea of the centrality of Brazil develops in the ambassador's mind as he gradually becomes aware of the growing importance of the products originating from that overseas conquest in the empire's economy, overshadowing the Far East. Sugar and tobacco, among other products, are seen from the standpoint of geopolitical strategies as trade currency with the other European economies. But among many other goods, gold takes on special importance, and because of it the ambassador begins attributing an immense importance to the *Minas Gerais* region as the location of that wealth. In 1719 he wrote to his old friend, the 2nd Count of Assumar, João de Almeida, and in his irreverent style declared that "that Brazil [and] those Minas are the old lady's goods."[10]

---

7 Cunha, *Instruções políticas*, pp. 279, 368–69.
8 Cunha, *Instruções políticas*, p. 371.
9 Cunha, *Instruções políticas*, pp. 369–73.
10 AHI, DEL, 343-1-9, f. 41, May 1719.

At first, the ambassador, like other members of the Portuguese elite, was skeptical of the discoveries' importance and duration, but he soon concluded that "the conquests, which I had believed to be an accessory of Portugal, I now consider to be its main asset, and even guarantors of its conservation, especially those of Brazil."[11] In 1720, confirming his most hopeful expectations, he exultantly states that "the fleet from Rio (…) comes laden with gold sent by the captaincy [and] totals six million and a half, twenty three thousand *cruzados* and nine hundred and twenty *réis*,"[12] perceiving the importance of that wealth to enable Portugal to recover its power and prestige among European nations.[13]

He saw in the wealth in the form of gold flowing from Brazil to Portugal the possibility to make up for Portugal's small geographical size and its meager economic resources. For that reason, according to his principles, that gold should not continue to fill the coffers of other European nations, especially England. He deplored the fact that Portugal had gone to "dig the mines so that foreigners could reap their precious production."[14] It was during his sojourn in England, from 1697 to 1709, that this idea started developing, upon perceiving the growing imbalance in Portugal's trade, aggravated by its involvement in the War of the Spanish Succession. He attentively observes that the way in which trade was carried out between those two nations created a continuous flow of Brazilian gold leaving Portugal, "since far more goods enter this kingdom [of Portugal] than products are extracted from it, the difference must constantly leave in cash."[15] The metal, transformed into coins, enriched England's coffers. After successively advising that trade between Portugal and England was becoming increasingly deficient year after year, he painfully recognized that his continuous warnings had done nothing to reverse this process and concluded that "we will always be dependent on England, which considers Portugal its best colony, since it gives it the gold and diamonds that its own colonies do not produce."[16]

The awareness of the growing importance of Portuguese gold circulating in Europe, which increasingly fueled the development of the economies of England, France, and the Netherlands, made the ambassador recognize the importance of Brazilian mines in this process. He became increasingly aware of the interest it aroused in Europe and the threat that interest represented

11 Cunha, "Carta de instruções a Marco António de Azevedo Coutinho," p. 279.
12 ANTT, MNE, livro 137, July 26, 1720. *Cruzados* and *réis* are old Portuguese currencies.
13 ANTT, MNE, livro 790, ff. 375–76, October 20, 1721.
14 Cunha, "Carta de instruções a Marco António de Azevedo Coutinho," p. 279.
15 ANTT, MNE, livro 781, f. 302, September 27, 1707.
16 Cunha, "Carta de instruções a Marco António de Azevedo Coutinho," p. 372.

to Portugal's empire, "since those dominions' riches are today the envy of all nations." He continuously alerts Portugal's authorities to the urgent need to protect them from foreign greed, maintaining that "all protection of the Americas is necessary because the wealth of our Minas makes increasingly more noise and causes greater envy."[17]

In the letters written by D. Luis da Cunha to Portugal and to other Portuguese ambassadors in other European courts, the idea of the importance of the fleets from Brazil and the riches they carried to keep the kingdom's economy afloat is a recurrent theme, and he expresses it constantly in his correspondence. Even from afar, he was always up to date on their flow, especially that of gold, to which he attributed the greatest importance. He both received news and sent information about the arrivals and departures of fleets, demonstrating his constant interest in the matter.[18]

At the same time, there is a constant fear that pirate attacks may deprive Portugal of such wealth. He paid special attention to monitoring expeditions, both English and French and even Russian, which could head toward Brazil. When he was in England he approached local authorities to spy on French pirates; when in Paris, he spied on the English. We see him constantly pressuring the authorities, demanding that the nations where he acted on behalf of the treaties of friendship established with Portugal respect the integrity of its merchant ships and the territories of its conquests. In 1710 and 1711, when he was in England, he monitored the French privateers of Le Clerc (1710) and Du Guay Trouin (1711), which were on their way to Rio de Janeiro.[19]

On February 3, 1711, D. Luis received disquieting news "from Bordeaux [which] brought word that M. Le Clerc had departed from Rocheford with six warships in the direction of Rio de Janeiro with the intention of looting the city, for which [purpose] he landed with 750 men." However, he was also informed of the heroic resistance of the local population, who fought off the enemy, which left him

> with great joy at seeing the able resistance of our soldiers, but with equal sorrow that our enemies consider it such an easy enterprise, and I do not doubt that they would attempt it again with greater forces if they were not informed of what our lord had done to defend such an important post.

---

17 ANTT, MNE, livro 790, f. 270, September 8, 1721.

18 BNP, R, cód.11.209, f. 324v, November 18, 1712; ANTT, MNE, caixa 926, ff. 4v–5, January 11, 1714.

19 BNP, R, cód.11.209, ff. 205–05v, September 1, 1711; 248, February 16, 1712.

He rejoiced at the expedition's failure, for according to him "all the joy that the news from Rio de Janeiro gave us is justified, and these costly experiences by our enemies take away their desire to repeat such an expedition."[20] Little did they know that their predictions were wrong, for at that time Du Guay Trouin's expedition, which was more successful, was already being planned.

Potential pirate attacks on the shores of Brazil required constant attention. Even invasions in the far south, on the Santa Catarina coast, could threaten Portugal's hold on Brazilian mines.[21] In particular, the years 1710 and 1711 and the danger they posed for Portugal's hold on Brazil made a strong impression on him. He feared the possible combination of an external threat (French pirates) and an internal uprising, for he was informed that the subjects from Minas Gerais had revolted—this was in reference to the disturbances resulting from the War of the *Emboabas*, when *Paulistas* and Portuguese immigrants vied for control over the mining region. Once the threat had lifted, D. Luis wrote: "All the joy that the news from Rio de Janeiro gave us is justified (...). Likewise I am glad that the disputes among His Majesty's subjects in Minas have been resolved, so that such a lucrative work is not interrupted."[22]

In the 1710s Dom Luis da Cunha established a connection between Colônia do Sacramento and the possession of Minas, proposing an increasingly assertive policy of the settlement and defense of the south of Brazil. He had already given similar advice in 1712, when he was informed of an Englishmen named Blau who was the captain of an expedition which planned to invade the south of Brazil. To prevent this, he suggested that Portugal "should immediately send, without delay, orders to Brazil to dispatch a ship to the islands of Santa Catarina and dos Patos with people taking some cabins, which would suffice to prove that we have possession over them." Effective colonization would be the strongest counterpoint to legally neutralize possible English colonization, since "in these terms other powers will not refrain from helping us uphold an unquestionable right." He concluded:

> The same should be done in any other port on the coast of Brazil from the Amazonas River to the Rio de la Plat, because it is a very inexpensive precaution that prevents many issues that could be inflicted upon us by the avarice of others awakened by the wealth of our mines.[23]

20 ANTT, MNE, livro 789, maço 7, carta 6, February 3, 1711; maço 9, carta 11, March 24, 1711.

21 ANTT, MNE, caixa 926, f. 56, June 21, 1714.

22 ANTT, MNE, livro 785, f. 75, March 17, 1711.

23 ANTT, MNE, livro 786, f. 403, Paris, September 2, 1712.

Early in 1714, he proposed some solutions to protect the Brazilian mines that could sometimes be "of little cost [and] would free us from many issues that could result from the avarice of others that the wealth of our mines is awakening." He was referring to protecting the coast of Brazil, encouraging the settlement of strategic areas such as Santa Catarina Island and various ports located between the mouth of the Amazon and the River Plate. Even if at that time he still wondered "whether the wealth of our mines (…) is greater in reputation than in effect."[24]

The governor of Colônia do Sacramento between 1722 and 1749, António Pedro de Vasconcelos, who opposed rendering the Colônia do Sacramento to Spain, echoed the geopolitics of Dom Luis da Cunha. He stated that it would open to Spain the "door to what until now was the key to our Colônia" and allow the Castilians to reach "the road which runs directly to the *sertão* of that hidden country," whose roads were "necessary to the traffic of Minas Gerais." According to him, surrendering Colônia do Sacramento would hand over "the key to our Minas."[25]

But a few years later, in 1719, he learned of further troubles afflicting Minas Gerais captaincy through his correspondence with his old friend, the 2nd Count of Assumar, the father of the governor of Minas at that time, D. Pedro de Almeida (1688–1756, later 3rd Count of Assumar) who suppressed the uprising. D. Luis writes:

> It was a most pleasurable occupation (…) to read the letters by D. Pedro de Almeida, [and] to learn that he was in good health and that he had uncovered the conspiracy of which I am informed, and for all of that I send infinite congratulations.[26]

But despite the young governor's precautions, in the early 1720s the ambassador was again assailed by the same fears.[27] Internal dissatisfaction could become unmanageable if it coincided with an external invasion motivated by the foreign powers' lust for Brazilian gold. In October, when the danger had passed, D. Luis da Cunha reassured that, according to the news arriving from Portugal, "the state of Brazil is calm and abundant."[28] But the peace was only apparent, since according to new letters from Lisbon the 1720 uprising could

---

24 ANTT, MNE, caixa 926, f. 56, June 21, 1714.
25 Cortesão, *Alexandre de Gusmão e o Tratado de Madrid*, vol. 2, pp. 386–87.
26 AHI, DEL, 343-1-9, f. 41, May 1719.
27 ANTT, MNE, caixa 816, November 5, 1720.
28 ANTT, MNE, livro 137, October 22, 1720.

have confirmed D. Luis's fears of the convergence of internal and external dangers as still in the context of the suppression of the uprising, the ambassador investigated in Paris the possibility of communications between some Frenchmen and Minas rebels during the revolt. Those communications could have been encouraged by the French crown with the intention of getting hold of Minas Gerais. His conclusion was that it was possible that "some *malheureux* [wretch] had corresponded with any of the dissenters from Minas, without the [French] court having any part in it," and that "all vigilance of Europe is needful, and all security of the Americas is necessary because the wealth of our Minas makes increasingly more noise and causes greater envy."[29] D. Luis da Cunha as early as 1711 had warned Portugal's authorities of the dangers that such a combination of internal and external threats represented. For D. Luis the gold mines were the answer to the fragility of the Portuguese economy at the time and worth the risks. By late 1730, after the arrival of so many fleets loaded with gold and diamonds, he had no more doubts about the centrality and importance of the region for the enrichment of the empire, a vision that culminated in his controversial proposal to transfer the Portuguese court to Brazil.

This geopolitical approach based on the centrality of the Brazilian mines is expressed in the cartography produced by d'Anville on the mining region, where Portuguese colonization of those territories is evident in the representation, leaving no room for doubt over their possession by the crown, as D. Luis da Cunha advocated. In the 1748 *Amérique méridionale* or in the copy made by Jacques-Nicolas Bellin in 1764 of part of d'Anville's map, entitled *Suite du Brésil, depuis la Baye de tous les Saints jusqu'a St. Paul* (Fig. 15), we can see that, between parallels 10° and 25° south, the eastern part of the São Francisco River, for which the cartographer mostly used documents provided by the ambassador, appears as a geographical entity whose space is traversed by the Portuguese presence. It contrasts with the western part of the same river, where we see a cartographic void, of which the cartographer wrote in *Amérique méridionale*: "deserted or little known extension of the country." This way, in the region that roughly corresponds to the Minas Gerais, whose capital was Vila Rica, and its surroundings captaincies (Rio de Janeiro, southern Bahia, São Paulo, and Espírito Santo), an entire set of graphical information drawn on the map—hierarchically organized roads, towns, villages, and resting places—bears witness to Portuguese colonization and penetration, leaving no doubt of their incontestable possession of that territory.

29 ANTT, MNE, livro 790, ff. 268–70, September 8, 1721.

FIGURE 15 *Suite du Brésil, depuis la Baye de tous les Saints jusqu'a St. Paul*, Jacques-Nicolas Bellin (Paris, printed by J.N. Bellin, 1764). Furtado Lins Collection

It was Portuguese settlers who revealed the area's geography to Europeans, which allowed the geographer to draw the networks of rivers and mountains that ran through the region, positioning them correctly and naming them the same way as the Portuguese did, revealing their sovereignty over the area expressed through language.[30] Naming places is a way to demonstrate ownership. It is thanks to those settlers that an entire set of geographical documents was produced, part of which reached d'Anville through D. Luis da Cunha. In the west, the absence of colonists means this geography remain unknown, which prevents the geographer from including it in the map, even though it was not unknown to Indigenous peoples who were able to continue their constant movements throughout the territory. To the ambassador, colonizing a territory is one of the legitimate ways of ensuring its possession when it is disputed by more than one nation that is *Uti possidetis*. In this sense, d'Anville's *Amérique méridionale* revealed the Portuguese presence in the mining region as undisputable, since "for a long time now the gold and diamond mines that Your Majesty gloriously possesses are the subject matter not only of discourses but much more so of envy."[31]

2 The State of Maranhão and the Mill Mule[32]

The late seventeenth century saw the beginning of the conflicts between the French and the Portuguese centered on the Cabo do Norte region (Figs. 12 and 16), which went from the northern bank of the Amazonas River to French Guyana. The dispute was due to the foundation of the French West India Company in 1663, to drive the occupation of the Guyana region under the French flag. In 1688 the French, commanded by Pierre-Eléonore de la Ville, Marquis de Ferrolles (?–1705), threatened the Portuguese fort of Araguari. "But it was only in 1697 that they decided to attack the forts of Macapá and Parú, which were destroyed, initiating the Portuguese–French conflict for control of the region."[33]

---

30 Furtado and Neil Safier, "O sertão das Minas como espaço vivido: Luis da Cunha e Jean-Baptiste Bourguignon D'Anville na construção da cartografia européia sobre o Brasil," in *Brasil-Portugal: sociedades, culturas e formas de governar no mundo português (séculos XVI–XVIII)*, ed. Eduardo França Paiva (São Paulo: Editora Annablume, 2006), pp. 263–77.

31 AUC, CLC, doc. 408, July 5, 1737.

32 Furtado, "O Tratado de Utrecht e a construção da cartografia como instrumento da diplomacia portuguesa."

33 Almeida, *A formação do espaço brasileiro*, p. 38.

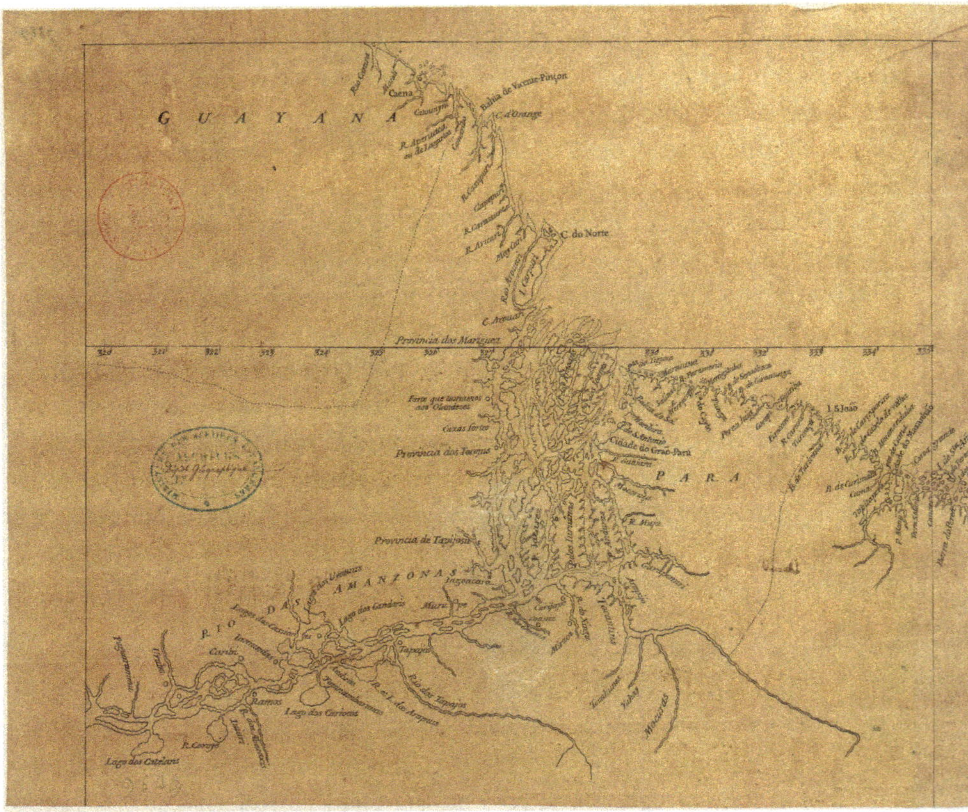

FIGURE 16 *Carte huilée des embouchures de la Rivière des Amazones et cotes voisines*, J.B.B. d'Anville (Paris by the author at the Louvre galleries, manuscript, 17??). BNF, DCP, GE DD-2987 (9549) REPRODUCED COURTESY OF FBN

In an attempt to resolve the conflict, in 1699 Louis XIV and D. Pedro II signed an agreement regarding the boundaries of the possessions of both crowns in the Amazonas region. The accord was negotiated by the 2nd Marquis of Cascais, Luis Álvaro de Castro (1644–1720), then ambassador extraordinary to the French court. In November, a Portuguese commission was appointed to determine the boundaries of Guyana and, on March 4, 1700 a Provisional Treaty was signed. In this treaty the Portuguese committed themselves to evacuating and demolishing the forts built north of the Amazonas River, such as the Araguari, Camaí, and especially Macapá (or Massapá). In turn, "France committed itself to provisionally refraining from creating any settlement on the northern bank of the river and to consider the lands situated between Macapá and the Oyapock (Oiapoque) or Vicente Pinzón River to be regulated in the

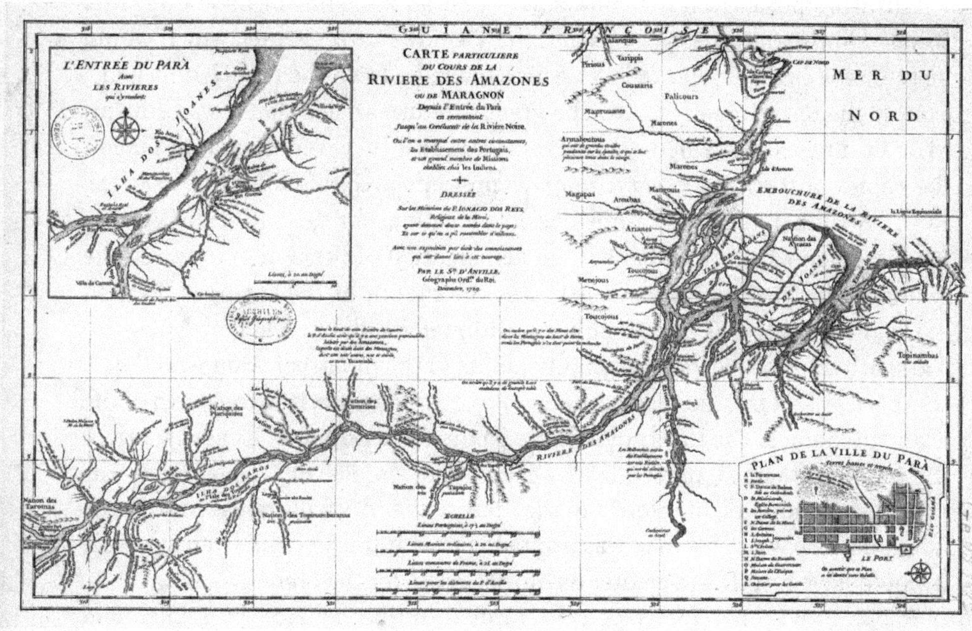

FIGURE 17 *Carte particulière du cours de la Rivière des Amazones ou de Maragnon, depuis l'entrée du Para en remontant jusqu'au confluent de la Rivière Noire où l'on a marqué entre autres circonstances, les établissements des Portugais, et un Grand nombre de missions établies chés les Indiens, dressée sur les mémoires du P. Ignacio dos Reys*, J.B.B. d'Anville (Paris, by the author at the Louvre galleries, printed, December 1729). BNF, DCP, GE DD-2987 (9550)
REPRODUCED COURTESY OF FBN

future."[34] The problem concerned the identification of the river that should regulate the border named the Vicente Pinzón River referring to the one discovered by this French navigator. The French insisted that it was the Araguari, farther south, and the Portuguese defended it was the Oyapock, which flowed into Cabo de Orange, expanding the Brazilian border farther north and giving them possession of the Cabo do Norte. The *Carte particulière du cours de la Rivière des Amazones ou de Maragnon*, drawn by d'Anville in 1729 based on a draft by the Mercedarian priest Ignácio dos Reis, shows the Portuguese forts on north bank of the mouth of Amazonas River as if they were still in operation, although by that time some of them were in ruins, as was the case with Parú and Desterro or Exílio, which had been demolished (Fig. 17).

34 E. Levasseur, *Le Brésil* (Paris: H. Lamirault, 1889), p. 2.

Portugal's position in the conflict was that all the lands in Cabo do Norte, situated between the Amazonas and the Vicente Pinzón Rivers, which would be the same as the Oyapock, belonged to them, and that the French only had a right to the territory north of it. The French in turn argued that it was difficult to determine exactly where these rivers were or if they were the same. One of the copies of the *Amérique méridionale* presents precisely this issue (Fig. 12). In it, d'Anville noted in red three possible locations for the bay of the Vicente Pinzón River, which would affect the demarcation of the border between the two crowns. The one farther north uses the mouth of the Oyapock, located at Cabo de Orange, as a landmark, a position defended by the Portuguese. To the south of this, a second possibility points to the mouth of what would be a southern arm of the Araguari (*Arawari*) River, which flows into the north of the island of Maracá. The third coincides with the Cabo do Norte, located south of the island of Maracá, named by the geographer the Bay of Calcuenne. It was at this last landmark, of interest to the French, that d'Anville chose to start the boundary line when his map was printed in 1748. If the first had been adopted, it would have resulted in greater territorial losses for the French.

In the Provisional Treaty of 1700, ratified on July 18, 1701, by France and Portugal due to the geographical ambiguities that persisted over the Cabo do Norte, it was decided that it would be the object of a future demarcation expedition that would identify, with precision, which was the river of Vicente Pinzón and only then could the region be occupied by the subjects of one of the two nations. In the agreement, Portugal managed to guarantee exclusive dominion over the left bank of the Amazonas River, but peace was short-lived in the region. In the context of the War of the Spanish Succession, fresh hostilities began in 1704, since Portugal abandoned the policy of neutrality, coming to the defense of the British position. The war, which ended up involving almost all the nations of western Europe, had significant repercussions in the Americas, since the battlefield between the Portuguese, the French, and the Spanish was not limited to Europe. In addition to the two French pirate invasions of Rio de Janeiro, France began to question the boundaries of its possessions in the Amazonas River region, along the borders between Brazil and Guyana, where it destroyed several Portuguese forts. The Congress of Utrecht was organized in an attempt to overcome a number of impasses that had arisen during the conflict. The Portuguese participated in preliminary negotiations in London in 1712 and in the congress itself, in the Netherlands, from 1713 to 1715. In 1712, D. Luis da Cunha was appointed Second Plenipotentiary, assisting the Count of Tarouca, who was First Plenipotentiary.[35]

35 Cluny, *D. Luis da Cunha e a ideia de diplomacia em Portugal*, pp. 74–86.

FIGURE 18 *Embouchure de la Rivière des Amazones*, unknown author (manuscript, ink and color on paper, 17??). BNF, DCP, GE DD-2987 (9552)
REPRODUCED COURTESY OF FBN

D. Luis believed that in the negotiation "the main point is that the northern bank of the Amazonas River remain with us freely with the entrance from the Cape" of the Vicente Pinzón.[36] Although an anonymous early eighteenth-century French map drew the Tordesillas Line, named "Demarcation Line by Pope Alexander VI", immediately to the left of the Amazonas River, thus denying Portugal the right to the body of the river (seen in *Embouchure de la Rivière des Amazones*, Fig. 18), since the treaties signed with France in the late seventeenth century, Portugal had considered its right to those territories assured. As a result, the position the ambassadors would maintain in Utrecht was to basically reaffirm what was already agreed between both courts in the Provisional Treaty of 1700, pointing that the Portuguese and the French had agreed to

36 BNP, R, cód. 11209, f. 310.

regulate the lands between the Oyapock and the Vicente Pinzón Rivers in the future. Portugal's position was that the boundaries should be established precisely at the Vicente Pinzón River, which would coincide with the Oyapock, and the ambassador considered that while obtaining the lands would not be that difficult, keeping them would be a different story. He called attention to the fact that mere possession would be of little use, for it would always be precarious and could be disputed if the lands were not effectively colonized.[37] That relationship between colonization and possession, as in the case of Cabo do Norte, started to become a constant theme as D. Luis led Portugal's successive negotiations throughout the century. Thus, in his geopolitical vision of the Americas, the concept of *Uti possidetis* begins to take shape; in other words, that effective colonization of the territory is more effective than discovery, conquest, or what previous treaties established.[38]

While the Portuguese ambassadors considered that the lands did not seem very valuable to the French, access to navigating the Amazonas River was a different story. The negotiations not only dealt with territorial possessions, but also with the right to trade in the region, for which opening the river and its tributaries was essential. D. Luis highlighted the lack of knowledge of the territory on the part of the French negotiators, whose information he believed came from only four travel books, and he once again called attention to the interest that Brazil awakened in the courts of Europe.[39] Negotiations with France regarding Maranhão began in Utrecht in early 1713 and, as the ambassador had predicted, they were relatively fast and efficient, unlike what happened with disagreements with Spain. The three points discussed by the representatives of both crowns dealt with "ceding [to France] the right it claimed over the lands of Cabo do Norte and that it give up navigation on the Amazonas River, with which it ultimately wanted to stay, and also agree not to use its entrance," the river's mouth.[40] D. Luis was aware that the most troublesome issue was "to gain free access (...) to the entrance and to navigate the Amazonas River."[41]

It was at this point that the Portuguese were impressed at how meticulous the instructions received by the French ambassadors were, quite unlike the orders they had received, which were generic and ambiguous. As supporting documentation for the Portuguese demands they had nothing more than a copy of the Provisional Treaty from the beginning of the century, although the

37 ANTT, ML, maço 638, carta 28, September 6, 1712.
38 Furtado, "Portuguese America under Foreign Threat."
39 ANTT, ML, maço 638, carta 64, February 24, 1713.
40 BNP, R, cód.11.209, f. 312, April 13, 1713.
41 ANTT, MNE, livro 787, f. 20, February 14, 1713.

ambassadors had insistently requested instructions and maps capable of orienting their actions.[42] Exasperated with the lack of direction from Portugal to establish a clear policy that could guide them, D. Luis wrote:

> I truly do not understand this ministry [of state] or the foreign ministry of our court. They all speak every day and are answered and it does no good, like a *mula de atafona* [mill mule].[43]

That metaphor—mill mule—used by the ambassador aptly illustrated what was going on in Portugal, where, as was typically the case with the politics of *ancien régime* monarchies, innumerable centers of power hindered decision-making, unlike Enlightenment criticism, which called for an increasingly cohesive, hierarchical, and rational administration. An *atafona* was a water mill used since Antiquity, pulled by a mule that walked around in circles all day long. His correspondent in this letter, ambassador José da Cunha Brochado (1631–1733), was, like him, a fierce critic of Portuguese internal politics, where

> everything is born of the few men the king has at his service, I mean to say that no one knows how to perform their duty because there is neither discipline nor doctrine, nor school; we know not how to command, nor how to obey.[44]

One of the great difficulties faced early on to establish the boundaries of the territory to be returned was the lack of maps of the area. When the congress began, D. Luis was anguished to see that the Count of Tarouca, who had come with instructions to guide Portugal's position, had arrived

> without a good map of that state [Maranhão] nor of the region of the Rio de la Plata, with an explanation of the lands that Your Majesty claims there with the Treaty of Alliance, as well as copies of what transpired regarding Colônia do Sacramento.

He sorrowfully confessed that "I always thought that the Count of Tarouca would have come furnished with all of these documents, for the promises

---

42 ANTT, MNE, livro 787, f. 20, February 14, 1713.

43 ANTT, ML, maço 638, carta 64, February 24, 1713.

44 Cf. Magalhães, "A acção diplomática no pensamento dos diplomatas portugueses dos séculos XVII e XVIII," p. 25.

made in treaties always have more strength when they are assisted by the same justice."[45] By "justice" the ambassador meant any document—maps, treaties, reports—that justified the historical possession of a territory and its unjust occupation by an enemy nation. Maps were beginning to be used in an increasingly systematic way to justify and define the borders negotiated among rival nations.[46]

D. Luis stresses that, to make things worse, the few instructions that had arrived from Portugal were confusing. The lack or ambiguity of the orders forced him and the Count of Tarouca to act with a certain amount of freedom, and that was precisely what they did regarding this issue. When both realized that it was impossible to reconcile confirming the Provisional Treat and forbidding French trade on the Amazonas River, they decided to stick to the latter, which seemed to them more important, since "if we requested confirmation of the treaty we would have had no room for action to demand the forbiddance of that trade, and to attempt the latter by revoking the treaty was against instructions." The solution to the impasse came when they perceived as early as the first conference "that the French had no intention of confirming the treaty, for it began with an offensive alliance against Castile, nor did they want to revoke it." It was then, as they inform that "we dared offer them an alternative, for them to answer us, as they did, that the treaty was not be discussed, and thus we were freed."[47]

A certain freedom of action was expected on the part of ambassadors during negotiations. French Minister of State Jean-Baptiste Colbert de Torcy (1665–1746) was informed by his ambassador in Utrecht that "the brave Tarouca said that the distance to Portugal and the uncertainty of the winds to send an express ship there made him decide on his own the demands of his court."[48] Tarouca was not quite truthful when he blamed the slowness of transportation on the wind, for certain freedom of action was inherent to the position of ambassador and kings had to trust their representatives. "This trust means granting ambassadors significant freedom of action regarding the best way to defend the specific interests of their government, and also listening to them regarding all important decisions affecting their mission."[49] The Count of Tarouca and D. Luis da Cunha enjoyed so much freedom of action in Utrecht

---

45 BNP, R, cód. 11.209, f. 282v, October 2, 1712.

46 David Buisseret, ed., *Monarchs, Ministers and Maps: The Emergence of Cartography as a Tool of Government in Early Modern Europe* (Chicago: Chicago University Press, 1992), p. 2.

47 BNP, R, cód.11.209, ff. 312–12v, April 13, 1713.

48 Cf. Bély, *Espions et ambassadeurs au temps de Louis XIV*, p. 501.

49 Magalhães, "A acção diplomática no pensamento dos diplomatas portugueses dos séculos XVII e XVIII," p. 20.

that at a certain moment they presented at the negotiation table "a blank piece of paper that their master signed and sent them to make the decision they considered appropriate to his service."[50]

By contrast, D. Luis da Cunha confessed that, as soon as the French sat down to negotiate, he and Tarouca were "surprised [at] the thoroughness [of their instructions] and the documents and maps with which they came authorized."[51] In other words, the reality of the French plenipotentiaries was quite different from the first impression he had of them. They were not inexperienced men, based only on four travel books as D. Luis said before the negotiations started. In fact, during the preceding months, the French representatives had prepared themselves extensively to negotiate with the Portuguese. Let us see.

In Utrecht, the English were the first to use cartography together with documents and reports, when in December 1712 they negotiated with the French the lands disputed by the two crowns in North America. The English representatives arrived very well prepared, provided with several documents including maps, and in their correspondence the French plenipotentiaries reported that this strategy was very successful to them. In contrast, the orders the latter had regarding which territories could be ceded were vague and imprecise, since "the instructions speak nothing of the limits that should divide the land of New France on one side and the English possessions on the other." This placed the French in a clearly disadvantageous position. To negotiate, both delegations therefore based themselves on a map representing New France, "communicated by the English, and where they themselves had drawn with a pink line the borders of the two states that they intended to establish from the banks of the lands of Labrador on the east to the coast of the continent on the west."[52]

The French ambassadors concluded that the strategy of using this map would "lead them securely," and that is what led them to organize in the same manner to face the Portuguese. From December 1712 to February 1713, they prepared themselves, like the English had done, gathering maps and documents to back up their positions. When they sat down with the Portuguese ambassadors at the beginning of 1713, the situation was quite different, now it was they who had detailed orders and maps of the area, which allowed

50 Cf. Bély, *Espions et ambassadeurs*, p. 501.

51 ANTT, MNE, livro 787, f. 20, February 14, 1713.

52 Cf. Bély, *Espions et ambassadeurs*, p. 718. My hypothesis is that the map would be the Beaver map (*A New Map of North America*) by Herman Moll (1654–1732), see Furtado, "O Tratado de Utrecht e a construção da cartografia como instrumento da diplomacia portuguesa."

them to insist on the primacy of cartography to configure the territory situated between the Amazonas and the Oyapock Rivers, and of which D. Luis da Cunha complained to Portugal's authorities, because they did not have a basic cartographic document that made Portugal's demands explicit. As the French ambassadors informed their king, Louis XIV, "we overlay the maps, we measure the land." Despite this initial advantage, the terms agreed were more favorable to Portugal's interests, and the French plenipotentiaries had to justify why they did not manage to overcome the Portuguese: "They never wanted to give up more than a fourth" of the disputed land.[53] Although as we will see the Portuguese won that "war" with the French—wars are really disputed more in the field of diplomacy than cartography—this first battle left a lifelong impression on D. Luis da Cunha. Since that moment he would incessantly advocate the use of maps to guide negotiations, insisting on their importance as diplomatic instruments and reaffirming Portugal's need to produce a cartography of Brazil to back up and justify their demands. For him, the Brazil's centrality for the empire and the need to precisely define its borders, thus guaranteeing Portugal's sovereignty in South America, marked the eighteenth century. That would be a war fought in peace congresses more than on battlefields, and precise maps were powerful weapons for convincing and persuading.

Despite their apparent disadvantage, having no maps of the region, the Portuguese representatives stuck to earlier treaties. This was another one of D. Luis's maxims: respect for what had been formerly accorded. Although he defended the principle that old treaties were powerful negotiation instruments, since they provided historical grounding for the arguments set forth, he knew that they were always provisional and generally respected or disrespected according to the present context. Regarding navigation on the Amazonas River, the Portuguese argued that it had been long forbidden to the French, since it was below the dividing line established by both nations, which used as its reference point the Vicente Pinzón River. They then presented some papers that had already been presented in 1710 to the French in The Hague. It was "the agreement made by Cardinal Richelieu (1585–1642) with the company, called Cabo do Norte," that stipulated the Vicente Pinzón River "marked the limit of both dominions."[54]

The strategy backfired. Basing their arguments on the measurements in degrees presented in this document without having a reliable map ended up

53 Cf. Bély, *Espions et ambassadeurs*, p. 718. About the probable maps presented by the diplomats of both courts see, Furtado, "O Tratado de Utrecht e a construção da cartografia como instrumento da diplomacia portuguesa."

54 ANTT, MNE, livro 787, f. 20, February 14, 1713.

turning against Portugal's intentions. Not without reason did D. Luis complain to Cardinal da Cunha that "we found ourselves without a good map of that state."[55] To make things worse, the French had, among many other maps, one that showed the Vicente Pinzón River positioned, according to those measurements in degrees, in France's possessions which "gave them a new argument to claim possession of part of those lands."[56] This was probably one by the engineer and hydrographer François Froger (1676–1715), who mapped the Guiana coast under the orders of the Marquis de Ferrolles. One of them is entitled *Carte du gouvernement de Cayenne ou France Equinotiale* and was printed in 1698. It depicts the Oyapock River (named *Oÿapoc*), flowing into Cabo de Orange, near the Aramichous Indians. At the top of the map, one can see the "terres des Portugais," which is limited to the southern part of the Amazonas River, where the fort of Corrupá is shown. Two Portuguese forts are also drawn on the northern bank on what was supposed to be the French side.

At the same time, D. Luis confesses "that we had no document with which we could show that our possession of that stretch always follows the course of the river" Amazonas. The only resources they had were diplomatic wiles and the expedient of pressuring the English as their allies to force the French to comply with their demands. They wisely resorted first to trying to regulate each crown's limits, which the French were more willing to accept, and then to approach the issue of navigation on the Amazonas River. To that end, they convinced the English that "the freedom of navigating in the river coveted by the French would be quite detrimental to trade with England." The strategy worked, and by order of their queen herself the English, who were in charge of coordinating the work, "suggested we first work at regulating the division and we would then deal with the other issue."[57] That is, navigation on the Amazonas River.

Portugal's position of negotiating the territory's division in the congress itself, without resorting to further expeditions to determine the limits, as had happened with the Provisional Treaty, turned out to be a good strategy, for it was an issue that D. Luis predicted would meet little resistance from the French. The English in fact presented Louis XIV with an ultimatum, which was essential for the problem to be resolved quickly and favorably to Portugal's interests. In this ultimatum, the Queen of England, Anne (1665–1714) declared

55 BNP, R, cód. 11.209, f. 282v, October 2, 1712.
56 ANTT, MNE, livro 787, ff. 20–21, February 14, 1713.
57 BNP, R, cód. 11.209, f. 310, April 13, 1713.

"that the lands on both sides of the river would belong to Portugal that could keep the disputed lands and rebuild the forts that were demolished."[58]

In March, D. Luis da Cunha triumphantly wrote to Portugal that regarding the issue of navigation on the Amazonas River, and despite the resistance of their opponents, they managed to forbid the French from navigating not only the river itself, but also its ports of entry.[59] According to the ambassador, for the

> truth [is] that if we wanted to trade through that river (...) it would be easy, with Portuguese troops reaching all the way to Quito and the vicinities of the South Sea, we can consider it of great importance to have that entrance to Peru unobstructed.

He prayed that "may God will we know how to make good use of them!"[60]

Years later, he told Marco António de Azevedo that the argument they had used to deny the French access to the Amazonas River at the Congress of Utrecht was a "plan by Father Acuña regarding navigation on the Amazonas River."[61] Father Cristóbal de Acuña (1597–1676), a Jesuit from Colegio de Cuenca, accompanied the Portuguese Pedro Teixeira's (*c.*1570/80–1641) expedition down the Amazonas River. From 1637 to 1638, Pedro Teixeira, at the head of a Luso-Brazilian expedition with about one thousand men, traveled up the Amazonas River from its mouth to its sources in Peru, going on to Quito before returning to Belém city at the mouth of the river.[62] In the context of the Iberian Union (*União Ibérica*, 1580–1640) between the crowns of Portugal and Spain, this expedition was part of a set of initiatives taken by the governor of Grão-Pará captaincy, aimed at guaranteeing Portugal's possession of the Amazonas region, neutralizing Dutch and French invaders, and delimiting the borders with Spain. According to D. Luis, the main idea behind that expedition was "that the King of Spain should take shelter in Maranhão and carry down the river all the silver from Peru, in order to ship out from there without risk." And it was therefore according to that same principle that "in our treaty, navigating that river was forbidden to the French, and we were granted both banks indefinitely so they could not carry Peru's silver down that same river." Both of them argued then that it was of "capital importance to have that door open

58 ANTT, MNE, livro 787, f. 46, March 12, 1713.

59 ANTT, MNE, caixa 562, M.E. III. A. Fr. C.3-nº.8, May 1, 1747.

60 BNP, R, cód. 11.209, f. 312, April 13, 1713.

61 ANTT, MNE, caixa 562, M.E. III. A. Fr. C.3-nº.8, May 1, 1747.

62 Cristóbal Acuña, *Nuevo descubrimento del Gran Rio de las Amazonas* (1641) (Madrid: Iberoamericana, 2009).

to Peru, which the King of France [finally recognized] it belongs to us and we closed it to them."[63]

A new Provisional Treaty was then signed with the French by both parties in Utrecht on April 11, 1713. With this treaty, "everything was obtained!", D. Luis da Cunha celebrated: "Your Majesty remains lord not only of the lands of Cabo do Norte, but also of both banks of the Amazonas River, indefinitely." Through Article 12, "all Frenchmen are excluded from trading in Maranhão." According to "Article 5, trade would be open [only] in the continent of both crowns, as it was previously done." He calls attention to the fact that he and Tarouca obtained much more than they were ordered, since "now we have been granted without limitation, and in fact to a much greater extent, the property of all the northern bank of the Amazonas River." He considered that all of the objectives had been accomplished much beyond expectations, and "if the French attempted to go to Brazil the same way the English or Dutch can, they could not make use of the Treaty of Utrecht."[64]

Although d'Anville's 1748 *Amérique méridionale* (Figs. 4 and 12) does not draw the boundaries as far as the Oyapock River, as the Portuguese achieved in Utrecht, otherwise it expresses quite clearly D. Luis da Cunha's geopolitics for the area, which did not give much value to the lands north of the Amazonas River but focused on securing the Portuguese a monopoly over its navigation. The dividing line, which was established by using Cabo do Norte equivalent to the mouth of the Araguari River as a geographical reference satisfying the French aspirations, initially runs parallel to and then up relative to the Amazonas River at a reasonable distance from it, guaranteeing to the Portuguese not only sovereignty over both banks, but a monopoly on its navigation as well, and grants it the right to rebuild the forts demolished or abandoned at the beginning of the eighteenth century.

3 The Colônia do Sacramento or the Touchstone

Colônia do Sacramento was established in 1680 by a Luso-Brazilian expedition commanded by D. Manoel Lobo (1635–1683), then governor of Rio de Janeiro captaincy. At the time, five ships landed there with the purpose of creating a settlement. The first Portuguese enclave situated on the Rio de la Plata, on its

63 ANTT, MNE, caixa 562, M.E. III. A. Fr. C.3-nº.8, May 1, 1747; ANTT, MNE, livro 787, f. 75, April 15, 1713.

64 ANTT, MNE, caixa 562, M.E. III. A. Fr. C.3-nº.8, May 1, 1747; ANTT, MNE, livro 787, f. 75, April 15, 1713.

northern bank opposite to Buenos Aires, it was of vital interest to the crown, since it was a strategic location for Portugal to take advantage of the contraband trade in the silver exploited by the Spanish in the Andean region. Its foundation provoked an immediate reaction from Spain, and a few months later Colônia do Sacramento was invaded and destroyed by an expedition sent by the governor of Buenos Aires.[65]

The issue of Colônia do Sacramento was the touchstone for Portuguese and Spanish negotiations in the first half of the eighteenth century. In 1681, through the Provisional Treaty signed by Portugal and Spain, the territory of Colônia do Sacramento was returned to the Portuguese, who sent a new expedition to restore the fort and repopulate it. This treaty established that both crowns' possessions in Rio de la Plata region would respect the principle of *al presente la tiene* (present possessions), "this way it was established that Colônia do Sacramento would belong to the Portuguese crown."[66] After recovering the territory, the area's colonization, which during the first occupation was limited to the fort, started expanding outside of it in the 1690s, constituting an agricultural belt to feed the residents.

In 1705, in the context of the War of Spanish Succession, Spain once again drove out the Portuguese troops stationed there. The negotiations between Portugal and Spain at the Congress of Utrecht, with the English as mediators, were costly and lengthy. They dealt not only with the issue of Colônia do Sacramento, but also with the restitution of Portuguese vessels that had been apprehended in Buenos Aires, accused of contraband.[67] As soon as D. Luis da Cunha arrived at the congress, he also complained of the brevity of instructions he received also on this matter, which practically consisted of a copy of the 1681 Provisional Treaty, and of the lack of maps to clearly define the desired limits of Colônia do Sacramento. He then requested that the royal cosmographer Manuel Pimentel be consulted.[68] After the consultation, the king's answer was that:

> Colônia do Sacramento was of no importance to Portugal and of much importance to my royal interests, but what would be most beneficial to that kingdom [is that] another colony be created in Portuguese

65 Almeida, "A Colónia do Sacramento e a Formação do Sul do Brasil," pp. 163–82.
66 ANTT, MNE, livro 793, ff. 285–86, 1724.
67 ANTT, MNE, livro 787, f. 7, January 6, 1713.
68 ANTT, MNE, livro 786, f. 403, September 2, 1712.

> territory to protect its own jurisdiction rather than being exposed to new contingencies.[69]

It is interesting to note that we learn of this resolution through the Spanish ambassador in Lisbon, since, as is known, diplomatic activities were accompanied by a good deal of spying on both sides.

The terms of the treaty went back and forth. Then the option of trading the territory of Colônia do Sacramento with the Spanish for an equivalent in lands was then mentioned for the first time. D. Luis da Cunha discussed this issue intensely with the Spanish ambassadors throughout 1713. But at that time he assumed an ambiguous position on the issue, claiming that he did not know which equivalent would be fair, and that he had no instructions on the matter and would not dare propose anything without specific orders from the king.[70] The truth, as he later confessed, was that the orders they had were "not to admit any equivalent for Colônia do Sacramento that was not in Europe."[71] Queen Anne of England pressured Portugal to accept an equivalent in the Americas and give up its pretensions in Europe—Albuquerque and Puebla—to which the ambassador countered that "what interests us in Brazil is too costly considering what we keep from acquiring at the borders" with Spain.[72] Thanks to the efforts of both plenipotentiaries, the negotiations ended up leaning not toward an exchange, but toward the restitution of the territory to the Portuguese. So, a second restitution of Colônia do Sacramento to Portugal was then agreed in the Treaty of Utrecht, finally signed between Portugal and Spain on February 6, 1715. The treaty established that "the Territory and Colônia do Sacramento" shall be restituted to the crown of Portugal.

The expression *al presente la tiene* (at present), that appears in the Provisional Treaty, had given rise to many controversies during the negotiations in Utrecht. After all, what lands did the Portuguese occupy then? The Spanish insisted that the territory of Colônia do Sacramento that should be restituted to the Portuguese should be limited to what they had up to the signing of the Provisional Treaty in 1681, which corresponded only to the existing fort, since that was what the Portuguese had at present. The Portuguese believed that, according to the Treaty of Tordesillas, they had a right to the whole northern bank of the Rio de la Plata and were opposed to the treaty clearly limiting their possessions. In this respect, since the Portuguese had expansionist intentions

69 AGS, E, Portugal, legajo 7084, f. 1, December 31, 1716.
70 ANTT, ML, maço 638, letters 71 to 74, February to April 1713.
71 ANTT, MNE, livro 789, f. 335, January 5, 1720.
72 ANTT, MNE, livro 787, ff. 47, 52, March 12, 1713.

in the region, the instructions that Tarouca and D. Luis received determined that "we could include the article on the restitution of Colônia do Sacramento, excluding the clause *como al presente la tiene* that was in the Provisional Treaty," given the differences in its interpretation by both parties.[73]

Having a certain amount of independence to maneuver in the negotiations, both ambassadors boasted of having succeeded in excluding the expression from the new agreement, replacing it with the "Territory and Colônia do Sacramento," the first word capitalized, meaning an extensive region beyond the fort. They reported that:

> for some time now we had decided not to speak of Colônia do Sacramento without saying together the word territory, and in effect we created a plural, as for example when we say, in Article Six: "That the Territory and Colônia do Sacramento are included.

D. Luis da Cunha later confirms that this new expression was their own independent creation, "because we put it in the peace treaty, without having an order to that effect." This way the ambassadors hoped to prove "that the restitution and concession by the King of Castile contains both things, not only Colônia do Sacramento, but [the] Territory that belongs to it."[74] From then on "the aforementioned Territory and Colônia do Sacramento are encompassed in the dominions of the Portuguese crown (Art. 14)."[75]

Initially, both plenipotentiaries rejoiced in the agreement: "It seemed to us that we had won much in this peace!"[76] Their hope was that, by including the term "Territory," Portugal's possessions would not be restricted to the fort nor only to the nearby surroundings, which after the second Portuguese occupation as of 1690 had been settled as an agricultural frontier to supply the residents, since both of them knew that Portugal had expansionist projects for the entire region of the Rio de la Plata, and the generic terms could encompass its whole northern bank of the river. But the expression was most likely carefully chosen by the ambassadors also because of the difficulty in defining the desired territory in more precise terms.[77] The lack of adequate maps hindered them from establishing a text that more precisely determined the limits

73 ANTT, MNE, livro 789, f. 286, December 15, 1719.

74 ANTT, MNE, livro 788, f. 281, February 12, 1715.

75 Treaty of Utrecht: http://www.info.lncc.br/utrech2.html.

76 Fabrício Prado, *Colônia do Sacramento: o extremo sul da América portuguesa* (Porto Alegre: Fabrício P. Prado, 2002), p. 92.

77 ANTT, MNE, livro 786, f. 403, September 2, 1712.

negotiated. In fact, D. Luis da Cunha believed that the agreement left no "doubt regarding [Portugal's] property, but only regarding the limits" of the territory.[78]

But contrary to what D. Luis da Cunha expected, its restitution to Portugal would still lead to many upheavals that demanded continuous diplomatic efforts for many years to come. The Spanish insisted that the term "Territory" inserted into the accord by the two ambassadors corresponded only to the countryside around the fort, and stubbornly refused to restitute more than the territory within the range of a cannonball,[79] which corresponded to about 3 km around the fort. What had seemed to be an astute political move soon became a nightmare. Consequently, shortly afterwards, trying to explain himself to the crown, D. Luis da Cunha regretted the use of the term "Territory," which "with the same confusion [as the term *al presente la tiene*] we inserted in the peace treaty."[80]

This idea of the range of a cannonball was suggested by the governor of Buenos Aires, D. Baltasar Garcia Ros (1670–1740), who wrote to king Filipe warning him of the harm that would result from the Portuguese reoccupation of Colônia do Sacramento. When that monarch finally decided not to restore the Colônia, as the governor suggested, it was already too late. The act of formal restitution had already taken place. So in his correspondence, Garcia Ros enumerates three possible ways of interpreting the term "Territory": the first would limit it to the range of a cannonball; the second would include the countryside, where the Portuguese raised cattle and exploited the leather; and the third would include all the lands claimed by Portugal, encompassing the whole northern bank of the Rio de la Plata. Therefore, there being no possibility of keeping Colônia do Sacramento without triggering a new armed conflict, the Spanish began obstinately insisting on interpreting the term "Territory" as meaning only a within the range of a cannonball around the fort.

Since Spain refused to adhere to what Portugal understood had been agreed in Utrecht, limiting Portugal's expansion to what was within the range of a cannonball, it was necessary to continue diplomatic efforts in order to put an end to the dispute. The first round of negotiations took place in Madrid, where D. Luis da Cunha went in 1719. His main mission consisted in pressuring Spain to comply with the agreement. In the ensuing negotiations, cartography once again became an indispensable tool for diplomatic bargaining. Those maps should clear up the impasses once and for all, and at the same time justify

---

78 ANTT, MNE, livro 788, f. 281, February 12, 1715.

79 Paulo Possamai, *A vida cotidiana na Colónia do Sacramento: um bastião português em terras do futuro Uruguai* (Lisbon: Bertrand, 2006), pp. 84–85.

80 ANTT, MNE, livro 788, pp. 264–66, January 18–20, 1715.

Portugal's claims. The ambassador once again insisted to the king that "regarding the territory of Colônia do Sacramento, it would be most convenient if I was sent a special map of that country, if such exists, declaring which are the boundaries desired by Your Majesty." Since nothing arrived from Portugal, once in Madrid he decided to look for the Spanish king's confessor, the Jesuit Daubenton, and

> telling him that although I understood that he did not get involved in political matters, I wished to speak to him of one, because it was of interest to the Society [of Jesus], and with that understanding I told him of the Territory of Colônia do Sacramento, since it should encompass the Jesuit missions.[81]

He then asked him whether he could obtain a map of the area among the many that belonged to the Jesuits' collection. We can see it is only a game of appearances (on the part of the confessor) and of rhetoric (on the part of the ambassador) and how the ambassador uses the Jesuits' interest in the region of the missions to favor the Portuguese. We can also see how he not only uses the word territory as it appeared in the treaty, but capitalizes it, establishing it as an independent entity from the fort itself. He also implies that the territory of Colônia do Sacramento would encompass the Jesuit missions of Paraguay, clearly assuming a much greater area than the fort itself or the 3 km around it.

It was during this mission that D. Luis learns from the Jesuit priests in Madrid of the peculiarities of Colônia do Sacramento region, and begins proposing that, given the lack of utility of that possession, together with the high political and economic costs of maintaining it, the best option to resolve the impasse with Spain was to exchange it for a territory in the Americas. Since he had received no orders to that effect, when he is chastised for discussing the matter with Filipe V's confessor, he justifies himself and disguises his viewpoint claiming that "I only said that we could give this hope to the Jesuits so they contribute as much as possible to adjust the limits, because, as I say, it would be most convenient to put an end to this matter." The information he obtains on the benefits derived from keeping Colônia do Sacramento under the Portuguese flag are not at all encouraging. One of the few activities in the region was "slaughtering cattle, since their leather is the only benefit obtained there."[82] In Portugal, one of the rare voices in this respect was that of the Duke of Cadaval, D. Nuno Álvares Pereira de Melo (1638–1727), Councilor

81 ANTT, MNE, livro 789, f. 286, December 15, 1719; ff. 632–33, August 1720.

82 ANTT, MNE, livro 789, f. 335, January 5, 1720; f. 294, December 18, 1719.

to D. João v, who maintained "that a king whose conquests produce so much gold had no need for a territory whose main wealth constitutes the extraction of leather."[83] Gradually, D. Luis da Cunha began advocating a similar position.

The issue of Colônia do Sacramento dragged on and nothing was resolved in Madrid. Preparations then began for the Congress of Cambrai (1720–1725). On September 4, 1720, D. Luis da Cunha and the Count of Tarouca received instructions to participate in the congress as Portugal's representatives. They hoped to include the issue of Colônia do Sacramento in the negotiations. However, they remained in Paris, for they were prevented from participating officially by the English, who claimed that the Portuguese had not taken part in the Quadruple Alliance, established between the Netherlands, England, France, and Spain to guarantee world peace.[84] Forced to stay in Paris, D. Luis da Cunha tried to obtain the French ministers and the English ambassador's mediation to carry forward the negotiations outside of Cambrai and put an end to the dispute regarding Colônia do Sacramento. To convince the French, in particular the Secretary of State, Cardinal Fleury, André-Hercule de Fleury (Bishop of Fréjus, Archbishop of Aix, 1653–1743), he wrote a memoir on Portugal's ambitions for the territory of Colônia do Sacramento to mediate the negotiations with the Spanish where he maintained "that the most Christian king [Filipe v] wants to reduce it to only those spaces covered by the fort's artillery" and attempted to demonstrate the act's illegality. He then wrote to his crown that, considering everything that the French and the Spanish ambassador told him, "this court's intention is not to allow Colônia do Sacramento to have any more territory than the aforementioned, so that, considering that the expenses required to maintain it would be greater than the use we derive from it, we would let it go for the equivalent they would be willing to offer us."[85] In this passage the ambassador clearly makes a point that he will henceforth insist on: that the lack of economic interest in the Colônia do Sacramento justified its exchange for an equivalent. Yet once again there was no tangible result.

In late 1735, the Spanish imposed a severe blockade on Colônia do Sacramento. This episode was called the Great Siege and "for almost two years (1735–1737) it led the residents of Colônia do Sacramento almost to starvation."[86] It was the outcome of a diplomatic incident caused by the incompetence of the then Portuguese ambassador in Madrid, the Lord of Belmonte, Pedro Álvares

83 Possamai, *A vida cotidiana na Colónia do Sacramento*, p. 74.

84 Ana Leal Faria, *Arquitetos da Paz: a diplomacia portuguesa de 1640 a 1815* (Lisbon: Tribuna, 2008), pp. 139–40.

85 ANTT, MNE, livro 789, f. 645, August 1720.

86 Prado, *Colônia do Sacramento*, p. 119.

Cabral. Breaking diplomatic protocol, his servants had freed and sheltered at the embassy an individual who had been arrested by the local authorities. As soon as D. Luis da Cunha heard of the incident, he feared with good reason that the episode could renew military hostilities between the two courts, which he suspected would not manifest themselves in the Iberian Peninsula, but in the Americas. He did not tire of warning that "the smallest of disagreements with Spain will have immediate effects not at our borders, but on the Rio de la Plata, as long as there is no resolution."[87] Despite promises that there would be no retaliation,[88] as the ambassador had expected the incident had direct repercussions in the Rio de la Plata region: the Spanish decreed the blockade of Colônia do Sacramento and fresh negotiations had to begin to try to resolve the impasse.

However, at that time, Spain's audacity and the danger that the siege represented to Portugal's plans for the Rio de la Plata basin led D. Luis da Cunha to abandon the almost invariably peaceable tone of his recommendations to the crown. In July 1736, he asked Portugal's authorities what their strategy for retaliating against Spain's military offensive would be: whether

> to pay no heed to the mediators' proposals as long as the Colônia do Sacramento is not restored, in case it is invaded, or to engage in negotiations [and] send to the Rio de la Plata whatever forces there are in Brazil to recuperate it and in addition attack Buenos Aires if possible.

He complained of the fact that the planned fort of Montevideo never made it past the blueprints, but argued that there was still time to build it, "for we will keep it while the dispute continues regarding how far the Territory of Colônia do Sacramento will extend toward the Uruguay River; and if perchance an equivalent in Europe is discussed, the more we have to give up, the better."[89] In fact, in 1736 the Portuguese would begin a period of more aggressive systematic occupation of southern Brazil, with the construction of forts (Jesus Maria José in 1737) in Rio Grande de São Pedro and the arrival of migrants from Azores to Santa Catarina Island (1740).

An examination of Spanish and Portuguese maps of the time points to the very different ways in which each crown understood the territory's possessions. Evidently, more than portraits of reality, these maps were intended to legitimate what each nation understood as its right. From this standpoint, the

---

87 ANTT, MNE, caixa 2, maço 2, doc. 48, July 3, 1741.
88 AUC, CLC, doc. 342, July 20, 1736.
89 AUC, CLC, doc. 359, August 17, 1736.

FIGURE 19 *O Grande Rio da Prata na América austral e portuguesa*, unknown author (manuscript, ink and color on paper, *c.*1740). BNF, DCP, GE DD-2987 (9451B) REPRODUCED COURTESY OF FBN

maps were once again, above all, instruments of power. Spanish maps represented the territory of Colônia do Sacramento that only covered the fort's surroundings, occupying the point of the peninsula that projects into the Rio de la Plata, limited to what can be estimated to correspond to the range of a cannonball. With an entirely different perspective, Portuguese cartography presents the territory extending far beyond the peninsula, going further into the interior, away from the fort, eventually reaching the entire northern bank of the Rio de la Plata. The maps in d'Anville's collection entitled *O Grande Rio da Prata na América austral e portuguesa* (The Greater *Rio de la Plata* in Southern and Portuguese America) (Fig. 19) and *Carte manuscrite de l'embouchure de Rio da Prata* (Fig. 20), both produced in the 1740s, portray the various ways the Portuguese represented these domains. In the first case, the title itself suggests that the greater Rio de la Plata extends into the southern part of Portuguese America. In other words, its entire northern bank belonged to Portugal, and did not establish a boundary around Colônia do Sacramento. The fort is drawn in red without any boundaries shown on the land around it. The second one proposes two boundary lines, drawn in yellow and red respectively. The first, located immediately to the west of the fort encompasses all the northern coast up to Ponta de Castillos, representing the Portuguese demands. The second enclosed the area in Spanish possession. It is interesting that there is a copy of

FIGURE 20 *Carte manuscrite de l'embouchure de Rio da Prata*, unknown author (manuscript, ink and color on paper, *c*.1740). BNF, DCP, GE DD-2987 (9450)
REPRODUCED COURTESY OF FBN

this map in Portugal that shows no lines.[90] This boundary, whose position corresponds approximately to the Tordesillas Line, extends northward and enters the territory of Rio Grande do Sul captaincy, encompassing the entire territory to the east, which would be in Portugal's possession.

In 1737 D. Luis da Cunha was again appointed Portuguese ambassador in Paris. From there he kept a close eye on events in the Rio de la Plata region, entreating other nations' diplomatic representatives (especially the British), and the French authorities to obtain their nations' military aid or mediation in negotiations with Spain. He tried to stay informed of the daily developments of the siege and of whether Spain had managed to take the Portuguese fortress or whether the attack had been repelled.[91] Aware that the garrison there was too small to defend the fortress, he sought France and England's military aid and advocated the displacement of Portuguese troops from other Brazilian posts

90 Almeida, *A formação do espaço brasileiro*, pp. 124–25.
91 AUC, CLC, doc. 387, November 6, 1736.

come to their assistance.[92] He presented to the 2nd Marquis d'Argenson, René Louis de Voyer de Paulmy, Louis XV's Secretary of State, Portugal's position regarding the Spanish blockade of Colônia do Sacramento, and requested that, once in Spain, Marshal Noailles, Adrien Maurice de Noailles (3rd Duke of Noailles, 1678–1766), make efforts "to impose the 1737 convention, whose meaning Spain intends to misrepresent."[93]

Once again the issue of exchanging territories was raised and D. Luis began speculating which could be bartered. He simultaneously attempted to influence French ministers to convince the Spanish and to obtain the Portuguese court's support to carry out the operation. He went to negotiate with the Secretary of State Cardinal Fleury, for the latter to serve as a mediator with the Spaniards. The minister replied that Colônia do Sacramento did not seem to him to be sufficiently important to either side to warrant a war. To which D. Luis replied: "allow me to send for the map of the Rio de la Plata that [the Portuguese Secretary of State] sent me when I was in Madrid, and it *will make you see* Colônia do Sacramento's importance to both nations." Once the map was brought in, it faithfully performed its intended role, and

> the recipient was quite impressed at seeing it because there is no other as distinct, and also because in the report it was not well understood what the Castilians possessed near Colônia do Sacramento nor the importance of their fortification in Montevideo, so that with my explanations he was convinced of how much interest we had in maintaining its possession as well as the Castilians' interest in depriving us of it.[94]

The map thus played the role of the perfect teaching tool in the context of diplomacy that could make a territory visible from afar. As the ambassador underscores, because of its accuracy, which set it apart from the others, it worked as a supplement to the written report he presented to the minister. Its effect was such that, in Paris, the minister "saw" and perfectly understood what was going on on the far-off banks of the Rio de la Plata. As D. Luis advocated, "although [territorial disputes] will not be decided with geographical maps," they were useful to make them visible "if we compare them to others and we are prepared for any argument that may arise."[95]

92 ANTT, MNE, livro 16, f. 20, October 26, 1737.

93 AUC, CLC, doc. 790, April 5, 1746.

94 BACL, SA, 608, f. 56v, April 6, 1736 (italics are the author's).

95 ANTT, MNE, caixa 560, ME.III-A-Fr.C1.nº. 111, April 8, 1737.

For D. Luis da Cunha, Portuguese sovereignty over Colônia do Sacramento only made sense if maintaining the right to the gold mines that depended on keeping the frontier secured. The perception of Brazil's centrality with regard to both Minas Gerais and the Amazon River basin was part of a broader geopolitical plan to ensure the Portuguese empire's sovereignty over the Americas. According to D. Luis's vision, as d'Anville precisely pointed out in 1737 in his *Discussion géographique*,[96] three main issues were in the way of negotiations between Spain and Portugal regarding their territories in the Americas. The first had to do with the post-Utrecht indecisiveness of the two great powers, France and England, about get involved as mediators in the negotiations with Spain on the issue of Colônia do Sacramento, which led him to conclude that it should be negotiated directly and separately by the two crowns and that it would not be resolved in the context of the great European peace treaties. The second one had to do with the imprecision of the true measurement of the Tordesillas Meridian, which left many doubts regarding the exact limits to be respected by both nations, which led the ambassador to insist on mapping Brazil more precisely and commissioning d'Anville with a geographic argumentation to support Portugal's claims. The third one was Portugal's settlement of territories in the mid-west and the Amazonas, which up to then belonged by right to Spain, since the differences with France had been resolved at Utrecht and they did not return to the negotiating table, which made him propose that the issue of the establishment of boundaries should go beyond the case of Colônia do Sacramento.

He then began advocating that the issue of setting boundaries with Spain should be addressed as a whole, taking into account not only all of South America, but also the region of the Moluccas Islands in the South Sea. Portugal's invasion of territories in the middle of the Americas and the purchase in 1529, through the Treaty of Zaragoza, of the Moluccas Islands from Spain—a territory to which Portugal already had a right according to the new measurements of the Treaty of Tordesillas—in addition to the inaccuracy of the location of that meridian, required, in his opinion, the establishment of new and more precise boundaries. His efforts in Utrecht, Cambrai, and the main European courts convinced the ambassador that any negotiation with Spain would not be limited to the borders in Europe, but would necessarily encompass an entire geopolitical reorientation of the boundaries between the two crowns. He was aware of the role that Brazil had come to occupy in the Portuguese empire, and that it was necessary to include as much territory as possible in

---

96 NL, VB, Ayer MS 1918, doc. 552.

the mid-western and Amazonas regions, while maintaining the security of the Amazonas in the south. He then began insistently arguing that Portugal's greatest wealth at that moment was located in Minas Gerais, and that the region of Colônia do Sacramento, at the mouth of the Rio de la Plata, given its complete lack of economic interest, would be the key bargaining chip with Spain during the negotiations. The unjust purchase of the Moluccas Islands could also be called on to ensure possession of Colônia do Sacramento if necessary. With regard to the Amazonas region, he believed that what was essential for Portugal was to keep control over both banks of the river, thus guaranteeing the monopoly over its navigation. These geographical ideas were advocated by d'Anville in his *Discussion géographique* made on the ambassador's orders.

4 The Fernando de Noronha Island, the War of Jenkins's Ear, and the English Threat

In October 1736, when the first negotiations regarding Colônia do Sacramento with Spain still dragged on, D. Luis da Cunha's most dismal predictions seemed about to come true. At that time he received a letter from Portugal with "the news of the occupation of the Island of Fernando de Noronha [which] was confirmed by the news communicated by the viceroy of Bahia, delivered by a ship of ours that landed there." He was instructed that "some Frenchmen were established who seemed to be pirates" and ordered "to investigate whether those men are pirates or not, and, depending on his findings, take the necessary actions in that court."[97]

The ambassador set his French information networks in motion and learned from one of his informants[98] in the government that a certain "Captain Lesquelin, by order of the [French] Company of the Indies, disembarked with nine men in the Island of Fernando de Noronha." From there, the captain wrote a detailed report of the island. While the first news was not so threatening—it was only eight men—the action's future consequences were quite harmful, since after this "information the company made the decision to create on that island, which it called Isle Dauphine, a large and useful establishment."[99]

97 ANTT, MNE, caixa 559, M.E. III-A-Fr-C.2-nº. 5, November 26, 1736.

98 One was *chevalier* Jean-Baptiste Yves Garnier du Fougeray (1689–1747), former director of the French Indian Company. Mediated by ambassadors, royal graces given to such men an important mechanism to obtain loyalties in foreign courts, as they served as spies. AUC, CLC, doc. 408, July 5, 1737.

99 AUC, CLC, doc. 408, July 5, 1737 (italics are the author's).

D. Luis then calls attention to the fact that, as part of the act of possession, the French had renamed the island. This act of naming together with the act of possession is part of the whole process of maritime expansion of the modern era, with Europeans renaming geographical elements as they learn of their existence. During his trip,

> Columbus knows perfectly well that the islands already have a name, in a way natural names (but with a different meaning of the term); other people's names, however, do not interest him much, and he wants to baptize the places in terms of their place in his discoveries (…). To give them the right names (…) is equivalent to taking possession.[100]

He also tries to demonstrate that, although there were initially only a few men, they planned "a large and useful establishment," in other words, something more long-lasting, and "for that purpose [the company] decides to send on board the ships, which in September should leave for India, 300 men with all materials and artillery necessary to fortify themselves well, leaving at that port two frigates to better ensure its security."[101]

He then suggested to the king that, in the face of the threat of French colonization, the Portuguese should get ahead of them and effectively colonize the island themselves. This insistence on settling the region was directly linked to the idea of being able to later justify the possession of a territory through the colonization established by a given crown there, extending the right of sovereignty of the monarch to that territory. It was no longer a case of building stone landmarks or crosses, or of claiming first arrival. It was a matter of instituting a society under the aegis of Portugal, of "interiorizing the metropolis"[102] as "the expansion of Iberian monarchies implied the dissemination, at immense distances, of [their] forms of social organization, institutions, and representations."[103]

This notion of the right of possession through settlement, which he had expressed in other cases and at other moments, appears clearly in the reflections he made at that time in regard to Fernando de Noronha. He informs that the French, after rebaptizing the island Isle Dauphine, built two fountains and

---

100 Tzvetan Todorov, *A conquista da América: a questão do outro* (São Paulo: Martins Fontes, 1983), p. 27.

101 AUC, CLC, doc. 408, July 5, 1737.

102 Furtado, *Homens de negócio: a interiorização da metrópole e do comércio nas Minas setecentistas* (São Paulo: Hucitec, 1999).

103 Cardim, "La jurisdicción real y su afirmación em la corona portuguesa," p. 364.

planned to build a fortress and leave two frigates. He suggests to the king that, before more Frenchmen arrive at the island, it was best that "there was a given settlement [there] so it cannot be said that it belongs to no nation, [since] it must belong to the one which takes possession of it and thus renewing ours." He warns that "we are to fear that, upon arrival of the ships with the more than 300 men and supplies, they will claim the act of possession that others made."[104] These two excerpts illustrate the importance of colonization to renew the possession obtained by the simple primacy of arrival. Revalidating the (Portuguese) possession through discovery or the (French) possession through conquest would depend on which of the two nations had the primacy of colonization, creating in the territory a stable settlement. For the second time in his writings he clearly expresses, albeit without naming it, the right of *Uti possidetis*.

It is interesting to note that d'Anville had in his collection a 1737 map by Philippe Buache, entitled *Carte de la partie de l'ocean vers l'équateur*, where beside the Archipelago of Fernando de Noronha appears the text: "Island of Fernando de Noronha, named Dauphine in 1734 by a French navigator" (Fig. 21). That map shows the island in detail and, next to the ruin of the old fort, can be seen the wooden cross planted by the French that year. "Naming the island and planting a wooden cross" were performances that were part of the ceremonies with which the French established their possessions in the New World.[105] Establishing a landmark, renaming a geographical feature, and inserting it in a map under the dominion of another nation were also important forms of possession, especially because maps were very much used as repositories of memory in diplomatic negotiations on the right, anchored in the past, which the different nations had over the disputed territories.[220] On the other hand, the anonymous *Isle de Fernão de Noronha, carte portuguaise manuscrite* (Fig. 22), also in d'Anville's collection, presented the legend: "no. 10—church and settlement made by the French" on the island, but the verb in the past tense suggests the enterprise's failure. In this context, Isle Dauphine was not a simple curiosity or cartographic aberration. It was a political act of declared possession by France that worried D. Luis da Cunha.

Although England was a traditional ally of Portugal, D. Luis da Cunha believed that since the reports of the Blau expedition in 1712 it would be interested in taking over Brazil, due to its wealth in gold and because the establishment in the Rio de la Plata region could be promising. In 1737, António Guedes

---

104 AUC, CLC, doc. 408, July 5, 1737.

105 Patrícia Seed, *Cerimônias de posse na conquista européia do Novo Mundo (1492–1640)* (São Paulo: Unesp, 1999), pp. 63–99.

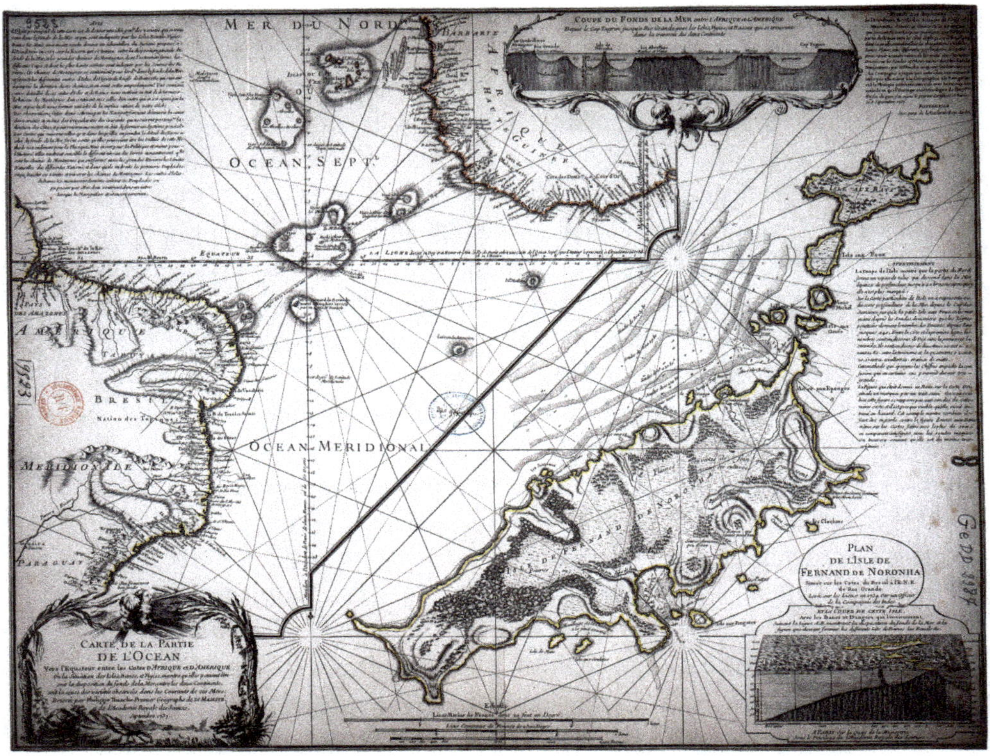

FIGURE 21 *Carte de la partie de l'ocean vers l'équateur, entre les côtes d'Afrique et d'Amérique où la situation des isles, bancs, et vigies montre qu'elles peuvent être soit à la disposition du fonds de la mer, entre les deux continents soit la cause des variétés observées dans les courants de ces mers*, Philippe Buache (Paris, Sur le Quay de la Mégisserie, printed, 1737). BNF, DCP, GE DD-2987 (9523)
REPRODUCED COURTESY OF FBN

Pereira, his secretary in the embassy in Paris, warned the Secretary of State that a certain Mr. Zambone had informed him

> that an Englishman and an Italian, who had been in Rio Grande de São Pedro [and] had probed its bar and examined its *campanhas*,[106] and had proposed to the court in London that it take possession of this area due to the great use that could be made from this establishment.

He says that the English were not interested in the project, but that the two adventurers offered it to Russia. In exchange, "the Italian had been given a

---

106 *Campanha* was the name given to the vast plains covered with grass where cattle could graze extensively, typical of the extreme south of Brazil and Uruguay.

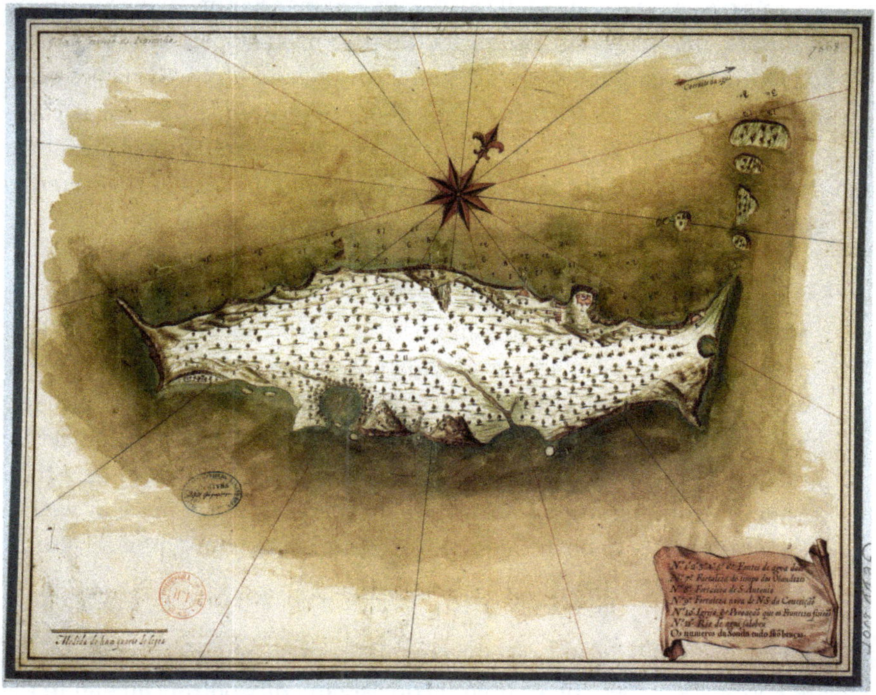

FIGURE 22 *Isle de Fernão de Noronha, carte portugaise manuscrite*, unknown author (manuscript, ink and color on paper, *c.*1740). BNF, DCP, GE DD-2987 (9525) REPRODUCED COURTESY OF FBN

place on the Council of Trade and the Englishman who returned to London where he currently still is, was given the rank of captain, while the court in Russia resolved to take possession of that place." Afterwards, Pereira described the Russian plans. He also revealed that the English officer had then presented the plan to London merchants

> who offered to establish a company, and recently, giving great support to this project, the Duke of Brijoater had resolved to the same court to send a ship on [this] mission, Captain Hervor Bogge, with 60 men and a garrison and the Nassau sloop *Captain David Burne*, who left on April 26 of last year, while the sloop is expected within two months with the intention to take possession of the said place to then send greater forces.[107]

107 ANTT, MNE, caixa 560, doc. ME.III-A-Fr.C1.nº.102, Paris, January 11, 1737.

In 1740, this time in the context of the War of Jenkins's Ear involving Spain and Great Britain, the latter sent two expeditions to invade Spain's overseas possessions. One was commanded by Admiral Anson, and it passed by the coast of Brazil. From Paris, D. Luis da Cunha monitored events. As always, there was uncertainty regarding the results of the enterprise, and it was considered that Portugal could benefit from conflicts between the British—their eternal allies—and Spain. For that reason, the ambassador speculated that

> the uninhabited coast south of the Rio de la Plata, which extends up to Magellan's Straits, is not an object [of the expedition] for reasons that are known to all who have read and know of the travels that were made in that region. And passing through that strait to the coasts of Chile is an idea of more work and difficulty than hope. Which led me to conjecture that no more advantageous acquisition could be conceived of here than Buenos Aires, because of the natural advantages of the place as well as the opportunity to navigate the River Plate: whose consequences are far from overlooked here.[108]

D. Luis believed that England's establishment in the Buenos Aires region would ensure Portugal's possession of Colônia do Sacramento, putting an end to the conflicts with Spain in that matter. However, that would be of little use, since to him maintaining that post was costly and resulted in little or no advantage. On the other hand, the consequences for Portugal's control of the rest of Brazil, especially Minas, which he considered vital for Portugal's enrichment, would be disastrous, because

> if the English become lords of Buenos Aires, they will soon become lords of the River Plate, taking Montevideo, as well as the Paraguay River and the Jesuit missions, from where they will confront the *Paulistas*, depriving them through their goods of the gold of our Minas, and as a consequence of the trade with our merchants, in addition to turning the ports of Brazil into their entrepôts, where their ships would enter alleging difficulties, or repairs, or provisions, [unless] they find it more convenient to establish themselves on Santa Catarina Island and Ilha Grande [on the Rio de Janeiro coast] to negotiate from there with Rio de Janeiro.[109]

108 BNP, R, maço 62, n. 2, doc. 133.
109 BNP, R, maço 62, n. 2. doc. 137, f. 3v.

It can thus be seen that Portuguese sovereignty over Colônia do Sacramento only made sense if maintaining the right to the gold mines dependent on it. To D. Luis da Cunha, the perception of Brazil's centrality with regard to Minas Gerais, the captaincies of Goiás and Mato Grosso, and the Amazonas was part of a broader geopolitical plan to ensure the sovereignty of the Portuguese empire over the Americas.

While the simple invasion of Fernando de Noronha no longer seemed to forecast the loss of Brazil as a whole or of Minas Gerais in particular, as he had feared in the 1720s, the same could not be said of control over Colônia do Sacramento. Due to its connections to the mining areas in Minas Gerais and the Goiás and Mato Grosso captaincies, he continued to warn that

> seafaring and trading nations are not engaged in wanting nor can they usurp them, [and] are always studying the ways to take advantage of them, and therefore it behooves us infinitely to hinder them, at least those who trade clandestinely with Brazil.[110]

5 The Uti possidetis or the Colonization as an Act of Possession

As a result of the disputes over Portugal's sovereignty over the Amazonas and Colônia do Sacramento regions and his preoccupation with maintaining control over gold production in Minas Gerais, D. Luis da Cunha began to articulate some of the issues that would become cornerstones of his thinking on how negotiations with Spain should be managed: the use of natural landmarks to establish limits, the respect of previous treaties, and the idea of *Uti possidetis*, that is, to "consider valid the possession each one currently has," according to each colonization process.[111] We will examine the last here, considering that it was not by chance that these three premises first appeared as a whole in arguments in favor of Portugal's territorial interests in the Americas in the Treaty of Madrid of 1750, clearly as a result of the influence of the geopolitical thinking outlined by the ambassador in the elaboration of Portugal's negotiation strategies.[112]

The issue of the definition of borders is directly related to the issue of the exercise of sovereignty by *ancien régime* monarchies during the modern era.[195]

110 AUC, CLC, doc. 408, Paris, July 5, 1737.

111 Cf. Cortesão, *Alexandre de Gusmão e o Tratado de Madrid*, vol. 4, ff. 92–93; Furtado, "Portuguese America under Foreign Threat," pp. 128–19.

112 Cluny, *D. Luis da Cunha e a ideia de diplomacia em Portugal*, pp. 99–102.

In fact, "almost all monarchies of western Europe began an intense process of incorporating territories (through inheritance, conquest, or agreement)."[113] During the Age of Discovery, Portugal and Spain's right to exercise dominion over their overseas territories was based primarily on the principle of discovery as recognized by Pope Alexander VI's bull. D. João II (1445–1495), fearing the consequences that such prerogatives could have on his claims to the Indies, signed the Treaty of Tordesillas with the Spanish monarchs and attached to the text of the treaty that, once the boundaries had been demarcated, all "seas, islands, lands, which remain with us and our successors, are ours and of our lordship and conquest, and thus of our kingdoms and their successors." The right of conquest, which had justified the formation of the Portuguese kingdom, taken over in the war against the Moors, became, alongside that of discovery, a justification for the possession of overseas territories, which constituted a hereditary right and had to be respected by the other European monarchies.[114]

Until the eighteenth century, discovery and conquest were rights unrelated to the act of colonization, which may or may not have been a consequence of the first two. Discovery and conquest (*jus belli*) alongside what had been recognized in previous negotiations (*jus tractum*) were the recognized possessory mechanisms for claiming territories in America, instrumentalized in diplomatic negotiations (*jus legations*), and disconnected from the process of colonization, which could or could not be carried out at the discretion of the possessor's interests. The concept of *jus de facto* that started to be used was a revival of *Uti possidetis*, derived from Roman law from during the process of the Roman empire's expansion and justifying the permanent establishment of settlers in colonies. But until the Treaty of Madrid colonization did not delegitimize possession justified by discovery, conquest, or previous treaties, it did not override them and what officially belonged to each Iberian crown overseas followed what was originally established in the papal bulls and the Treaty of Tordesillas.

113 Pedro Cardim, "La jurisdicción real y su afirmación en la corona portuguesa y sus territórios ultramarinos (siglo XVI–XVIII): reflexiones sobre la historiografia," in Francisco José Aranda Pérez and José Damião Rodrigues, eds., *De re publica Hispaniae: uma vindicación de la cultura política en los reinos ibéricos en la primera modernidad* (Madrid: Sílex, 2008), p. 362.

114 Furtado, "Da conquista à dicotomia colônia-metrópole: uma cartografia dos vocábulos políticos no contexto da independência," in Furtado and Andréa Slemian, eds., *Uma cartografia dos brasis: poderes, disputas e sociabilidades na independência* (Belo Horizonte: Fino Traço, 2022), pp. 41–68.

While history and historical cartography were systematically invoked during negotiations of the territories in dispute, it was the actual colonization of the space which in the final analysis justified the maintenance of its possession and served as an argument to deny the right to the territory to the rival power, even if it previously had rights over it. In other words, *jus de facto* overpowered *jus de jure*. It is no coincidence that at the turn of the seventeenth and eighteenth centuries, the practice of demarcation based on maps branched out as an important tool of territorialization of the European empires. This is especially evident from the European wars with the Ottoman empire, in which the *Uti possidetis* principle of delimitation was applied, and the use of maps was developed. For example, the Peace Treaty of Carlowitz was the first in which "the Ottoman interpretation of the Uti possidetis (as you possess) principle was accepted,"[115] and to establish a complete methodology for the application of this principle and elaborate in detail the practice of mapping and marking borders for diplomatic purposes.[116]

D. Luis da Cunha didn't comment on the use of *Uti possidetis* in the Treaty of Carlowitz, but it can be assumed that he knew the details of the negotiation, not only because it was part of his diplomatic culture to study the terms of the various treaties, but also because the ambassador had closely followed the wars against the Ottoman empire. His interest was piqued after the Infante Manuel, the younger brother of D. João v, left Portugal in 1715 against the king's wishes. Two years later, together with the son of the Count of Tarouca, he fought in the Battle of Belgrade as part of the army commanded by Prince Eugenio. As the Infante was wounded in the battle, the two men went on to Utrecht and The Hague, taking shelter in the Portuguese embassy where D. Luis and the Count of Tarouca were serving.[117] It was in 1714, in the context of the Treaty of Utrecht, that D. Luis da Cunha reflected on the artifice of occupation (*Uti possidetis*) as a guarantor of possession for the first time, recalling what happened when the Scottish were allowed to colonize the territory of the Gulf of Darien (located on the border between present-day Colombia and Panama). They requested that the right be extended to all the "lands not possessed by any European power, to [all] uninhabited lands." In other words, uninhabited lands could not be claimed by European powers, because they had not been previously

115 Zsolt G. Török, "Karlowitz, Treaty of (1699)," in Gábor Ágoston and Bruce Masters, eds., *Encyclopedia of the Ottoman Empire* (Chicago: Chicago University Press, 2008), p. 709.

116 Török, "Karlowitz, Treaty of (1699)," pp. 709–10; Rifaat A. Abou-el-Haj, "The Formal Closure of the Ottoman Frontier in Europe: 1699–1703," *Journal of the American Oriental Society*, 89, 3 (July–September 1969), pp. 467–75; Colin Heywood and Ivan Parvev, eds., *The Treaties of Carlowitz (1699): Antecedents, Course and Consequences* (Leiden: Brill, 2019).

117 ANTT. MNE. Livro 137; BNP. R. AT 26[8].

colonized. They were lands without an owner (excluding Native Peoples from that right) which could now be claimed by a nation that was only beginning its colonization in the region. Although the Scottish Company was denied that right, because "it was not convenient to England," D. Luis da Cunha worried at the validity of the argument that a lack of previous colonization resulted in the absence of the right of possession, hitherto assured by discovery or conquest, and that an intention to colonize could extend the possession to contiguous areas. Therefore, regarding Brazil, in the context of several foreign invasions, especially French and Spanish, he warned that "it would be good to take precautions with acts of possession of all places still unoccupied that could seem propitious for such settlements."[118]

It was from this perspective that he began gradually developing the notion of *Uti possidetis*, the right to the land by those who had occupied and populated the territory, first as a complementary and later as the sovereign element to what was previously agreed in treaties, taking precedence over them. He begins expressing this concept more clearly in 1736, an ill-fated year as his fears of a possible loss of Portuguese hegemony in Brazil became a reality. Colônia do Sacramento was in danger, surrounded by the Spanish, and the French had invaded Brazilian territory, taking hold of Fernando de Noronha Island.

As mentioned, it was precisely in 1736, in the context of the Spanish siege of Colônia do Sacramento and the invasion of Fernando de Noronha Island, that D. Luis da Cunha radically changed his discourse and started to push the crown to adopt a more active and openly expansionist policy in southern Brazil. It was a quite different position from the one he had held in 1720–1721. He put aside his policies of relative inaction, since up to that point he had feared Spain's reaction to an increased Portuguese presence on the northern bank of the Rio de la Plata estuary. He then suggested that the king should: 1) "send to Rio de la Plata the forces we have in Brazil to recuperate and in addition attack Buenos Aires if possible"; 2) "in the surrounding waters let us not only displace our enemies, but also build the planned fort of Montevideo, because we will keep it while the extension of the territory of Colônia do Sacramento toward the Uruguay River is disputed; if perchance an equivalent in Europe is discussed, the more we have to give up, the better"; 3) "build [the fort] in Montevideo, settling it 'under the cover of the people and ships sailing from Bahia and Rio de Janeiro'"; 4) "begin fortifying and settling the islands that may be apt for it, as for example Santa Catarina, etc."[119]

118 BNP, R, cód. 11.209, ff. 435–35v, June 21, 1714.

119 AUC, CLC, doc. 342, July 20, 1736; doc 359, August 17, 1736; doc. 346, August 24, 1736; doc. 408, July 5, 1737.

Immediately after, D. Luis wrote to the king making suggestions about reinforcing the Portuguese presence at the mouth of the Rio de la Plata. Among the various measures to keep control of the territory's strategic areas he suggested building a fortress in Montevideo, on the septentrional border of the river, settling it "under the cover of people and ships sailing from Bahia and Rio de Janeiro, because it is easier to negotiate when it is done based on *Uti possidetis* than otherwise."[120]

It is significant and it cannot be perceived as a mere coincidence that it was precisely in 1737 that, as the ambassador suggested the previous year, the Portuguese crown began settling and more effectively colonizing the region of Rio Grande do Sul, building the fort of Jesus Maria José. Bringing together his reflections on Colônia do Sacramento and Fernando de Noronha Island, D. Luis da Cunha finally articulates the concept of *Uti possidetis*, naming it explicitly, which then goes from his correspondence to the text of the *Dissertation qui determine tant geografiquement que par les traités faits entre la Courone de Portugal et celle D'Espagne*,[121] which he had been writing together with Alexandre de Gusmão, private secretary of D. João v. The document was supposed to synthesize Portugal's view of the differences with Spain over Colônia do Sacramento, summarizing the terms of previous negotiations and expounding the way in which future ones should be guided. D. Luis da Cunha corrected and disagreed with some of the positions defended by Gusmão. His corrections and comments would be inserted in the body or placed in the margins of the document. The events of 1736, together with all of his previous experience, had a direct impact on D. Luis's intervention in the process of drafting this document.[122]

D. Luis's most important contribution to the original text of the *Dissertation* appeared in the part of the text where Gusmão began to discuss the meaning of the term "Colônia do Sacramento and its Territory", which the Portuguese ambassadors had put into the Treaty of Utrecht. In the margin of the document where he discusses the term colony and *colonus* under the Roman empire, evoking *Uti possidetes* although not naming it, he added that it is

120 AUC, CLC, doc. 346, August 24, 1736.

121 Alexandre de Gusmão, "Dissertation qui Determine Tant Geografiquement que par les Traités Faits Entre la Courone de Portugal et Celle D'Espagne Quels Sont les Limites de Leurs Dominations en Amérique c'est a dire du Coté de la Riviere de la Plate (avec la collaboration de Dom Luis da Cunha)," in Cortesão, *Alexandre de Gusmão e o Tratado de Madrid*, pt. 3, vol. 3, pp. 367–417.

122 Furtado, "Portuguese America under Foreign Threat," p. 127.

> known that the Romans gave the name of colony, which comes from *colonus*, to a certain number of people sent to cultivate and settle land they possess or which they intend to get revenue from, so that when His Catholic Majesty [the Spanish king] in the treaties states that he cedes Colônia do Sacramento, it does not mean that he cedes the people who have settled there and use the land, but the lands themselves.[123]

Two important transformations can be seen here in relation to the mechanisms of frontier negotiations: one semantic and the other juridical. The principles of discovery and conquest were no longer invoked but of colonizing the land. The name *Colônia* do Sacramento became a clear reference to the colonizing act and created a direct link between its residents as permanent settlers of the *colonus* and between them and the process of colonization. Following Roman law from the period when Roman colonies were being expanded, the legal concept that the colonist and the colony were inseparable entities guaranteed *de facto Uti possidetis* over a territory. In D. Luis da Cunha's view, colonization was thus configured as a juridical condition in itself and the Roman principle of *Uti possidetis* no longer had to refer to possession based on previous rights, instead becoming based on *de facto* possession.

In his *Instruções políticas*, aimed at Marco António de Azevedo Coutinho, D. Luis does not use the term *Uti possidetis* but mentions the colonization strategy as a form of making "the Castilians lose, which must always be our main object." For this the encouragement of migration was suggested, even of Jews and foreigners, since "it is not advisable to depopulate Portugal to populate Brazil." The intention was to settle them on those lands to "penetrate that vast country," cultivating its fields and discovering new mines, including silver. Once they arrived, they were prohibited from returning, "there they would marry, and their children would be very good Portuguese and Roman Catholics," and they could "bear fruit in some part of Brazil, because being so vast, it seems that any canton can be found." This would increase its revenue and expand its frontiers.[124]

For D. Luis da Cunha, as is clear in his comments on the *Dissertation*, it was a judicial principle to be used diplomatically in a long-term strategy. As can be seen in his interventions in the original text, for him it was not only to be used occasionally in campaigns in an intermittent and shared form, but was a *de facto* settlement made by individuals who held them as possessions, and this was one of the proofs of strength which assured Portuguese dominion over

---

123 Cortesão, *Alexandre de Gusmão e o Tratado de Madrid*, vol. 3, p. 395.

124 Cunha, *Instruções políticas*, pp. 342–43, 346–47.

the extensive northern bank of the Rio de La Plata, far beyond the Portuguese Colônia do Sacramento fortress.[125] Mirroring D. Luis da Cunha's geographic and diplomatic strategies, d'Anville's *Amérique méridionale* shows the Portuguese territories in South America filled with colonized settlements in a way that *Uti possidetis* could be used to justify their rights over Brazil's continental territory. Strategically, according to the ambassador's instructions, the map prepared in Paris to subsidize the diplomatic negotiations between the two crowns would be a documentary proof of Luso-Brazilian dominium and at the same time would guide the new demarcation line of the frontier with the Spanish colonies.

125 Cortesão, *Alexandre de Gusmão e o Tratado de Madrid*, vol. 3, p. 404.

PART 3

Maps

CHAPTER 5

Amérique méridionale, a Modern Knowledge

1 An Enlightened Map

D'Anville placed the South American continent in a unique position of being able to advance the geographical knowledge of his time. He emphasized the importance of his *Amérique méridionale* map given the relative lack of knowledge of an extensive part of inland Brazil, which he considered "an object of great speculation (…) in one of the vastest continents in geography."[1] Hoping to make his own contribution, he affirms that his "map of *Amérique méridionale* lays bare before the eyes, with details of the place, many new circumstances, given the empty state of our knowledge of these totally bare spaces." In opposition to this ignorance as to the interior geography of the continent, d'Anville holds up his map as the best poised to further understanding of "that which previous maps attempted to represent as vague things and in figures based on pure imagination, which was before then an absolute void."[2] In his works he would make this point several times, employing terms like "void," "bare," and "unknown" to refer to territorial spaces still unmapped by geographers. According to him, the great geographic discoveries of that century would be observed on the South American continent, precisely "in the interior of that grand swathe of land."[3] A "continent so vast in geography, will give us the opportunity to add to these new acquisitions a great improvement in the detail of what was believed to be known."[4]

The memoir he wrote about his *Carte de l'Inde*, published some years later in 1753, presents yet another justification for the important role the geographer attributed to the *Amérique méridionale*. He makes it clear that in making the map of India he had to rely on a cartographical foundation inherited from Antiquity. Only through the modern geographer's critical reading of ancient cartography—with improvements, additions, and clarifications—could a new map be created. He argued that this mechanism could be used in mapping the three major continents—Asia, Europe, and even Africa—seeing as they

1 RBC, n. 539 (1), f. 1.
2 RBC, n. 539 (1), f. 2.
3 Anville, *Considérations générales sur l'étude*, p. 64.
4 Anville, *Considérations générales sur l'étude*, p. 1.

 | DOI:10.1163/9789004710764_007

had been represented on maps since Antiquity.[5] However, the *Amérique méridionale* demanded a radically different attitude on the cartographer's part, given that no similar tradition existed for that region. This was an unknown world—the New World. And so America, as a geographic entity, would be a modern creation, the only continental space to be entirely developed by the light of a new, equally modern geography. One thus sees that the *Amérique méridionale* could only be constructed using a range of modern sources, produced after the discovery of America—and, given that reconnaissance of the interior of the continent was even more recent, most of these sources would be practically contemporaneous with the mapmaking process. Because of that he would reassert: "If we shall bear witness to a considerable increase in geographical knowledge, it will be found in America, precisely in the interior of this great swathe of land."[6]

Over time, d'Anville produced three different versions of *Amérique méridionale*: 1737, *c.*1742, and 1748 (Figs. 1, 2, 3). A glance at the 1737 map produced for the private use of the Duke of Chartres, at the map attributed to *c.*1742, and the 1748 map, reveals the advance the geographer managed to effect in the knowledge of South America. The first two are manuscript and the third is printed, the map attributed to 1742 is undated. I work from the assumption that this is a copy of the map made at the request of D. Luis da Cunha at that point.[7] It is very similar to the 1737 version, but it is not identical. A few toponyms have disappeared, such as the Rio de la Plata, or Castillos Point; others were corrected. In the two first maps (1737 and *c.*1742), the entire central region of the continent is depicted as an enormous void, labeled "unknown country." The volume of geographical information grows exponentially in the 1748 map, visible in the detailing of the hydrography of the Andes, Amazonas, southeast and south of Brazil, revealing progressive European control of the South American interior. D'Anville's access to the Franco-Spanish expedition's documents was fundamental in this respect, as was his access to Portuguese sources, provided by D. Luis da Cunha. But at the same time the interior void was also enlarged.

This being d'Anville's largest map to date, the printed 1748 *Amérique méridionale* demanded "three (separate) large sheets," each a separate map itself.[8] Together they made up approximately 127 cm long by 80 cm. The north, sheet number 1, was titled *Carte du Pérou et Brésil septentrionale de Tierra-Firma, de Guayana et de la Rivière des Amazones* (Fig. 4). The center of the continent,

5 Anville, *Éclaircissemens géographiques sur la carte de l'Inde.*
6 Anville, *Considérations générales sur l'étude*, p. 64.
7 Cortesão, *Alexandre de Gusmão e o Tratado de Madrid*, vol. 3, p. 787; vol. 4, p. 908.
8 *Mercure de France* (November 17, 1749); *Mercure de France* (December 1749), p. 328.

sheet 2, was taken up by the *Carte qui represente la partie méridionale du Brésil, et du Pérou, le Chili Septentrional, et le Paraguay*. The south, sheet 3, was represented in the *Carte du Chili Méridional, du Rio de La Plata, des Patagons et du Détroit de Magellan*. Brazil occupies sheets 1 and 2. The strategy of separating the continent into three pieces would make it easier to revise specific areas, as would happen over the coming years. Subsequent editions of the map, dated 1754, 1760, 1765, 1772, and 1779, respectively, would allow D'Anville to alter small areas of the map.[9] Strange as it may seem, however, these various printings with their posterior corrections and additions were all stamped with the date of 1748. The repetition of the year of the first edition, which makes it difficult to correctly identify the various other printings, was not common at the time; normally, when modifications were made in a given map, geographers would emphasize it to boost sales of the new edition, highlighting the new map's degree of precision in relation to the previous one. It was common to update the year on the cartouche and include observations such as "revised and expanded in many places in accordance with the most recent memoirs," or "according to the newest precise observations."

Alexander von Humboldt (1769–1859) was among those who complained that d'Anville, "in making important corrections on the copperplates of his maps, did not mark the period in which these changes were made." He pointed out that among the consequences of this lapse was the fact that "geographers who do not know the circumstances could be inclined to err about the dates of many discoveries."[10] However, d'Anville's manuscript and printed memoirs, in which he explains the revision of each edition of the map, aid in revealing when and why specific changes were made. The English geographer John Green (b.1688–17?) was another who took issue with d'Anville, as he, despite his promises, did not publish his memoirs on the *Amérique méridionale* in book form.[11] Though he did not release a book, as he did with other maps, he did publish two memoirs, as well as articles, in the *Journal des sçavans* in March

9 In the 1754 version, he added mountains around Lima; in the 1760, he altered the course of the Orinoco and its tributaries; in the 1765 version he corrected the position of Lake Xarais, in the interior of Brazil; in 1772 he modified the shape of the Falkland Islands; and finally, in 1779, he reconfigured the southern part of Brazil, around the region of Paraná, as well as Paraguay. Dacier, *Notice des ouvrages de M. D'Anville*, pp. 98–99.

10 Cf. Neil Safier, *Measuring the New World: Enlightenment Science and South America* (Chicago: Chicago University Press, 2008), p. 123.

11 John Green, *Remarks in Support of the New Chart of North and South America* (London: J. Jefferys, 1753), p. 6.

and April of 1750, shortly after the printing of the map.[12] This journal, tied to the *Académie Royale des Sciences de Paris*, constituted the primary vehicle for the publications of French scientists. Later on, in 1779, d'Anville drew up two more manuscript memoirs in which he refers to posterior changes made to the map.[13]

D'Anville made note of the fact that the continent took in a vast span not only from east to west, but also from north to south, permitting observation of the progressive contraction of the meridians as they approached the poles.[14] Aware that any projection meant the deformation of the object being represented, throughout his production, he did not limit himself to a single form, varying in accordance with the location on the Globe of the area represented. Because of this he chose the pseudocylindrical projection for his 1748 *Amérique méridionale*.[15] This projection, frequently used to represent the hemispheres, represents them as cones that unfurl across a plane, thus accentuating the reduction in distance between meridians as one approaches the poles—a sign of d'Anville's acceptance, at this point, of Newton's theory.[16] The map is oriented around the Ferro Meridian, cutting through one of the Canary Islands. One of the interesting aspects of this meridian is that it would become a symbolic mark positioned as it was halfway between the New World and the Old.

The *Amérique méridionale* was printed by Guillaume Nicolas Delahaye (1725–1802), a descendant of a traditional family of printers. By this period his workshop was established on the Rue de l'Arbre Sec, just past the Louvre Palace. Delahaye had already printed the *Amérique septentrionale* for d'Anville in 1746. Since the geographer envisioned—and advertised in the *Mercure de France*—uniformity across the two maps, down to questions of scale, this was yet another reason beside his skill at the trade to choose Delahaye as printer. Later on, and probably for the same reasons, Delahaye would print d'Anville's *Carte de Canada, Louisiane et Terres Angloises*, in 1755, and the *Carta de la provincia de Quito*.

In 1751, D'Anville explained to the president of London's Royal Society, Martin Folkes (1690–1754), that *Afrique*, published in 1748, was part of "a series of very extensive general letters in several sheets, [produced] for the benefit of M. duke d'Orleans" and, in fact, the map of the African continent was

12 Anville, *Journal des sçavans* (March 1750), pp. 522–33; Anville, *Journal des sçavans* (April 1750), pp. 625–73.

13 RBC, 229, ns. 539 (1) and 539 (2).

14 NL, VB, Ayer MS 1918, doc. 552, f. D-3.

15 Aguilar, *Historia da Cartografia*, p. 218.

16 Furtado, "Embates na terra de papel."

published following the *Amérique septentrionale* of 1746 and the *Amérique méridionale* of 1748, and before the *Asie*, on which he reckons he was already working, and the *Europe* that follows it.[17] In the announcement that D'Anville posted about *Afrique* in the *Mercure de France*, he states that his maps from the other continents "add, to the precision we have tried to put into them, a great elegance of execution." As he sought a more accurate representation of the continents' geography, D'Anville developed his methodology and improved the techniques of composition of cartographic elements, seeking increasingly universal and rational symbols of representation. His charts of the continents produced from the 1740s onwards reveal that, by this time, he was capable of masterfully manipulating not only the critical methodology to produce them, but also the visual refinement in the way of representing the territory. About *Amérique méridionale* he stated, "he devoted a part of his time to forming his personal tastes and style through the study of our great writers."[18]

As art, cartography was also being modified by the new demands of a more naturalistic taste, without exaggerations or rococo and its transformations were already moving "toward the classicism, advocating a *Good Taste* that is defined in proportion, order and unity, factors that should apply as much to a written work as to a painting."[19] When announcing his new *Amérique méridionale* in 1748, d'Anville wrote to the *Mercure de France* communicating his achievement and asking the editors to inform the reading public since "your taste and your sense become more interesting day by day." He then describes the aspects of the new map that were in accordance with the new cultural codes, providing "the understanding of the subject," since "the knowledge I have introduced there and the uniformity of scale, (…) and the degree of accuracy is provided almost necessarily to the detail of knowledge. To these advantages is added a great elegance in the execution of the engraving." He also informs us that it was produced "by the desire of a great prince," the Duke of Orleans, "who encourages my studies by his benedictions." The patron's indication pointed to the quality of the work, produced for consumption by the nobility, according to their new demands of taste, as "the magnificence of the prince gives way to nothing to economize."[20] D'Anville also credits the duke with the good taste employed in the map, not only because "his lights & his

17 Anville, "Lettre de M. d'Anville, à M. Folkes, Président de la Société Royale de Londres, sur une copie faite à Londres de la *Carte de l'Amérique septentrionale*," *Mercure de France* (Paris, November 15, 1750), p. 151.

18 Dacier, "Éloge de M. D'Anville," p. 42.

19 Nuno Saldanha, *Poéticas da imagem* (Lisbon: Caminho Editorial, 1995), p. 280.

20 Anville, "Lettre de M. d'Anville, à M. Remond de Sainte Albine, sur une nouvelle *Carte d'Amérique Méridionale*," *Mercure de France* (Paris, November 17–December 1749), p. 169.

taste in the sciences commit me to composing works of this kind," but because "the magnificence of the prince gave way to sparing nothing in this respect."[21]

It is noted that taste "appeared both as a criterion for evaluating the quality of maps but also as a general rule in their production, applicable to every cartographic mode."[22] This meant not only the pursuit of geographical accuracy, but also harmony of form, expressed in restraint, proportion, and hierarchy of both graphic and textual elements. In the cartographic style, this meant banishing from the map all decorative excesses. This is what can be observed from the *Amérique méridionale* cartouche. So in order to get to know the map better, we must start by analyzing its cartouche (Fig. 23). Designed by his brother Gravelot, it expresses both figurative contention and rare beauty, in keeping with the new taste of the triumphant rationalism of which d'Anville was a part. The cartouche depicted the victory of European Christianity: a woman who holds a cross in one hand and a chalice in the other, the infidel natives of the region at her feet. In this allegory of the triumph of civilization over barbarism, the Americans are represented by three male figures. The two on the left—a wild American Indian with a feather loincloth and a native of the Andes, perhaps an Inca, dressed in the fabric of the region—are kneeling, both in submission and in adoration of the cross. The figure on the right, holding a bow and arrows, sprawls languidly across the frame, revealing the title of the map with a flourish of his hand. Around the frame are elements of the local flora and fauna (an armadillo, a bird, and a sloth), suggesting the lush tropical nature of the continent.

Apart from the title, the first section of the cartouche is blazoned with a dedication to the Duke of Orleans, the First Prince of the Blood of France, a great patron and sponsor of d'Anville's cartographical work, and whom he was currently serving as secretary. This reference not only made the homage to the duke public, it also gave the map status—connecting it to one of the most important authorities in the country, a man also known for his vast culture, taste, and for his sponsorship of the arts. Below the date appears the phrase "avec privilege." At this time there was no legislation in France guaranteeing the recognition of the authorship of an intellectual work, nor the corresponding rights. However, there was an institution—"the *Librairie*,[23] [which] supervised and controlled the book and print trade." According to the law, "the protection for an author or a graphic designer and/or engraver was found in

21 Anville, "Lettre de M. d'Anville à l'auteur du *Mercure* sur la publication d'une nouvelle *Carte de l'Afrique*," *Mercure de France* (Paris, March 1751), pp. 157–58.

22 Nicolas Verdier and Jean-Marc Besse, "Art and Design of Maps," p. 118.

23 Antoine Perrin, *Almanach de la librarie* (1781) (Paris: Aubel, 1984).

FIGURE 23 Cartouche, by Gravelot, in *Amérique méridionale*, J.B.B. d'Anville (Paris, printed by Gravelot, 1748). BNF, DCP, GE C-6149
REPRODUCED COURTESY OF FBN

the permission and the privilege" conceded by this organ,[24] without which the publication of any work was expressly prohibited. In each area, a censor examined titles to be approved; in the case of geography, that included both books and maps.[25] By paying a fee, which might be divided between the author, the engraver, and the publisher, and by agreeing to leave three copies of the map in the king's *Librairie*, such a permission or privilege would be conceded. There were two types of concessions: the *exclusif* or *général* and the *privilège* or *permission simple*. The first, which was more expensive, "granted exclusive rights to make, distribute, and sell a product throughout France for a period of from three to twenty years." The second, which was cheaper, allowed the author to sell his work to more than one publisher, covering a three- to six-year period.[26] By opting for a *privilège simple* for the *Amérique méridionale*, d'Anville was guaranteeing the controlled production, sale, and distribution of his map, but over a shorter period. As was written to the left of the cartouche and advertised in the *Mercure de France*,[27] the map could be acquired in the author's own studio. The concession of a *privilège* from the *Librarie* meant that the work was protected by the king and could not be copied by others within French territory. This had no bearing on the reproduction of the map in other countries, however, where an unauthorized version of the map appeared very early on, titled *Map of South America* and published by J. Harrison.

According to d'Anville, "one of the great advantages of the *Amérique méridionale* (…) is its having acquired, in a country occupied by the Spanish, a degree of perfection which cannot go unsung."[28] Without a doubt, the 1735 Franco-Spanish Geodesic Expedition to Quito, sent by the *Académie Royale des Sciences de Paris* to make a meridian arc measurement at the Equator and thus contribute to an understanding of the shape of the Earth, provided information fundamental for the refinement of d'Anville's maps in the region of the viceroyalty of Peru.[29] One by one, as the members of this expedition returned to France, they became the geographer's collaborators.

However, despite the affirmations of the cartouche, the expedition's contribution was not limited to the Hispanic part of the continent; their findings

24 Pedley, *The Commerce of Cartography*, p. 103.

25 Perrin, *Almanach de la librarie*, pp. 7, 14.

26 Pedley, *The Commerce of Cartography*, p. 104.

27 *Mercure de France* (November 17–December 1749), p. 169.

28 Anville, *Considérations générales sur l'étude*, p. 65. For all contributions that d'Anville used to draw the north of Brazil see Furtado, and Cintra, "Bourguignon D'Anville's *Carte de l'Amérique méridionale*."

29 Safier, *Measuring the New World*.

also helped to set the length of the course of the Amazonas River.[30] A number of mishaps forced a longer-than-planned stay in America, and the new geodesic observations from Peru were not crucial in revealing the shape of the Earth. For this reason, the Frenchman Charles Marie de La Condamine and the Hispanic-American Pedro Vicente Maldonado (1704–1748), decided to return to Europe by going down the Amazonas River; the river's course was of great interest to European scholars, given that their knowledge of its position was at this point fairly relative. La Condamine was a member of the *Académie Royale des Sciences de Paris* and Maldonado an enlightened Spanish *criollo*, governor of the province of Esmeraldas. Taking slightly different routes, the two went down the river from Peru to the city of Belém do Pará. The information they collected about the positioning of the riverbed was only partially incorporated into 1748 *Amérique méridionale*, although d'Anville hid this in the geographical memoirs he wrote about the map.[31]

On the lower part of the cartouche, the geographer paid tribute to the members of the expedition for their contributions, but made a point of differentiating the contributions of each one. He wrote that the establishment of the "course of the River of the Amazons and its tributaries" had been based on "collecting the documents" that La Condamine had brought back with him, in addition to direct observations made as he traveled down the river, as he had already "communicated all of interest to an active and clarifying curiosity" in regards to what he had seen of the region on his trip. D'Anville also declared himself "indebted for the use he had made of a few important pieces of information" provided by Pierre Bouguer, head of the expedition, without specifying what precisely this scholar provided him with; and he thanks Maldonado, whose contribution is limited to the geography of the "northern part of Peru."

When La Condamine returned to France, Bouguer—who had arrived before him—had already read his *Succinct Abridgment of a Voyage Made within the Inland Parts of South-America* before the assembly of the *Académie* on November 14, 1744.[32] His report "related his journey and the operations he had undertaken to determine the form of the Earth."[33] La Condamine thus resolved to simply publicize his deeds on the Amazonas River. With data that he then

30 Furtado and Jorge Pimentel Cintra, "A *Carte de l'Amérique méridionale* de Bourguignon D'Anville."

31 Safier, "Como era ardiloso o meu francês: Charles Marie de la Condamine e a Amazônia das Luzes," *Revista brasileira de história*, 29, 57 (2009), pp. 91–114.

32 Mary Louise Pratt, *Os olhos do império: relatos de viagem e transculturação* (Bauru: Edusc, 1999), p. 47.

33 Cf. Monteiro, ed., *Meu pai e meu senhor muito do meu coração*, p. 49.

FIGURE 24 *Carte du cours du Maragnon ou de la grande route des Amazones dans sa partie navigable depuis Jaen de Bracomoros jusqu'à son embouchure et qui comprend la province de Quito, et la côte de la Guiane depuis le Cap de Nord jusqu'à Essequebè levée en 1743 et 1744 et assujetie aux observations astronomiques par M. de La Condamine*, J.B.B. d'Anville (Paris, printed by G.N. Delahaye, 1743–1744). BNF, DCP, GE DD-2987 (9542) and (9543)
REPRODUCED COURTESY OF FBN

claimed had been collected from direct observation in the territory,[34] his texts, read in the *Académie* and then published in books, were destined to impress intellectual circles but also the general public; for this reason, his work would be published as a fantastical narrative.[35]

The story of his journey down the river was turned into a book, in two slightly different versions: the 1745 *Voyage sur l'Amazone* and the 1751 *Journal du voyage fait par ordre du roi a l'equateur*. In publishing the former, La Condamine partnered with d'Anville to produce a few maps illustrating points of the story. The *Carte du cours du Maragnon ou de la grande route des Amazones* (…) *levée en 1743 et 1744 et assujetie aux observations astronomiques par M. de La Condamine* (Fig. 24) was introduced in 1745 to illustrate this edition. Based on La Condamine's longitude measurements, this was the first map of the Amazonas River, worked up from a sketch provided to d'Anville—"whose ability," La Condamine writes, "is well known, [and] was of great advantage to me in the joining and composition of these sparse materials."[36]

34 Safier, "Como era ardiloso o meu francês," pp. 91–114.
35 Pratt, *Os olhos do império*, pp. 47–48.
36 Charles-Marie de La Condamine, *Viagem na América meridional descendo o Rio Amazonas* (1745) (Brasília: Senado Federal, 1984), p. 34.

In keeping with his style—with careful attention to the study and comparative establishment of measurements—*Amérique méridionale* has four different scales. The first, in French leagues (*lieues*), is used as a point of reference. The second scale works with French, English, and Spanish nautical leagues; these were measurements necessary for navigation, as these maps were important for guiding navigators, and the uncertainty of navigating through "a vast ocean" demanded "the prudence of having this stretch of ocean in greater detail."[37] Then came Spanish leagues, which, as d'Anville explains, were also used by the Dutch in Guyana. Finally came a travel measurement, which corresponded to the hours and distances a traveler could expect to traverse in a given range. In addition to maps, armchair geographers made constant use of travel accounts which described the geography of the place: these are so-called mental maps. D'Anville became a specialist in transforming the geographical information contained in these accounts into cartographical representations and time of travel into measurements of distance. This kind of document was used as a source in several parts of the *Amérique méridionale*.

2 Placing Names

An analysis of the three maps (1737, *c.*1742, and 1748—Figs. 1, 2, 3) reveals the development of the geographer's style, which attained "a sort of perfection," it being declared of the maps that "they were drawn with style and clarity."[38] If D. Luis da Cunha considered Delisle somewhat careless in the finishing of his maps, the opposite could be said of d'Anville. One of his greatest concerns was with the presentation of the information in each map, with the aim being absolute clarity. One proof of this was the care he lavished on the style and nomenclature of place names. In this regard, he set down strict rules: 1) the overall name should be in the center, running horizontally, in larger letters; 2) province names should be centralized, horizontal, in thick letters; 3) river names should follow their courses; and 4) place names should be horizontal when possible, but to avoid "a principal, tiring monotony, (which) gives no sense of the variety which nature bestowed upon the work," the artist was given certain license.[39]

Examining the maps from 1737 and *c.*1742, one sees the beginning of the application of d'Anville's place name style. The name of the Pampas, for

37 Anville, *Considérations générales sur l'étude*, p. 52.
38 Du Bocage, "Avant propos," p. 9.
39 Anville, *Considérations générales sur l'étude*, pp. 66–67.

example, just below that of Buenos Aires, which appears off-center and written vertically on the first map, is then made horizontal and centralized in relation to the territory. By the 1748 version, all the rules are perfectly applied. Take the name of the Tucumán Plain, which is mirrored in relation to the two previous maps to place it in the correct position for the reader, from left to right. The region of *Paraguay* is exemplary, with an abundance of information marshaled perfectly clearly given the mastery with which he arranges the various sizes, thickness, spacing, and shapes of the letters and the way he positions the toponyms. Region names are written in larger and thicker letters; the names of the rivers follow their courses; place names, mainly Jesuit missions, are frequently written horizontally, but when necessary for reasons of clarity, as is the case of Santa Maria do Iguaçu, the rule is broken to fit the full length of this name into the small space available. Comparing this region to Minas Gerais, one sees that in both the format of the letters' size allows one to rank population clusters and differentiate geographic features. In the two previous maps, this was not carried out in any evident sense. In 1737, populated areas are almost all represented by the same symbol—red circles—with a few rare exceptions, such as Buenos Aires, being topped by a cross, while in the *c.*1742 map they are simply signaled by colorless circles. Moreover, on these two maps, there is little notable difference in terms of letter sizing.

All this care was not limited to representations of physical geography. In keeping with his propensity to think in terms of human geography, d'Anville does not neglect to note many population centers that no longer exist. In this sense, the map becomes a living witness to the human presence and the transformation of the space over time, functioning like an encyclopedia in its collection of natural history, ethnographic, geographical, and historical information. This is the case for the missions in Paraguay. And so, in the 1748 map, he informs in the upper right-hand corner that "many places with this mark + designate the towns or missions that no longer exist." Other notations recall the process of discovery and colonization. In Chile, the port near the Chonos Archipelago "was found by the English in 1741"; the volcano on Fogo Island was "spotted in 1712"; the hamlet of Nuestra Señora de las Nieves de Yurimaguas, near the Amazonas River, had been founded from "an old Spanish mission"; the confluence of the Ene and Apurinac Rivers in Peru was "mapped in 1742." The dating of events reveals the map as history encyclopedia and reflects the present temporality of mapmaking.

A refinement of this taste, marked by illustrative rationality and contention, keeps d'Anville from putting practically any color in his maps. Mountain bases (always from left to right); coasts; and rivers and lakes, indicating the existence of water, are merely crosshatched. In the first two maps, colors are a much greater presence, used to signal the limits of provinces and distinguish possession: Portuguese land in blue-green, Spanish in red, French in yellow, and Dutch in green. In the 1748 version, only a few copies are colored. In this case they are used to demarcate the external borders of Spanish and Portuguese America, and the latter's various viceroyalties, with the Equator colored in brown. In the case of the 1737 map, produced for the edification of the Duke of Chartres, colors are given a clear pedagogical value and serve to emphasize the political divisions of the territory for the benefit of the keen student.

Another concern of d'Anville's had to do with place names.[40] In the first two maps (1737 and *c.*1742) names are written in French, as opposed to the 1748 version, where the places are written in Spanish, Portuguese, and even Indigenous languages, with rare exceptions (one being Brazil, which is labeled with the French spelling, *Brésil*). D'Anville believed that toponyms ought to represent the local language. Geographers were obliged to "know, among the dominant nations, the correct terms to designate the nature of the places, rivers, mountains, islands, etc." As far as place names, he argued for a few fairly strict rules, such as respect for the spelling of the name of the mapped area, not following the author of the map and obeying the antique spelling even when later changes had been made. This demanded a thorough study and knowledge of languages, which would, he said, allow for the formation of a dictionary of geographic toponyms—a volume of great utility and interest.[41] He argued that nomenclature could not be overlooked when drawing up a map. In the *Amérique méridionale*, his meticulous spirit led him to take the greatest care even with the smallest details. For example, he opts for the name "River of the Amazons," but notes that the same river, "below the Negro River, is called Solimoens or Rivière des Poisons by the Portuguese."

However, D. Luis's doctor in Paris, António Ribeiro Sanches, wrote in regards to the place names of the *Amérique méridionale* that

> many of the names of rivers and places that one reads in said treaty [of San Ildefonso] are not to be found in Mr. d'Anville's map, on the contrary, either because they are written by a foreigner or because this geographer had no knowledge of them.[42]

40 Íris Kantor, "Cartografia e diplomacia: usos geopolíticos da informação toponímica (1750–1850)," *Anais do Museu Paulista*, 17, 2 (2009), pp. 39–61; Kantor, "L'appropriation des cartes de d'Anville dans le monde luso-brésilien."

41 Anville, *Considérations générales sur l'étude*, pp. 68, 80–81.

42 Kantor, "Cartografia e diplomacia," pp. 44–45.

In fact, in spite of all the geographer's care, nuances of local spelling often escaped him. Solimões is written as Solimoens, Baixos as Baxos, Casa as Caza, and Boipeba as Buipeba. In the case of South America, Indigenous names ought to be included alongside those in Spanish and Portuguese. All this aside, d'Anville does display a certain linguistic command of these languages, even in terms of the sonority of the words.

His knowledge of the meaning of Indigenous words is made clear in his writing out the toponym "Pampas." Just below the name he writes that "the great plains are called Pampas, from an Indigenous word which in the Quichua language of Peru means, rather, valley." One sign that the copies of his maps made outside France were unauthorized is a certain carelessness in regard to the spelling of place names. In the British copies, for example, local languages are mixed with French and English. D'Anville was right to complain about this copyist, whose works, he said, were "of crude execution (…), far from elegant, as were the original maps."[43] A quick comparison of the same region of the map in the French and English editions, between the Tropic of Capricorn and the city of Salvador, reveals these discrepancies. For example, Cabo Frio becomes Cape Frio, Baxos de Abrolhos is written out as Banks of Abrolhos, St. Antonio becomes St. Anthony, and Bahia de Todos os Santos is written as B. of All Saints. In this case, the copyist was more concerned with making place names accessible to the English consuming public than respecting the local naming, as d'Anville firmly advocated. It is no wonder, then, that he felt doubly betrayed by these copies.

The push into the South American interior, which picked up speed over the course of the eighteenth century, put Europeans in contact with many Native peoples, many hitherto unknown; and the maps of the period, including the *Amérique méridionale*, are mirrors of this movement. Indigenous maps, or even the Natives themselves, became great sources of information about local geography, indispensable for the gradual penetration of these areas, and methods of appropriating Native geographical knowledge are often evident in European maps.[44] The 1748 *Amérique méridionale* has several examples of contributions from local populations. Next to the coast of the Argentine pampas, for example, d'Anville writes, "Indians report that the islands formed in the Sauces River are inhabited by whites."

43 Anville, *Considérations générales sur l'étude*, p. 111.
44 Black, *Visions of the World*, pp. 66–67.

Indigenous peoples were not merely informants on local geography, however.[45] In keeping with the Enlightenment spirit, a nascent ethnology converted the Natives into objects of study for European scholars. Cartography was one of the physical media for the crystallization of Europeans' information and vision of these peoples. Nevertheless, though these maps were useful in giving visibility to the knowledge gathered by European explorers, their fidelity had its limits. The difficulty in accurately registering the nomadic character of some of these peoples, for example, can be seen in several maps. In the *Amérique méridionale*, when writing about the Natives who lived on the Pampas, d'Anville writes that "they live wandering across the plains." As for the Tupinambás, these were "a great nation which extends from its first dwelling on the coasts of Rio de Janeiro to this point [the interior of the Brazilian northeast] and has even spread to the border with Peru."

The map also presents several ethnographic observations with regard to Indigenous people: in French Guyana are the "Amicouanes, a nation of large ears"; in Brazil, "whose interior is largely unknown, the Natives of the country are called by a common name—Tapuias"; in the Pampas one can find "the *Puelches*, called 'serranos,' or 'of the mountain,' as they live near a great *Sierra*"; while in Patagonia "are established the Argueles and the Cesares, who are said to have mixed with the Spanish who left Chile in 1554." Intriguing enough, in Suriname, D'Anville placed at the headwaters of the river of that name—its mouth being in Paramaribo—a colony of "Muslim Negroes," referring to a community of runaway slaves.

3 Map Sources

As an armchair geographer, d'Anville used several old and new maps to make his own. Let us begin with the maps he had available to him at the time. The enormous *D'Anville Collection* now in the BNF includes both the geographer's published maps and those he consulted. Referencing previous maps was a part of the geographer's method, always taking care to introduce improvements and not simply copy the work of others, which would be detrimental to one's reputation. D'Anville had 462 maps of the South American continent, a total of 4.4 per cent of the *Collection* of 10,502 cartographical items estimated for 1780[46]

45 About Natives in eighteenth-century cartography see Furtado, and Neil Safier, "Indigenous Peoples and European Cartography," in Edney and Pedley, eds., *The History of Cartography*, pp. 663–72.

46 Du Bus, "Les Collections D'Anville à la Bibliothèque Nationale."

and 4.8 per cent of the 9556 estimated in 2018 (Tables 1 and 2).[47] Setting aside those by d'Anville and those produced after 1748, mostly by the royal hydrographer, Jacques-Nicolas Bellin, we find a total of 420 maps used as sources for the 1748 *Amérique méridionale*. Here are a few figures:

One notes from Table 2 that the subcontinent as a whole—including maps of America as a whole, or simply the southern part, or even world maps—take up 9.3 per cent of the maps representing the subcontinent, 39 maps in all. Following the tendency of the *Collection* as a whole, the cartography of subregions, which takes up 90.7 per cent of the 420 maps, is more numerous than the depiction of the subcontinent as a whole. Brazil is the province with the greatest number of maps: 78 in all, 18.6 per cent of the whole. Its size and geographic diversity explain this predominance. But Spanish America as a whole is better represented, with 274 maps, 65.2 per cent of the total, distributed among the various provinces and viceroyalties. Of these, Chile and the Audience of Lima have the greatest number of maps. To the former we might add the 27 maps of the Terre Magellanique, since, as d'Anville pointed out, this was an erroneous name, the region actually constituting part of the Audience of Chile.

The large number of maps of Chile can be explained by the length of the territory and the fact that knowledge of its coastline was crucial for navigating the Pacific; Quito's predominance, meanwhile, was due to the presence of the Franco-Spanish Geodesic Expedition, which gathered and produced a significant number of maps of the area, which the members then passed on to d'Anville. The Audience of Panama has 38 maps, but only the lower part of the isthmus is represented in the *Amérique méridionale*. Notably, one map of Panama was drawn up by Bouguer and brought to France, where D'Anville would use it to trace this portion of his map. Comparing it with his *Carte réduite de la Mer du Sud* (Fig. 10), he managed to establish the correct position of the isthmus, between the Gulf of Darien in the South Sea and Escudo de Veraguas, in the North Sea (Atlantic Ocean). D'Anville also used a map of the coast and nearby islands drawn up by La Condamine, who for his part used a map by the Jesuit priest Juan/Jean Magnin (1701–1753), acquired on his journey.[48]

Paraguay, with 18 maps, provides all cartography relating to the Rio de la Plata and the missions' region, objects of keen interest for D. Luis da Cunha. We cannot forget, of course, that d'Anville had already published a *Carte du Paraguay* (Fig. 9), in 1733. Since then he had emphasized that access to a few manuscript maps produced by the Jesuits had been fundamental for his knowledge of the area, as the priests had established several catechizing

47 Hofmann, "La 'Collection d'Anville' au Ministère des Affaires Étrangeres."

48 Anville, *Journal des sçavans* (March 1750), pp. 530–31.

TABLE 2 Places depicted on the maps of South America[a]

Place	Number	%
South America as a whole*	39	9.3
Paraguay	18	4.3
Province of Tierra Firme	4	1.0
Audience of Panama	38	9.0
Audience of Nova Granada	33	7.9
Curaçao	4	1.0
Islands of Cuba, Marguerite, and La Trinité	9	2.1
Orinoco (Spanish Guyana)	14	3.3
Peru in general	7	1.7
Audience of Quito	18	4.3
Audience of Lima	44	10.5
Audience of Les Charcal	6	1.4
Chile (including Chiloé Island)	46	10.9
Juan Fernandes Islands	6	1.4
Chile—Terre Magellanique	27	6.4
Subtotal for Spanish South America	274	65.2
Brazil	78	18.6
French Guyana	21	5.0
Dutch Guyana	8	1.9
TOTAL	420	100

* This category includes world maps that included the American continent and South America.

a This classification is based on that adopted by the catalogue of the collection at the BNF. The organization which produced this classification was begun by D'Anville himself, and carried on after his death by Barbié du Bocage. Du Bus, "Les Collections D'Anville à la Bibliothèque Nationale," p. 21.

missions in the region. In his own words, and in step with the ambassador's view, "the priests' cavalry"[49] would shut off access to the area, thus monopolizing the local geographical knowledge.[50] There are six Jesuit-produced maps in the *D'Anville Collection*, a highlight being Matthias Seutter's (1678–1757),

49 NL, VB, Ayer MS 1918, doc. 549.

50 Anville, "Observations geographiques sur la *Carte du Paraguai*," in *Lettres édifiantes et curieuses, écrites des missions étrangéres*, vol. 8. (Toulouse: Noel-Etienne Sens/Auguste Gaude, 1810–1811), pp. 254–66.

FIGURE 25 *Amérique méridionale*, Gerard Mercator (printed, J. Hondius, 16??). BNF, DCP, GE DD-2987 (9153)
REPRODUCED COURTESY OF FBN

titled *Paraguaia Provincia Sotietatis Jesu com adjacentitus novissima descriptis*. It was published in 1732, based on an earlier Jesuit map from 1726, attributed to the missionary Juan Francisco d'Ávila (1682–1733).[51] D'Anville had this map as well; it varies slightly from Seutter's, especially in terms of the course of the Paraguay River and its tributaries, and also in terms of Lake Titicaca, which appears a bit farther north.

The chronology of these maps reflects fluctuating tendencies in European mapmaking, which grew exponentially over time and peaked in the eighteenth century, as seen in the production of general maps of South America. In terms of portraits of America as a whole, d'Anville had five maps from the sixteenth century: two versions of an ancient map by Ptolemy (*c*.100–*c*.170), one titled *Amérique méridionale* (Fig. 25), by Gerardus Mercator (1512–1594), one by Paolo Forlani (1560–1571), and one by Arnold Florent van Langren (1580–1644). In this century, maps produced in the Netherlands and Amsterdam in particular

51 Mário Clemente Ferreira, "O *Mapa das cortes* e o Tratado de Madrid: a cartografia a serviço da diplomacia," *Varia Historia*, 23, 37 (2007), p. 56.

dominated the world market, which is reflected in the geographer's *Collection*. There are maps by Joan Blaeu (1596–1573), Nicolas Visscher II (1649–1702), Jodocus Hondius (1563–1612), and a reprinting of one of Mercator's. French cartography, which was beginning to distinguish itself on the European market, is represented by Nicolas Sanson and Nicolas de Fer for example. As the eighteenth century had barely begun, the period is underrepresented in this particular selection; but even so it outnumbers all the maps from the previous two centuries. These are 22 maps, 56.4 per cent of the whole. Having access to "new things" was crucial for assuring the quality of the *Amérique méridionale*.

France was the place of production of the majority of the general maps of South America—unsurprisingly, as d'Anville had the greatest access to French cartography and it had become central in the eighteenth century, the predominant period of his *Collection*. Many maps were duplicated from the Navy's archives, which since 1720 had accumulated an important cartographical collection to be used on expeditions it sponsored.[52] Of the French authors, Guillaume Delisle has eight maps, such as his *Carte d'Amérique* (Fig. 26), made in 1722 according to the new longitude measurements, followed by Nicolas Sanson with three. Amsterdam, another important center for mapmaking in Europe, comes next. However, when one cross-references places and dates, one sees that this city dominates both the sixteenth and seventeenth centuries with ten maps out of 17, almost 59 per cent of the total.[53] In contrast, of the 22 maps from the eighteenth century, 14–63.6 per cent—were produced in France, reflecting the shifting axis of printed cartography from Holland to France.[54] Another phenomenon we might observe from reading this table is an increase in the diversity of the sites of mapmaking in the eighteenth century, as opposed to previous periods. In the sixteenth century, for example, Italy is only represented by Forlani, and by Vicenzo Maria Coronelli for the seventeenth century, the latter living for a time in Paris in the service of Louis XIV. In the eighteenth century, we see England, with Hermann Moll; and Spain, with Vicente de La Fuente (?–17?).

Examining the 77 maps of Brazil, one notes a few gaps. Difficulties in accessing Portuguese maps, generally produced in manuscript form for the private use of the crown, meant that fewer maps of Brazil were made elsewhere. For this reason, the Dutch maps of Brazil—above all, those resulting from the

52 Crone, *Maps and Their Makers*, p. 70.

53 Crone, *Maps and Their Makers*, p. 81.

54 Josef Konvitz, *Cartography in France, 1660–1848: Science, Engineering, and Statecraft* (Chicago: Chicago University Press, 1987), p. XVIII.

FIGURE 26 *Carte d'Amérique dressée pour l'usage du Roy*, Guillaume Delisle (Paris, by the author, printed, 1722)
REPRODUCED COURTESY OF FBN

Dutch invasion (1630–1654)—have a privileged position in the *Collection*, possibly because they collected information *in loco*. Under the administration of Count Maurice of Nassau (1637–1644), a representative of the Dutch West Indies, important studies were carried out on the medicine, natural history, and geography of Dutch Brazil. With him, the count brought a group of scholars, tasked with studying and illustrating Brazilian nature and geography. Upon his return the count published the *History of the Deeds Recently Undertaken over the Course of Eight Years in Brazil and Other Parts under the Government of the Illustrious João Maurício, Prince of Nassau*, by Gaspar Barleo (i.e. Barléu, 1584–1648), printed in Amsterdam, 1647, and *Natural History of Brazil*, by Willem Piso (1611–1678), Georg Marggraf (1610–1644), and Johannes de Laet (1581–1649), printed in Leiden in 1648. Besides containing several paintings, such as those by Albert Eckhout (*c.*1610–1665) and Franz Post (1612–1680), these two works provided the European public with the cartographical conformation of the Brazilian northeast. In terms of this region, d'Anville had a significant *Collection* of maps and prints, 16 copies in all. The prints, with their apparently three-dimensional city views, were another form of cartographical representation.

The combination of the Portuguese monopoly on maps of Brazil and the visibility of Dutch cartography around the period of the invasion meant a great number of maps of Brazil in the seventeenth century (42 items or 54.5 per cent), with a decline in the eighteenth century (35 items or 45.5 per cent)—a different trend than that of general maps of America, or even in terms of *D'Anville Collection* as a whole. And so Amsterdam, not France, was the greatest producer of maps of Brazil, producing 29 (37.6 per cent) and 22 (28.6 per cent) maps respectively for a total of 77. However, one sees a decline in Dutch cartographical production, consisting of 27 maps in the seventeenth century and just two in the eighteenth. Highlights of Dutch production include the prints and maps by Blaeu, Post, and Marggraf, who portrayed Dutch Brazil. France's production, mirroring the gradual rise of its cartography, sees the opposite trend: seven maps in the seventeenth century and 15 in the eighteenth. In the first case, we see Nicolas Sanson and André Thevet (1516–1590), who produced a few maps around the time of the French invasion of Rio de Janeiro as *L'Isle Henri*. In the second, we see Antoine Coquart (16?–17?), who accompanied the expedition of the French pirate Duguay-Trouin, who attacked Rio de Janeiro in 1711; Philippe Buache, from a traditional family of geographers; Amédée François Frézier (1682–1773), member of a 1714 expedition to Chile and Peru who gathered information about Bahia and Santa Catarina Island and produced his own map; and La Condamine. This group also includes a map by Delisle, published in Nuremberg, in Germany. This is the *Typus geographicus Chili Paraguay Freti Magellanici*, from 1733, a map demonstrating a process typical of this period's cartography: a geographer updates a preceding map, using another's copperplate. These copperplates sustained a profitable market, being hotly fought over by dynasties of cartographers. Originally drawn up by the Jesuit Alfonso d'Ovalle (1601–1651), the map updated by Delisle and printed by Nicolas Techo (1611–1680) includes Frézier's information about South America.

Many of these maps were produced in the wake of pirate attacks or foreign journeys to Brazil. D'Anville's configuration of the bay of Rio de Janeiro is mainly derived from the maps created after the 1711 invasion of René Duguay-Trouin, given that through Coquart he owned *A plan de la baye et de la ville de Rio de Janeiro, prise par l'escadre commandé par M. Duguay Troüin, apres 1711*, and another based on this expedition, made by the royal hydrographer, Jacques-Nicolas Bellin; he had one more map by the latter as well, which was in turn copied by Nicolas Denis d'Mannevillette's (1707–1780) *Plan de la baye et du port de Rio de Janeiro*. His *Collection* also includes maps by François Froger, who accompanied Jean-Baptiste de Gennes (*c.*1656–1705) on

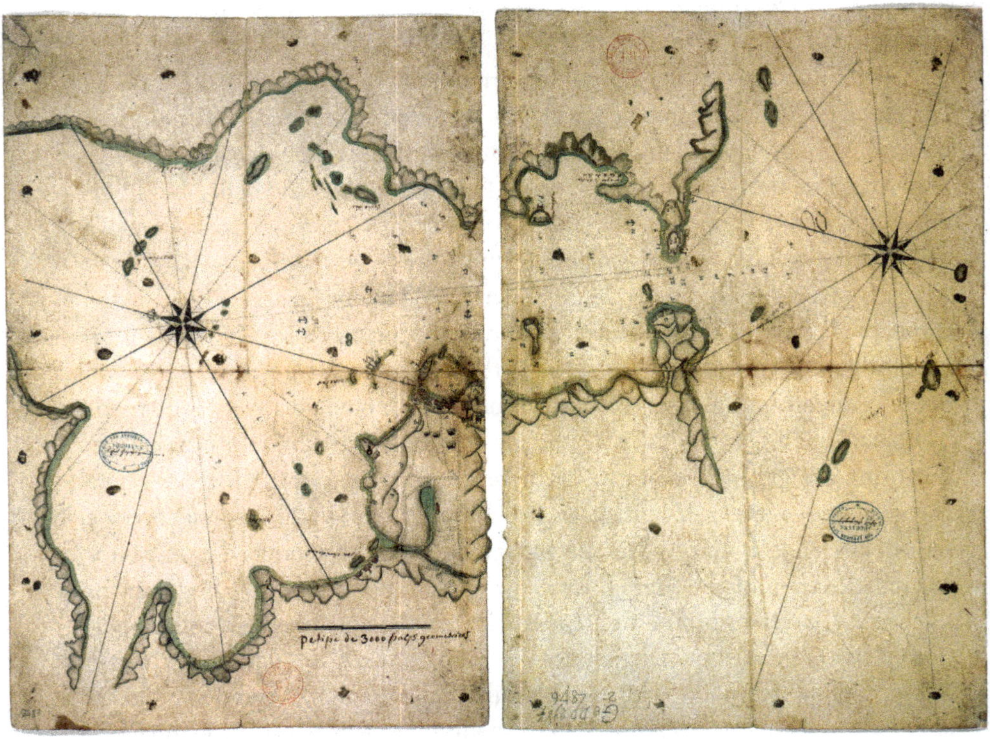

FIGURE 27 *Plan de la ville de St Sebastien à la coste du Bresil ou autrement la Baye de Rio de Janeiro*, unknown author (manuscript, ink and color on paper, 17??). BNP, DCP, GE DD-2987 (9487) REPRODUCED COURTESY OF FBN

a journey to Africa and Brazil, all the way to the Strait of Magellan. He also had a map of the island conquered during the first French invasion between 1555–1570, by Andre Thevet. All this French cartography was checked against a magnificent Portuguese manuscript map entitled *Plan de la ville de St Sebastien à la coste du Bresil ou autrement la baye de Rio de Janeiro* (Fig. 27); on a separate sheet of this map, d'Anville sketched out a French-language legend of the principal points of the city and its surroundings. As for the Bahia de Todos os Santos, d'Anville mainly relied on the configuration established by Amédée François Frézier when this traveler passed through the city of Salvador but also had a manuscript map of the Bahia of Todos os Santos. A nautical map by W.H. Smith—based, according to the author, on the observations of Dutch and English sailors—also appears in the *D'Anville Collection*.

As for the other sites of cartographical production on Brazil, Italy shows up with seven maps, all by Andreas Antonius Horatius (16?–17?), a geographer who adapted preexisting maps by Marggraf or Portuguese mapmakers and published them in Rome during the early eighteenth century. Nearly all

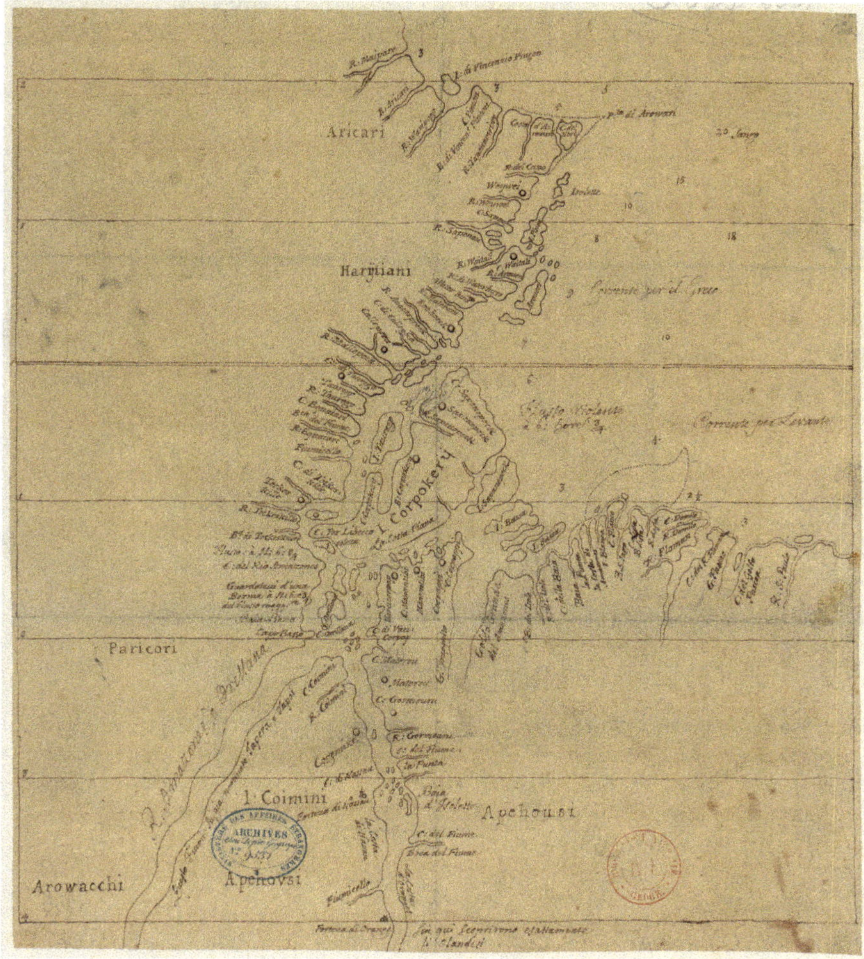

FIGURE 28 *Carte de l'embouchure de l'Amazone, d'après la Carte de Robert Dudley*, unknown author (manuscript, brown ink on paper, 17??). BNF, DCP, GE DD-2987 (9551) REPRODUCED COURTESY OF FBN

the maps have the Portuguese coat of arms, and appear to be part of an atlas, like *Provincie de S. Vicenzo e del Rio de Janeiro*. England is represented by a seventeenth-century map named *Carte de l'embouchure de l'Amazone* (Fig. 28), based on one by Robert Dudley (1574–1649), a geographer who was sponsored by the Duke of Tuscany and produced *Del Arcano Mare*, published in Florence.[55] This is an eighteenth-century map copied from an original map of the author's, of the headwaters of the Amazonas River. D'Anville also had two

55 Geoffrey C. Gunn, *First Globalization: The Eurasian Exchange 1500–1800* (Lanham/Oxford: Rowman & Littlefield, 2003), p. 121.

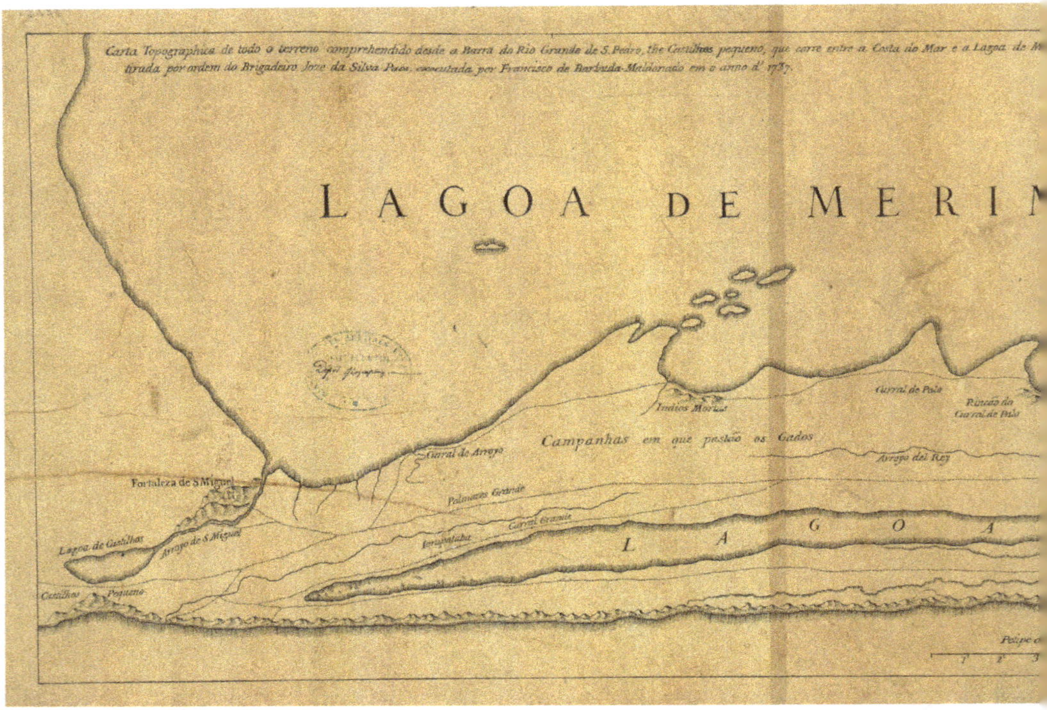

FIGURE 29 *Carta topográfica de todo o terreno compreendido desde a Barra do Rio Grande de São Pedro, thé Castilhos Pequeno, que corre entre a costa do mar e a Lagoa de Merim, tirada por ordem do Brigadeiro Joze da Silva Paos, executada por Francisco de Barbuda Maldonado em o anno 1737* (Brazil, manuscript, brown ink on paper, 1737). BNF, DCP, GE DD-2987 (9471 B)
REPRODUCED COURTESY OF FBN

maps from George Anson's (1697–1762) expedition headed to the South Sea in 1749, from which he gathered measurements to be used in various parts of his *Amérique méridionale*'s later editions. One was of Chiloé Island (off the coast of Chile), the configuration of which changed in the 1760 edition.

Though the Portuguese strove to keep geographical knowledge of their territories secret, d'Anville had access to Portuguese manuscript maps which were essential for the accuracy of his map's configuration of Brazil. There were 13 of these maps, of critical importance. Most of them were provided by D. Luis da Cunha and his inner circle. Not by chance, the better part of the most recent Portuguese maps d'Anville had access to depict the mouth of the Rio de la Plata, Amazonas and Tocantins Rivers, and Minas Gerais captaincy; the ambassador wanted to ensure a precise mapping of these areas, as they took

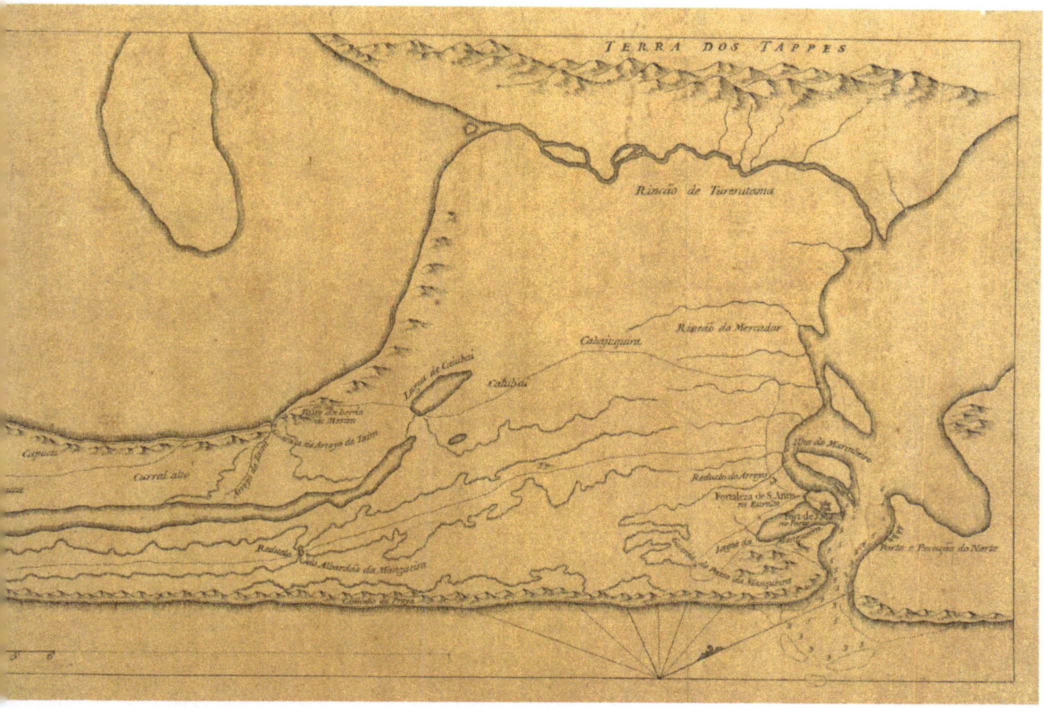

on an important role in his geopolitical vision of the borders of South America. The area around the Rio de la Plata basin could potentially be exchanged in order to guarantee possession of Minas, Tocantins and the Amazonas Rivers. The configuration of Brazil was ultimately determined by access to a collection of maps and travel reports that D. Luis da Cunha provided to d'Anville, having gathered them over the course of his many diplomatic postings. Among this collection, one notes the *Carta topográfica de todo o terreno compreendido desde a Barra do Rio Grande de São Pedro* (Fig. 29), drawn up by Francisco de Barbuda-Maldonado (b.1651–*c.*1700) at the orders of the brigadier and military engineer José da Silva Paes; a manuscript map of the Tocantins River (Fig. 40), a work of rare beauty; and the *Esquisse au crayon de la region de Santos au Brésil* (Fig. 30), drawn using information from Manuel Gonçalo de Lacerda,

FIGURE 30 *Esquisse au crayon de la région de Santos au Brésil, Informed by Manuel Gonçalo de Lacerda*, J.B.B. d'Anville (manuscript, pencil on paper, *c*.1740). BNF, DCP, GE D-10691
REPRODUCED COURTESY OF FBN

secretary to D. Luis da Cunha, after two other maps of the borders between Minas Gerais and São Paulo captaincies both named *Mapa manuscrito da região de São Paulo* (Figs. 31 and 32).[56]

D'Anville also had access to and continually used the longitude and latitude measurements determined by the Royal Cosmographer Manuel Pimentel in

56 RBC, n. 229.

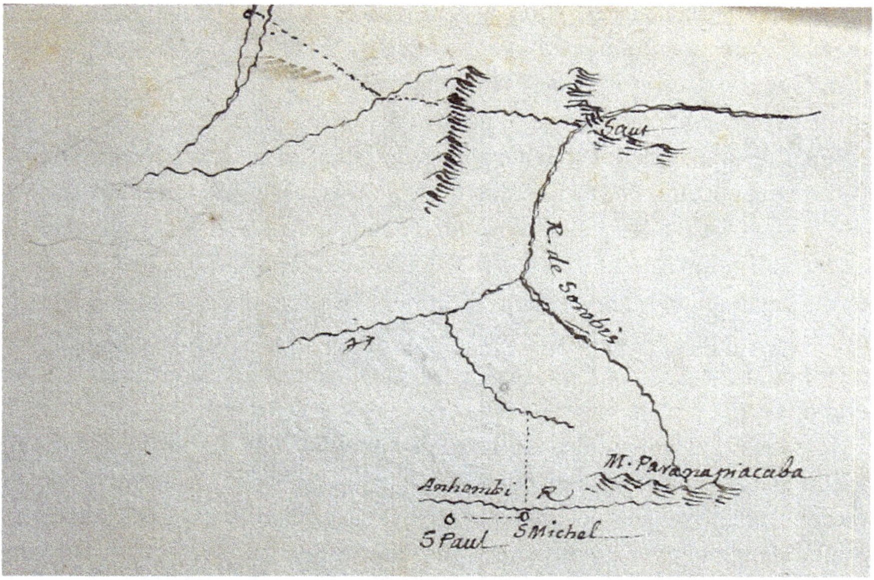

FIGURE 31 *Mapa manuscrito da região de São Paulo*, Manuel Gonçalo de Lacerda (manuscript, ink on paper, 17??). RBC, 229 (1)
REPRODUCED COURTESY OF BOSCH COMPANY

FIGURE 32 *Mapa manuscrito da região de São Paulo*, Manuel Gonçalo de Lacerda (manuscript, pencil on paper, 17??). RBC, 229 (2)
REPRODUCED COURTESY OF BOSCH COMPANY

his book *Arte de navegar*, published in Lisbon in 1746. D'Anville considered them very accurate, repeatedly declaring that "I can find nothing to contradict Pimentel's authority here."[57] Interesting is the fact that he had been using the measurements of "Manuel Pimentel, royal cosmographer" since 1736, far before they were published in book form, to establish geographic points of the Portuguese empire, as well as in his study on the Tordesillas Meridian, carried out at the request of D. Luis da Cunha.[58] Access to Pimentel's tables before their publication can only be due to the ambassador, who certainly gave them to the geographer. Many of these documents were made available to D. Luis da Cunha thanks to his circle of friends. Manuel Pimentel had also been a member of the *Academia dos Generosos*, in which D. Luis and his father were also involved.

It seems that it was also via Manuel Pimentel that the ambassador had access to the map of José da Silva Paes,[59] given that Pimentel's nephew was married to Silva Paes's daughter (Fig. 29). Silva Paes was a brigadier and military engineer who took part in the War of Spanish Succession, working on the repair and construction of Portuguese fortifications under the order of the *engenheiro mor do reino* (royal military engineer) Francisco Pimentel (1652–1706), Manuel's brother. From 1722 onward he served on the *Conselho Ultramarino* (Overseas Council), working particularly on issues regarding military engineering and cartography.[60] Silva Paes was a *savant* of his time, with a library "made up of 437 volumes, several bundles of loose papers and ongoing series of national and foreign magazines." His library mirrors the circulation of Enlightenment culture among the members of the European Republic of Letters, presenting a character "based in rationalism and firmly linked to Cartesian and Newtonian currents."[61]

In Brazil, he went on two expeditions to the Colônia do Sacramento with an eye to repairing and reinforcing its fortifications, shoring up against a possible invasion just after the Spaniards surrounded the fortress, Silva Paes directing the disembarkation and ground operations of the large squadron sent from Lisbon in 1736 and directed to free the fortress. After the ceasefire he was sent to Rio Grande de São Pedro captaincy, where "he applied himself with (…) zeal and prudence to the fortifications and the execution of their borders." He was

---

57 Anville, *Journal des sçavans* (April 1750), p. 650.

58 NL, VB, Ayer MS 1918, doc. 552.

59 Walter F. Piazza, *O Brigadeiro José da Silva Paes, estruturador do Brasil Meridional* (Florianópolis: Editora da UFSC, 1988).

60 ANTT, MR, DS, "José da Silva Pais," maço 118, 19, f. 1.

61 Ana Cristina Araújo, "Livros de uma vida: critérios e modalidades de uma livraria particular no século XVIII," *História das ideias*, 20 (1999), pp. 157, 162.

even twice governor of Santa Catarina Island, "applying himself with the same zeal in its fortification and establishment."[62] With a superior knowledge of cartography, Silva Paes supervised the production of the *Carta topográfica de todo o terreno compreendido desde a Barra do Rio Grande de São Pedro* (Fig. 29) in 1737, which came into D. Luis da Cunha's hands, who gave it to d'Anville, who praised it very much.[63] What is certain is that he obtained the map shortly after its production, given that he provided it to d'Anville while the latter was still working on his study of the position of the Tordesillas Meridian, at precisely this time.[64]

D'Anville used Silva Paes's map to glean more accurate information about southern Brazil—particularly as to the existence and character of the Merim (now Mirim) Lagoon, as well as the geography of this region up to Castillos Point. As the geographer attests, "it is to this map which I owe my knowledge of a great lake, until know unknown in this region of America,"[65] which "gave me no fewer than 50 maritime leagues for the length of this lake, [and] which my *Amérique méridionale* was the first to bring to light."[66] In fact, one of the great revelations of his map in this region rested on the inclusion of Patos and Merim Lagoons. The geographer prides himself on this improvement, affirming that "I was able to draw this extension with much more accuracy compared to the greatest [geographers] who first represented this continent."[67] D'Anville held the brigadier and his map in high esteem. This document, he believed, had been drawn up with the greatest accuracy, based on direct measurements taken onsite using the most modern methods and instruments.[68] During his tenure as governor of Santa Catarina Island, Silva Paes took in the squadron of the English privateer, Admiral Anson,[69] after disease broke out on board.[70] In his book *A Voyage around the World*, Anson referred to Silva Paes as "an expert engineer,"[71] a passage cited by d'Anville in order to attest to the credibility of the man and his map.[72]

62 ANTT, MR, DS, maço 118, doc. 19, f. 1.
63 Anville, *Journal des sçavans* (April 1750), p. 644.
64 NL, VB, Ayer MS 1918, doc. 552, f. G6.
65 Anville, *Journal des sçavans* (April 1750), p. 645.
66 RBC. n. 539 (1), f. 4.
67 RBC. n. 539 (1), f. 4.
68 NL, VB, Ayer MS 1918, doc. 552, f. G6.
69 ANTT, MR, DS, maço 118, doc. 19, f. 1.
70 George Anson, *A Voyage Round the World, in The Years 1740, 41, 42, 43, 44* (London: John and Paul Knapton, 1749), pp. 42–56.
71 Anson, *A Voyage Round the World*, p. 44.
72 Anville, *Journal des sçavans* (April 1750), p. 645.

4 Oracles of a Living Geography

The close collaboration between D. Luis da Cunha and d'Anville for the creation of the 1748 *Amérique méridionale* map developed around that set of documents (maps, latitude and longitude tables, books, and chronicles), most of them Portuguese-Brazilian, today dispersed in several archives, which the Portuguese ambassador gave the geographer and which served as the bases, together or in contrast with the current cartographic tradition, for creating a new profile of Portuguese America on the map, thus constituting sources of a lived geography. From a distant study in France, these maps and memoirs, many of them given by D. Luis da Cunha, provided a window from Paris for d'Anville to take a closer look at the South American space to be represented in his map, written or drawn by people who travelled through the region—oracles of a living geography. By examining this documentation, it is possible to see how the geographer chose his sources, selecting some pieces of information over others, subjecting them to validation criteria that ensured their reliability, which he tried to explain in his written memoirs on the map. Once chosen, he consolidated them in a map drawn with an Enlightenment perspective. This documentation also helps unveil the information networks created by the ambassador to provide himself with precise information not only on Brazilian territory, but also on what transpired in the region, issues that became relevant given the role that Brazil and especially Minas Gerais captaincy played in his geopolitical strategy.

Some of those documents were intentionally sought out by D. Luis da Cunha. Others, he obtained almost by chance in the exercise of his everyday diplomatic activities, which included a certain amount of spying, not only on the activities of enemy crowns, but also on citizens who circulated through foreign countries and who could provide vital information on the empire, whether they were Portuguese or not. Aware of the attention that Brazil's mines had begun to attract and preoccupied with ensuring that the possession of the mining region remain undisputable, he began to collect an intriguing set of reports on the mining region and its surroundings[73] and provided them to d'Anville.

Without direct access to the measurements that were being made directly on the territory on the king's orders, establishing the correct position of Minas Gerais captaincy was one of the challenges that d'Anville faced in his *Amérique méridionale*. The captaincy was divided into four *comarcas* (counties): Ouro

73 The first he gathered were three memoirs written by French pirates about Rio de Janeiro and its surroundings. ANTT, MNE, livro 790, ff. 269, 291–306, 311–13, 332.

Preto, Rio das Velhas, Rio das Mortes, and Serro do Frio, the last was where the diamond mines were located. According to him, "the main area of Minas de Ouro county is determined in latitude by the position of Vila Rica town,"[74] since he knew that by establishing the position of this urban center he would be able to position the main geographical features of its surroundings. Far away in his Parisian studio, without ever setting foot in Minas Gerais, d'Anville was nonetheless able to establish quite precisely the position of its capital, Vila Rica, which he locates between 20° and 21° in latitude and very close to the Rio de Janeiro Meridian establish by the Mathematical Priests in the 1730s. To do so he made use of "the best maps, and some of them were given to me by D. Gonzalo de La Cerda" (*sic*).[75] He was referring to Manuel Gonçalo de Lacerda, who held the position of envoy at the Paris embassy and served as secretary to D. Luis da Cunha. Lacerda had been a member of the *Conselho Ultramarino*, a position that probably gave him access to those documents, and it was he who shared with the ambassador a manuscript map "that represents the surroundings of Vila Rica."[76] Lacerda gave him the three manuscript maps that represented the border with São Paulo captaincy that helped d'Anville position Minas Gerais (Figs. 30, 31, and 32).

To determine the location of Vila Rica a memoir was also essential, entitled *Descrição das Minas Geraes do Brasil*. In it there was a description of the town with its exact location at

> 70 leagues to the west of the coast of Brazil (…), behind some mountains and sierras that extend from the coast to that location, which is also monstrous. At its most rugged point, and at the latitude of 20° and 25′ south, Vila Rica was founded, because it is the place where gold is most abundant.

The document also located other villages of the captaincy and their distances from Vila Rica, which made it possible to locate them on the map as Ribeirão do Carmo to the east; S. Sebastião, S. Bartolomeu, and Catas Altas to the north; and Forquim, Gualachos, and Guaripiranga to the southeast.[77]

D'Anville states that this memoir on the mines was given to him by Pedro Nolasco Convay (or Couvay), and it was in fact with that title that it was catalogued in his collection. Convay was one of the greatest Parisian financiers in

---

74 Anville, *Considération générales sur l'étude*, p. 65.
75 Anville, *Journal des sçavans* (April 1750), p. 648.
76 RBC, 229, n. 539 (1), f. 3; Anville, *Journal des sçavans* (April 1750), p. 661.
77 RBC, 229, n. 555 (3).

the city. In 1724 Convay went to Portugal to make contacts, thinking of maybe taking up residence there or establishing a firm, which D. Luis da Cunha highly recommended.[78] While the ambassador lived in Paris, Convay was part of his social and intellectual circle. In addition to being a financier, Convay was a *savant*,[79] "recognized for his taste in literature, which led him to create a library" where he had a "considerable [collection] of Spanish and Portuguese books,"[80] "the most curious manuscripts," and maps that made it "unique in Europe."[81] It was part of that collection which he made available to d'Anville, undoubtedly at the ambassador's request, whom he knew because he was the great financier of his expenses and those of other Portuguese diplomats in Paris.[82] In gratitude for his services in favor of Portuguese interests in France, the king granted him membership in the Order of Christ,[83] which made him a noble, and the ambassador recommended that he be named general consul of Portugal,[84] which happened.

In his memoirs of the *Amérique méridionale*'s mapmaking process d'Anville states that the document's date is 1732, and in one of them he states that the author was António Blem.[85] Little is known about António or Antoine Blem, or sometimes Belém, only that he was a French merchant, a native of Marseille, who first lived in Lisbon[86] and was later in Rio de Janeiro with Francisco Xavier da Távora (1687–?), governor of Rio de Janeiro who arrived in the captaincy in 1713.[87] It is noteworthy that Rio de Janeiro city had just survived the two French invasions (1710 and 1711) and the new governor arrived with a Frenchman in his retinue, which was probably not well received by the residents. Not much is known of Blem's stay in Brazil, returning to Portugal in 1722, but he was undoubtedly in Minas Gerais, which included a stay in the diamond region around Tejuco village in the Diamond District. This voyage allowed him to gather the information he recorded in his memoir.[88] D. Luis da Cunha monitored his arrival in France from Portugal in 1723. He informed the Portuguese

---

78 ANTT, MNE, livro 790, f. 395.

79 RBC, 229, n. 539 (2), f. v.

80 Anville, *Journal des sçavans* (April 1750), p. 661.

81 *Catalogue des livres de la bibliothèque de Feu Mr. de Convay, Chevalier des Ordres du Roi de Portugal* (Paris: Chez Damonneyville, 175), f. v.

82 ANTT, MNE, livro 793, f. 407, August 6, 1724.

83 ANTT, MNE, livro 59, f. 74v, August 19, 1721.

84 ANTT, MNE, livro 790, f. 394.

85 Anville, *Journal des sçavans* (April 1750), p. 662.

86 Jean-François Labourdette, *La nation française a Lisbonne de 1669 a 1790: entre colbertisme et liberalisme* (Paris: Fondation Calouste Gulbenkian, 1988), p. 146.

87 ANTT, MNE, livro 14, f. 48v, October 18, 1723.

88 RBC, ns. 229, 555 (2).

court of Blem's arrival in Saint-Malo describing him as "a good-looking young man and good to go to India, where the turbulent are tamed, sometimes becoming great men."[89]

When in Lisbon, Blem was a correspondent of the firm that Pierre de PeyreLongue, a native of Bayonne, France, owned in that city.[90] PeyreLongue, once head of the firm in Lisbon returned to France and was living in Paris in 1723, when Blem arrived. PeyreLongue was certainly in contact with Convay as both were great local merchants and the latter was planning to move to Lisbon, where Blem returned in 1724 probably in his company as d'Anville later states that the two men were friends.[91] Business brought PeyreLongue, Convay, and Blem together, and the last two became very close, which explains why when Convay returned to Paris they continued corresponding with each other. In one of those letters Convay received news about Minas Gerais from Blem that he passed on to d'Anville, probably through D. Luis da Cunha who was acquainted with them all.

Exactly in 1723, when Blem arrived in Paris, the Swiss Charles Frédéric Merveilleux arrived in Lisbon, returning from a trip to Louisiana, eager to put himself at the service of the Portuguese king as a naturalist.[92] He also indicated a German specialist in mining called Blumenstein, who lived in France.[93] Around the same time, PeyreLonge wrote from Paris to D. Luis supporting Blumenstein. At first, the ambassador said, "I paid no attention, for I was tired of hearing from enterprisers who just cared to feed themselves as long as their hoaxes were not discovered." But, since some weeks later the German was also recommended to him by Robert-Alexandre d'Hermand, mathematician to the French king, and he received orders to that effect, the ambassador decided to write to PeyreLongue to obtain more information on the man.[94]

Blumenstein, who was exploring a lead mine near Lyon, gave D. Luis da Cunha a memoir on how to modernize mining technology for Brazilian gold. Blem, who was at the time in contact with the ambassador undoubtedly through Convay and PeyreLongue, read that memoir, certainly at his request, and "wrote a response on Brazil's mines." In 1723 Blem wrote one manuscript on the Brazilian mines and offered it to D. Luis, and there he stated that "Mr. Blumenstein's foundry will not improve their quality." Informed of the

89 ANTT, MNE, livro 792, f. 416, October 18, 1723.

90 ANTT, MNE, livro 792, f. 407, October 7, 1723.

91 ANTT, MNE, livro 793, f. 447, August 28, 1724; Anville, *Journal des sçavans* (April 1750), p. 661.

92 Cluny, *D. Luis da Cunha e a ideia de diplomacia em Portugal*, p. 109.

93 ANTT, MNE, livro 14, f. 27, July 27, 1723.

94 ANTT, MNE, livro 792, f. 333, August 23, 1723.

criticisms, Blumenstein in turn "respond[ed] to him." Based on that debate, the ambassador felt that "from his own [Blumenstein's] information, the usefulness of his intervention is strengthened, as your excellency shall see in the letter he writes to PeyreLongue and his response to Blem's doubts."[95]

The Count of Tarouca warned D. Luis da Cunha that "António Belém [Blem] is a Frenchman grafted as a Portuguese who pretends to know it all, (...) and therefore no attention should be paid to what he wrote." He also told him that he and the crown gave much more credit to the "memoir that Blumenstein gave you." Regarding both memoirs, Blumenstein's in French and Blem's in Portuguese, the Secretary of State Diogo Mendonça Corte Real said to the ambassador that "I will communicate the papers in French and Portuguese about our mines in Brazil to knowledgeable people to determine whether what is said in them is correct,"[96] deciding to resort to a third opinion.

In reality the memoir on the Brazilian mines that is in d'Anville's archives, which he claims António Blem sent to Convay in a letter in 1732, was probably not very different from the one he gave to D. Luis da Cunha in 1723, since most of the information in it derived from his experience in the region in the decade of 1711. Everything indicates that they are almost the same text, updated with a little more recent data accessible to any Lisbon resident, such as the establishment of a foundry in Minas Gerais in 1720 and the payment of a donation at the time of the Prince of Portugal (future D. José I) and the Infanta's (Maria Barbara) marriages with the Spanish heirs in 1728.[97] Convay undoubtedly obtained that new memoir with the ambassador's backing and by his request, to provide d'Anville the necessary documents for his cartographic production. In the manuscript, Blem reveals that he is aware of the mining techniques used in the captaincy at the beginning of the century that he described in negative terms.

A good part of this negative description of the technology used for mining is not quite accurate, but it was the result of the Portuguese view of Minas, whereby the *Paulistas* were accused of all the problems and blunders, which constituted a black legend about them that was first spawned by the Jesuits.[98] In tune with this view, Blem states that the region "is governed by men who never saw gold extracted except by *Paulistas*, who have always been very ignorant in this art," and because of the techniques they used "the water washes

95 ANTT, MNE, livro 14, f. 27, June 27, 1723; livro 792, f. 407, October 7, 1723.
96 ANTT, MNE, livro 14, f. 48v, October 18, 1723; f. 27, June 27, 1723; f. 40, September 21, 1723.
97 RBC, ns. 229, 555 (3), ff. 4–5.
98 Furtado, "As Índias do conhecimento ou a geografia imaginária da conquista do ouro," *Anais de história de Além-Mar*, 4 (2003), pp. 155–212.

away more gold than it collects." He thus suggests that this "loss could be overcome with the various instruments that European mining masters developed to separate mountain gold, but since there is great vigilance here to keep foreigners from entering into that state, the residents remain in their ignorance."[99]

Intriguingly, Blem did not share with D. Luis da Cunha his view of the mining region's wealth and its economic centrality in the Portuguese empire and was not able to make up his mind on the subject. Actually, his moralist discourse perceives such wealth as ephemeral and likely to bring more evil than good because of the moral disintegration it creates. In accordance with that view, he writes that

> in conclusion I shall say that Brazil's wealth should be considered temporary and not perennial, because most of the gold that was near the water has been already removed, and in a few years whatever is left will be exhausted. Diamonds are finished; topazes are few and of little value; emeralds are false; and silver only rarely appears. We can therefore say that this and other forms of wealth in the world are all apparent, and that most of them only serve to divert men from the path to salvation, which is the only happiness that does not expire.[100]

D'Anville extracted from Blem's text several pieces of information of a geographical nature, referenced from Vila Rica: the location of the captaincy's main towns and districts, the distance to São Paulo, Espírito Santo, and other adjacent captaincies, and the location of some rivers, such as one that

> runs [in] the county of Rio das Mortes, which after going through several fields and hills, flows into the renowned Paraguay River, which after a long voyage forms the Rio de la Plata, which passes by Buenos Aires and the New Colony [Colônia do Sacramento].[101]

This river basin connection between the Colônia do Sacramento and Minas Gerais fed D. Luis da Cunha's fear that an invasion of the former could result in the loss of the latter. But d'Anville extracts other pieces of information from this text which are less evident. When speaking of the discovery of silver mines in Rio Grande de São Pedro, Blem refers to the Society of Jesus and

99 RBC, 229, n. 555 (3), ff. 2–3.
100 RBC, 229, n. 555 (3), f. 7.
101 RBC, 229, n. 555 (3), f. 2.

uses the expression "the cavalry of the company's priests,"[102] the same expression used by d'Anville shortly afterwards in his letter to Jean-Baptiste Du Halde (1674–1743), who was coordinating the publication of *Letters Édifiantes*, to refer to the Jesuit presence in Paraguay.[103] In his mistrust of the Jesuits, Blem shared the ambassador's view of that order.[104] Regardless of Blem's opinion, once the Portuguese court was convinced of the German's value, the crown sent orders to D. Luis da Cunha to "arrange for the aforementioned Blumenstein [to] come to this kingdom."[105] Blumenstein and Merveilleux were then hired, and the former went on to work in Lisbon,[106] where he saw the forgotten eyepiece that Hermand had sent as a present for D. João v, and assembled it.[107]

It is intriguing that D. Luis da Cunha gave the French geographer "a manuscript map drawn by D. António Alv[ar]es da Cunha, his nephew." This map was drawn "according to the notions that a sojourn of many years in that part of Brazil (…) allowed him to acquire."[108] António Álvares or Alves da Cunha (1700–1791), 1st Count da Cunha, was later governor of Mazagão (1745–1753) and Angola (1753–1758). He also became a member of the *Conselho Ultramarino*, deputy of the *Junta dos Três Estados* (Council of the Three States, since 1761), vice-king of Brazil (1763–1767), and finally president of the *Conselho Ultramarino* (1768–1791). He was the firstborn son of D. Luis da Cunha's oldest brother, and it was the ambassador who obtained the title of Count for his nephew, negotiating it as a reward for his and his father's services.[109] However, all of those positions were occupied in the second half of the eighteenth century, and after the *Amérique méridionale* was produced in the 1740s.

It is d'Anville who reports the origin of that document and the biography of its provider, and his information cannot be doubted, for despite a slight confusion he seems to know precisely of whom he speaks. In his second memoir on *Amérique méridionale*, written in 1750, he says that the count was at the time "governor of Mazagão (now El Jadida, Morocco), on the coast of Africa."[110] Some years later, in the 1779 memoir, he updates some biographical information and states that he "believed that he currently served as ambassador in

102 RBC, 229, n. 555 (3), f. 6.
103 NL. VB, Ayer MS 1918, doc. 549.
104 Cunha, *Instruções políticas*, pp. 220–23.
105 ANTT, MNE, livro 14, f. 27, July 27, 1723.
106 ANTT, MNE, livro 14, ff. 27, 40, 41, 66v; Cluny, *D. Luis da Cunha e a ideia de diplomacia em Portugal*, pp. 105–10.
107 ANTT, MNE, livro 14, f. 186, July 26, 1725.
108 Anville, *Journal des sçavans* (April 1750), p. 659.
109 BNP, R, maço 62, doc. 7, f. 2, February 13, 1728; ANTT, MR, D, maço 6, July 22, 1738.
110 Anville, *Journal des sçavans* (April 1750), p. 659.

the States General of the United Provinces."[111] However, there is a confusion here, because it was actually his half-brother, D. José Vasques Álvares da Cunha (1734–1812), who inherited the title and was become the 2nd Count da Cunha, which explains the mistake, and was plenipotentiary minister of Portugal in the Netherlands from 1766 to 1776.

António Álvares da Cunha's information about Brazil may have come through his father, Pedro Álvares da Cunha (1658–1728), 18th Lord of Tábua, who was governor of Madeira Island; or through his uncle, D. João Lourenço da Cunha (1652–?), who served in India and stopped in Rio de Janeiro on his way back to Portugal. D. Luis da Cunha himself mentions this. When he suggests to the king that he should move to Brazil, he states that "the most appropriate place for his residence would be the city of Rio de Janeiro." He justifies his choice with the information that his brother gathered *in loco* and "assured me that it was very healthy and similar to ours." Among the advantages claimed by the ambassador to justify his choice, based on his brother's observations, was the city's geographic position, which was "so close to the gold and diamond mines."[112] We can thus see that D. João brought news not only of the climate and the fertility of the land, but also of the geographic location of the city and the distances from it to Minas Gerais, and it is not improbable that his son accompanied him on that occasion or that he drew the map based on information provided by his father. "It was an ideal map drawn by António Alves da Cunha, his nephew (...) which extends from Vila Rica, capital of the district of Minas do Ouro, to the Paraguay River."[113]

The only map available in the *D'Anville Collection* that corresponds to that description is the *Carte manuscrite des côtes du Brésil*, which is a handmade copy by d'Anville of a Portuguese map (Fig. 33). It extends from Maranhão to the far south of Brazil, and it is the only one in the *Collection* that portrays the region of Vila Rica. This map was effectively used by d'Anville to draw the urban network of Minas Gerais around that city in *Amérique méridionale*. It also served to delineate the São Francisco River and its confluence with the das Velhas River, as well as to establish the contour and the connection between the Paraná, Grande, Tietê, and das Mortes Rivers.

If one compares the configuration of the São Francisco River on the *Manuscrite des côtes du Brésil* with the *Amérique méridionale*, the similarity between the two is striking. On both maps, the Verde and Paracatu Rivers appear as tributaries on the left bank of the São Francisco. On the course just

111 RBC, ns. 229, 539 (1) (2).
112 Cunha, *Instruções políticas*, p. 366.
113 Anville, *Journal des sçavans* (April 1750), p. 659.

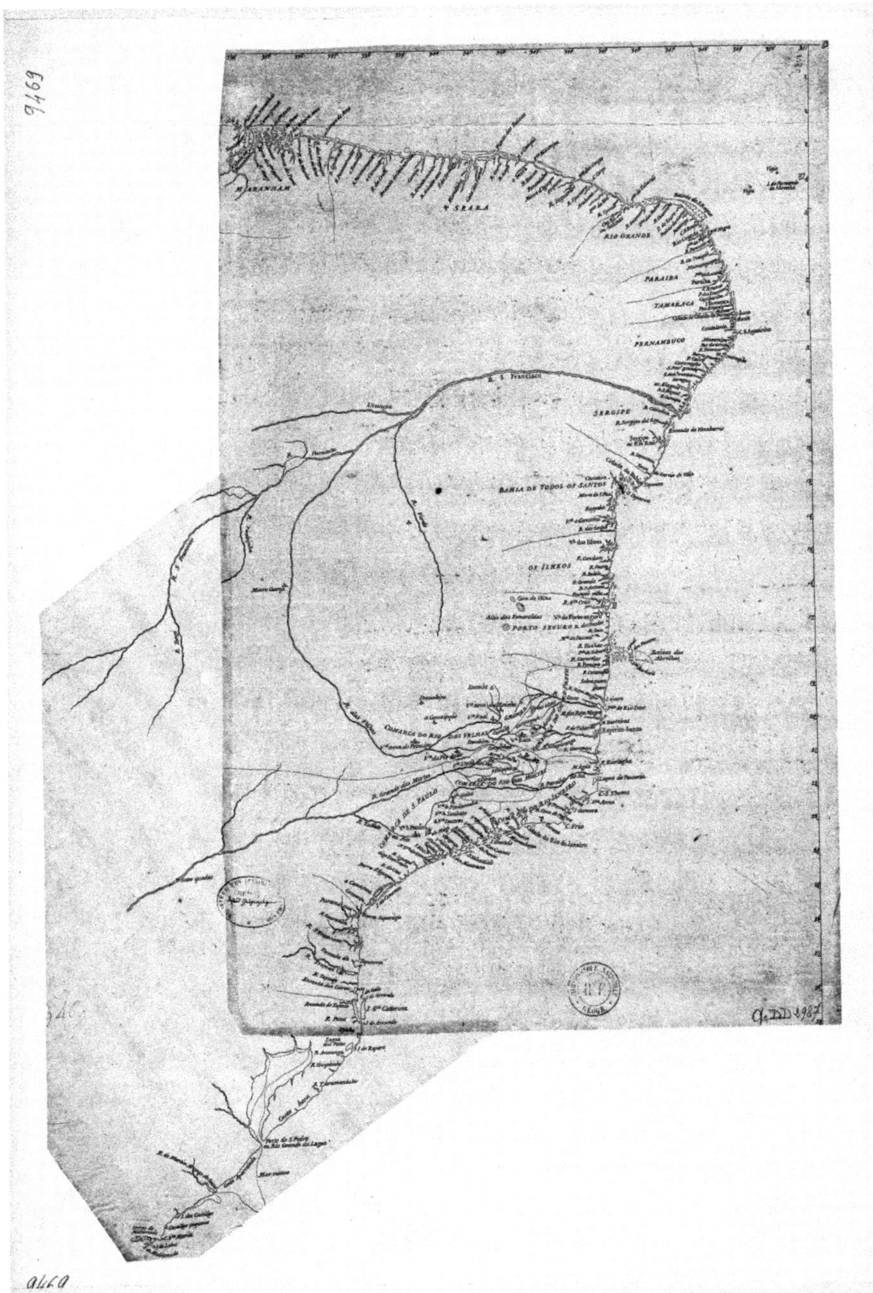

FIGURE 33 *Carte manuscrite des côtes du Brésil* (António Álvares da Cunha?), copy by J.B.B. d'Anville (manuscript, ink on brown paper, 17??). BNF, DCP, GE DD-2987 (9469 B)
REPRODUCED COURTESY OF FBN

above the Paracatu, both depict a lake, which d'Anville labels Parapitinga, and the Tangi appears as a tributary of the Paracatu River. The use of this map to fix the location of Minas Gerais meant that d'Anville did not draw the limits of the district, incorporating the counties of Rio das Mortes and Ouro Preto into the neighboring districts of São Paulo and Espírito Santo, respectively. We should not be surprised by this absence of borders, however; in the 1720s, the decade in which D. Luis collected most of the documents he gave to d'Anville, Minas was not an autonomous captaincy but rather joined to São Paulo and both maps reflect this political configuration. In a text positioned at the top left of the map, d'Anville explains to his reader that he has represented "the captaincies of Brazil, namely, Bahia de Todos os Santos, Sergipe, Olinda, Paraíba, Rio Grande, Ceará, Maranhão e Pará, Ilheus, Porto Seguro, Espírito Santo, Rio de Janeiro, being separate, with the distinction of the comarcas of São Paulo and Minas do Ouro" which is not exactly clear at first sight.

5 The Nunes Brothers

D. Luis da Cunha also provided the geographer with an intriguing itinerary guide produced by the Nunes brothers. The document, describing the *News of the Mines of the Americas Called Geraes Belonging to the King of Portugal*, is divided into four parts, describing 1) the roads by which one could reach the São Francisco River from Salvador da Bahia, along two different routes—the Travessia de João Amaro and the Travessia de Dona Joana—and from there, hugging the right bank, go up the Rio das Velhas and arrive at Vila Rica, in Minas Gerais; 2) an account of the War of the *Emboabas*, a struggle for control of the area waged between 1708 and 1711 between *Paulistas* and newer Portuguese arrivals, generically called *Emboabas*, meaning "chicken with boots" that differentiated them from those who were barefoot; 3) the path to the Emerald Mountain; and 4) a succinct record of the Indigenous groups to be found along the Doce River.[114]

When one compares the information in this document and the geographical conformation of the same territory on the map, the correspondence becomes clear. In his memoirs on the process of producing the map, the cartographer himself declared that he had based his work on the Nunes brothers' account to lay out the stretch between Salvador da Bahia and the backcountry of Minas

114 RBC, ns. 229, 555 (1). About the Nunes brothers see Furtado and Safier, "O Sertão das Minas como Espaço Vivido"; Furtado, "Trajetórias carto-geográficas de uma família de Cristãos-Novos dos sertões das Gerais."

Gerais;[115] and, in the book he published in 1777 to refine the study of geography, he would state once again that "an itinerary of close to 250 leagues guided me, going from the Bahia de Todos os Santos to Vila Rica."[116]

The Nunes brothers were called Sebastião, Diogo, and João. They were Portuguese New Christians from Vila de São Vicente da Beira, in the Bishopric of Guarda, near the border with Spain, and descended from Jewish families who had fled persecution by the Inquisition in Spain and were forced to convert to Catholicism in Portugal in the late fifteenth century, but their exact dates of birth are uncertain.[117] The Nunes family, as eighteenth-century New Christians, saw the expansion of frontiers in Brazil as a place to find shelter, to emigrate to, and as a source of wealth.[118] Diogo followed in the steps of his brother João, who already lived in Bahia and was a *tratante* (trafficker) in Minas Gerais, and Sebastião, who practiced the same trade in that captaincy.[119] Traffickers worked as agents, making deals, transporting merchandize, or collecting debts, working always on commission for a greater merchant. Because of that they established business and friendship networks that covered great areas,[120] as was the case with the Nunes brothers, whose connections extended from Bahia to the mining region, passing through the hinterland, until reaching Rio de Janeiro. In Minas, Diogo became rich, and when he was detained by the Inquisition he identified himself as a *homem de negócio* (businessman).[121] He could not read or write, but knew how to sign his name.[122]

The discovery of gold and the significant ensuing population flow, including New Christians, resulted in an intense production of itinerary guides and maps that described the ways to enter the region. It is believed that in that broad community of New Christians, whose information networks are still little known, a number of these itineraries circulated, which served to indicate the roads to be taken by the new waves of migrants to the mining region. These networks extended not only to various captaincies of Brazil, Portugal, and Spain, but also connected Jewish communities in the Netherlands and especially Great Britain. Until now, the only known survivor of what is believed to have been an extremely common source of information at the time was the *Itinerário geográfico: com a verdadeira descripção sos caminhos,*

115 Anville, *Journal des sçavans* (April 1750), pp. 661–62.
116 Anville, *Considérations générales sur l'étude*, p. 67.
117 RBC, 229, n. 555 (1), ff. 1, 5; ANTT, IL, processo 7488, September 7, 1729.
118 Neusa Fernandes, *A Inquisição em Minas Gerais* (Rio de Janeiro: Eduerj, 2000), p. 97.
119 ANTT, IL, processo 7488, f. 3–3v.
120 Furtado, *Homens de negócio*.
121 About the different types of merchants, see Furtado, *Homens de negócio*, pp. 197–272.
122 ANTT, IL, processo 7488.

estradas, rossas, cítios, povoaçõens, lugares, villas, rios, montes, e serras, que há da cidade de S. Sebastião do Rio de Janeiro, Até as Minas do Ouro, a guide written by Francisco Tavares de Brito (1700–?) and published in Seville in 1732. The Nunes brothers took up residence in Curralinho, in the Campos de Itaubira, which was one of the districts of the parish of Cachoeira do Campo, part of the area of Vila Rica, the captaincy city capital. Cachoeira do Campo was one of the oldest settlements in Minas Gerais. From there, they constantly traded along the road that connected Bahia to Minas, one of the branches of which led to the region of Serro do Frio, to the east, where diamonds were found in 1720. Though he was constantly traveling, Diogo Nunes considered this his place of residence where he had a farm. Since they often traveled through the territory, it was not difficult for the Nunes brothers to write the guide, and it is not unlikely that they sent for other members of their own family to seek refuge in Brazil. The three brothers stayed in Minas Gerais for close to 15 years, between 1709 and 1724, circulating between Minas Gerais and Bahia captaincies every year and eventually to Rio de Janeiro. In that year they decided to leave Brazil; and they returned to Salvador, where they caught a ship back to Portugal. It is very likely that they resolved to flee when the long arm of the Inquisition reached as far as Curralinho as they secretly followed the Judaic rites. In 1726 they were settling on plans for leaving Portugal and by the next year, 1727, they were all living in London. In 1728, however, the three Nunes separated, and for the first time their fates diverged. Sebastião, by this time a married man, decided to stay in London with his wife; and Diogo and João set off for France. It is likely that, as was the custom among families involved in commerce, and as the three had done before, they decided to distribute themselves in different markets. Diogo and João went to live in Dunkirk, north of Calais, in France. There, after the unexpected death of João and the news that his wife had died, Diogo decided to return to Portugal. He arrived in Lisbon in 1729 and, on September 7, he knocked at the doors of the Estaus Palace, the building which housed the Inquisition in Lisbon as he had decided to present himself "spontaneously," a common strategy among those hoping to minimize their punishments. His move to collaborate paid off; he received a light punishment and did not have his goods confiscated.[123]

We do not know the exact date when the Nunes brothers wrote their marvelously detailed account of the route between Salvador and Vila Rica. The text itself drops some hints, even as it opens itself up to further questioning. (As we know that Diogo was illiterate and could only sign his name—and the

123 ANTT, IL, processo 7488.

same was likely true of his brothers—they may have used a scrivener.) The text opens by saying that the brothers found themselves in Salvador, in 1709, when they decided to go to Minas—suggesting to the reader the journey about to be described. However, as the text unfolds, one observes that many other possible trajectories between the two extremes of the route are described with a richness and acuity of detail. This knowledge could only be the product of innumerable trips between Minas and Salvador, such as those the Nunes' undertook annually. It is likely, however, that the text was written in a later period, with information collected on several occasions.

If it is beyond doubt that the document served as a source for d'Anville to establish the geography of Minas Gerais, Bahia, and Rio de Janeiro in his *Amérique méridionale*, it is less evident how it fell into the hands of the French geographer. What we do know is that the Portuguese ambassador, D. Luis da Cunha, provided him with this document, but we don't know the path that this interesting document took between the three brothers' quill, the ambassador's hands, and the geographer's study in Paris. There are a number of ways in which D. Luis da Cunha might have obtained that document, since in his various postings he constantly interacted with immigrant Portuguese Jews or New Christians. What is more, the three brothers had relatives in those countries, and they themselves lived a nomadic life. Back in Portugal, they were in Lisbon and Almada and in some towns in Alentejo, where they met several members of their family and other New Christians. In England they met with more members of their family and other Portuguese who openly practiced Judaism, such as their cousin the doctor António Nunes Ribeiro Sanches who fled Portugal with the brothers years before. Also in Dunkirk, France, João and Diogo were in touch with the local Portuguese community, many of them openly practicing Judaism.[124] In any of those places the brothers may have offered the Minas itinerary guide to serve other New Christians willing to follow it, and the ambassador may have had access to it. First in England, where D. Luis had close contact with the Jewish community through the doctor Fernão Mendes da Costa and Jacob de Castro Sarmento, who arrived there to become a rabbi. Then during his stay in The Hague, where Ana Nunes (maybe António Nunes Ribeiro Sanches's sister) was his servant and Álvaro Nunes da Costa, another financier of the Portuguese court, became his friend—through whose family (Nunes) he made contacts with the local Portuguese Jewish community.[125] Finally, during his first stays in Paris, when he was extremely mindful of collecting documents and maps of Brazil in several parts of France, even through Convay and his

124 ANTT, IL, processo 7488.

125 Cluny, *D. Luis da Cunha e a ideia de diplomacia em Portugal*, pp. 149–52.

networks, and was alert to the movements of the Portuguese Jewish communities in Bordeaux and Dunkirk, and was attended by António Nunes Ribeiro Sanches who became his personal doctor until his death.

This itinerary was essential for d'Anville to map the region of Minas Gerais, situated between the coast and the São Francisco River, as an area marked by the presence of Portuguese colonizers. Following the brothers' guide and comparing it to the line that depicts the same road in the *Amérique méridionale* or in *Suite du Brésil, depuis la Baye de tous les Saints jusqu'a St. Paul* that highlight that region (Fig. 15), one can see the similarity between them, which demonstrates beyond doubt that the former was a source for the latter. But unlike the text in the brothers' account, however, someone looking over D'Anville's map would follow the path clearly traced on the map, just one of the many paths that could be taken from Salvador to Minas Gerais. D'Anville states that this "road, which measures about 250 leagues, led me from Bahia de Todos os Santos to Vila Rica."[126] Thus, the viewer who studies the map travels along the same road, with its population centers and geographical features, that the three brothers did in their adventure toward Minas. One of the salient aspects of the Nunes brothers' guide is the recurring positive description of nature and the praise of the mining region—with its abundance and diversity of fruits and foodstuffs—which made it Paradise on Earth. In the brothers' narrative as well as D. Luis da Cunha's view, the central role played by the area in the cosmology of many Portuguese of the time stands out, as a space destined to be occupied by them in opposition to claims by *Paulistas*, who were the true discoverers, constituting an *Emboaba*'s view of the captaincy. However, in this narrative we can also see the different nuances that characterized this perception. The brothers shared an Edenic view of Minas, while for the ambassador it was a matter of geopolitics based on pragmatic issues.

The brothers' itinerary guide is concerned with ordering in precise journeys the 25 to 30 days facing travelers when they began the voyage. Travelers are thus advised of the places they will go through and the various rest points they will find along the way. The issue of water supplies for such a long journey was important, and because of that they suggest that the voyage should be begun in October, at the beginning of the rainy season, and the guide specifies the locations of wells and other water supplies. To d'Anville, who was an expert in transforming time itineraries into distance measurements, this type of detail was vital information for calculating, based on the brothers' travel narrative, the distances between the points traveled. A correct idea of distances

126 Anville, *Considérations générales sur l'étude*, p. 67.

left permanent marks on tenacious travelers. Hence, when Diogo Nunes was pressured by the Inquisition into denouncing other New Christians, he had no difficulty remembering the exact distances between the places where he had met them, sometimes referring to them in terms of days travelled. We can see in the trial records that he often forgot the names or mistook the family relations of the Jews he met in his travels and with whom he shared Jewish ceremonies and rites (although these oversights could of course be intentional). Yet he was able to state with precision, even though some of the voyages made had happened over 20 years earlier, that Vila Rica was 6 leagues away from the site of Curralinho, that Itaubira was in turn another 6 leagues from that same town, and that Salgado was 3 leagues away from the city of Bahia, which was in turn 14 or 15 leagues away from Santo Amaro. Distances could also take the form of time periods, as was the case with Guarapiranga, which according to him was three days away from Vila do Carmo. D'Anville also used the data from the brothers' text to locate the main points of the mining region in the *Amérique méridionale* and concluded that the distance between Salvador and Minas Gerais was about 250 leagues.[127]

Pedro Nolasco Convay gave d'Anville another document written by Blem. It was "a particular description of the diamond district."[128] The document was used to establish the location of Vila do Príncipe, the capital of Serro do Frio county where the diamond area was located.[129] The geographer compared its information with "a general map of Brazil" provided by D. Gonçalo de Lacerda,[130] secretary of D. Luis da Cunha. Unfortunately, this map is not in the *D'Anville Collection* but it is not impossible that he was referring to António Álvares da Cunha's map, although this doesn't show the village, but the locations of Itambé village and Vila Real do Sabará were used to locate the Emerald Mountains (Fig. 33). D'Anville declares it was also with the help of the Nunes brothers' itinerary, about the location of the emerald discoveries,[131] that he fixed the position of the Emerald Mountains.[132] In the 1748 *Amérique méridionale*, this mountain range is found along a tributary of the Doce River, whose source lies in Porto Seguro captaincy, north of Minas Gerais (Fig. 34). On the other side of the mountain, "there emerges another river called the Preto River, where this Nunes believes there to be much gold, due to circumstances having occurred there. Following this River, one comes to another, called

127 Anville, *Journal des sçavans* (April 1750), pp. 661–62.
128 Anville, *Journal des sçavans* (April 1750), p. 661.
129 RBC, n. 555 (2).
130 RBC, n. 555 (2), f. 2.
131 RBC, n. 555 (1), ff. 18–19.
132 Anville, *Journal des sçavans* (April 1750), p. 662.

FIGURE 34 Detail of Emerald Mountain in *Amérique méridionale*, J.B.B. d'Anville (Paris, printed by Gravelot, 1779). Furtado Lins Collection
REPRODUCED COURTESY OF FURTADO LINS COLLECTION

Jequitinhonha."[133] It is on the bank of this Jequitinhonha or "Jutiquinhonha" (*sic*) River that Blem affirms that Vila do Príncipe is located, around the parallels 16° to 17° south.[134] It is the exact position that d'Anville put it on the 1748 *Amérique méridionale*, far north of its real position, but also according to da Cunha's map as it was placed on the northwest of the Emerald Mountain.

By comparing the maps and memoirs that D. Luis da Cunha and Convay gave him of Minas Gerais, d'Anville was able to locate the position of the captaincy relative to adjacent captaincies. The three manuscript maps of his authorship based on drafts made by Gonçalo Lacerda consolidated the notes he gathered, establishing the position of the mining region relative to São Paulo, which he believed was one of his map's great contributions. He states that all of these Portuguese sources were used as a foundation and defined "the

133 RBC, n. 555 (1), f. 19.
134 RBC, n. 555 (2), f. 1.

way of representing São Paulo's surroundings and the distance to Vila Rica, [which] are new things in the *Amérique méridionale*."[135]

In the catalog of books owned by Louis Charles-Joseph de Manne (1773–1832), there was a section on manuscripts and maps from d'Anville's studio, inherited from the geographer. Among them is a *Nouvelle relation de la Rivière des Amazones, et description de l'interieur du Brésil qui renferme des mines d'or et des pierres précieuses*. It was a quarto manuscript containing 120 sheets. Its description explains that it is a "very interesting manuscript composed from private memoirs furnished by Mr. Convay, Knight of the Order of Christ, in 1729, which speaks of the missions of the Capuchins and the Jesuits in that country, as well as its language."[136] As we can see, d'Anville collected in a manuscript the information he had on Brazil provided by Convay, regarding both the mining region that produced gold, diamonds, and precious stones, and the Amazonas region. Unfortunately, that manuscript has not been located, but based on the documents provided by Convay on those regions that have been found up to now it is possible to reconstruct its general terms. However, on September 9, 10, and 11, 1729, d'Anville met in Convay's house in Paris the Mercedarian Father Ignácio dos Reis/Reys who worked as a missionary in the region. Around the upper Urubu and Guatuma Rivers, on the north bank of Amazonas River, close to the mouth of Madeira River, the Mercedarians had established their missions, such as those of São Lourenço, São Pedro Nolasco, and São Raimundo, where Father Reis had served; he frequently went back and forth between them. After the priest's detailed description, d'Anville drew three manuscript maps of the region where both rivers flow into the Amazonas. One is titled *Confluent de l'Amazone, de l'Urubu et du Guatuma* (Fig. 35, see also 45). In it, d'Anville wrote that he "should have taken care to situate the course of the Urubu River farther north from Concepção (Mission) to São Pedro Nolasco." He drew this stretch of the river in ink, but added its course as far as another mission in pencil. He was able to conclude "also, that there would be 10 *lieues* from the tip of the great island to the entrance of Madeira." On the same day, with the priest's information, d'Anville drew another sketch of the Abacaxi River at the point where it flowed into the Amazonas at its south bank close to the Madeira titled *Confluent de l'Amazone et du Madeira, confluent de l'Amazone et du Rio Abacaxis*.

---

135 Anville, *Journal des sçavans* (April 1750), p. 663.

136 De Manne, *Catalogue des livres de feu M. de Manne: suivi de manuscrits, lettres autographes, et autres documents provenant du cabinet de M. D'Anville* (Paris: François, 1863), p. 246.

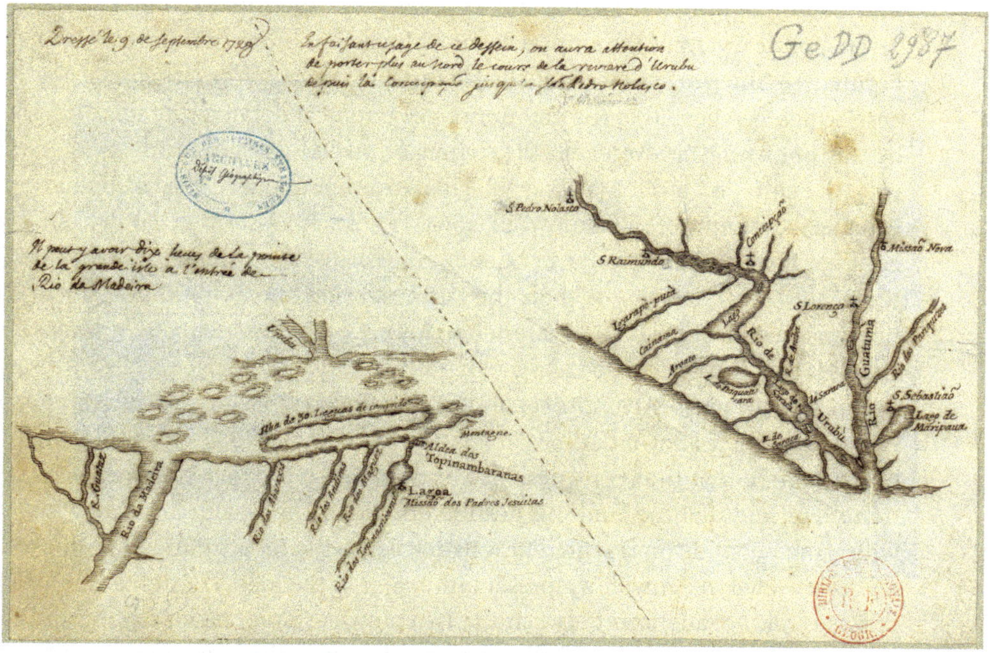

FIGURE 35 *Confluent de l'Amazone, de l'Urubu et du Guatuma, dressée chez Monsieur Couvay sur les deseins et instructions du R.P. Ignacio dos Reys*, J.B.B. d'Anville (Paris, manuscript, brown ink on paper, September 1729). BNF, DCP, GE DD-2987 (9555)
REPRODUCED COURTESY OF FBN

The manuscript that de Manne refers to has connections with or was in part written by Father Reis. In the map that d'Anville published in December 1729 entitled *Carte du cours de la Rivière des Amazones ou de Maragnon* (Fig. 17), he states in the subtitle that it was "drawn according to the memoirs of Father Inacio dos Reys, religious of Mercy, after living 12 years in the country," plus what he can gather from other sources, without specifying them. He also states that the map was accompanied by "a written exposition of the knowledge that gave rise to this work," but it was not located. The relationship between the *Nouvelle relation* cited by de Manne, Father Reis's memoir, and d'Anville's map, on the one hand, and Minas Gerais and the Amazonas region, on the other, can be established by a text written in the left margin (north) of the Amazonas River on the map, which states that "some assure that there are gold mines in the mountains of Alto do Parou,"—mountains located near the mouth of the Amazon, where the Portuguese built a fort of the same name. D'Anville regrets that the "Portuguese don't look for it."

The documents furnished by D. Luis da Cunha or, through him, by Convay, were essential to portraying the mining region. This region's predominance over the rest of Brazil in the documents provided by the ambassador reflected his geopolitical view of the empire: to him, Minas Gerais was the vital center of Portuguese America, and it should be kept within the crown's dominion at any cost, which is reflected in the amount of geographical detail in the map. And he was not wrong, as even after the signing of the Treaty of Madrid (1750), in 1759, D. Pedro de Cevallos (1715–1778), later governor and vice-king of Buenos Aires, sent a map to D. Ricardo Wall (1694–1777), a Spanish-Irish cavalry officer then Spanish ambassador in London, in which Minas Gerais was within the Spanish limits.[137] But by then, the Portuguese had no doubt that its location was east of the Tordesillas Meridian. In the *Amérique méridionale*, Minas Gerais is graphically depicted as a geographical entity whose space is filled with Portuguese presence. An entire set of cartographic symbols—rivers, mountain ranges with their known outlines, roads, towns, and rest stops—demonstrate beyond doubt that the Portuguese had created and reproduced their civilization there, and that the land was destined to them. The wealth of detail with which the region is represented unveils the centrality he attributed to Minas and is reflected in d'Anville's representation of it in the map. D. Luis da Cunha insisted that negotiations between the two crowns should be based on solid cartographic foundations and that, without ignoring the geographic reality, they should be guided by what had already been agreed on in previous treaties, respecting the land's natural limits and the principle of *Uti possidetis*. In the ambassador's view, Portuguese colonization of the mining region, which was clearly visible in d'Anville's map, made the maintenance of its possession indisputable as its possession was guaranteed by the Portuguese effectively colonizing of the area.

But the different ways in which d'Anville apprehends the documents he collected to reconfigure South America and the different ways he makes that interpretation public raise interesting issues regarding the various levels of truth he confers to those sources—whether cartographic or not—which were essential to develop his worldview. Hence in the map's cartouche he openly declares that he bases his work on the textual and cartographic sources by La Condamine and other members of the Franco-Spanish Geodesic Expedition, and diminishes those who, traversing the interior scrublands of Minas Gerais, the Pampas of the Rio de la Plata, and the Amazonas jungle, were also veritable eyes for the cartographer and thus became oracles of a lived geography. By

137 "Mapa Anexo á un Oficio Dirigido por D. Pedro de Cevallos al Ministro de Estado D. Ricardo Wall, fecha 20 de Enero de 1759." https://collections.library.yale.edu/catalog/15821086 (accessed December 2023).

granting different statuses to the various narratives and information he used to configure his geography of the Americas, d'Anville does not make it evident in the map, but only in the debate that followed its publication, that these men of the hinterland were essential to give a new dimension to the Americas, even if their narratives were not written with that purpose in mind. This way it is in the *Amérique méridionale*, and only in it, that three different oracles converge and reveal the substrata and the various languages that sustain the map: cartographic—the dominion of the French sage—political—the Portuguese diplomat's field—and the lived space—on the trails traveled by anonymous travelers or naturalists in the Americas.

CHAPTER 6

Amérique méridionale, a Distorting Mirror

1 The Blank and the Void

While European cartography produced in the sixteenth and seventeenth centuries seemed to abhor the void, eighteenth-century cartography slowly alters this outlook and abandons its fear of white space as geographical voids come to take up vast swathes of the lands being represented, revealing Europeans' relative ignorance of the American, African, and Asian interiors.[1] From 1700 onward, Guillaume Delisle—in his world map and others representing the four parts of the world—would be the first to move away from this obsession with filling in all the geographical space of a given map. In his map of "North America, he left 'blank' the whole northwest of the continent, as it was absolutely unknown." Like no other geographer, however, d'Anville "banished decorative marvels from all his maps," not flinching before "leaving 'blanks.'"[2] As J.B. Harley (1932–1991) wrote, "that which is absent from maps is as much a proper field for enquiry as that which is present."[3] Following this suggestion, issues related to the white spaces in the *Amérique méridionale* will be the principal object of this chapter.

If we compare the first, simplest version of the *Amérique méridionale* from 1737 (Fig. 1) with the two following it (Figs. 2 and 3) we see that while in some areas blanks disappear, in others they grow significantly. The region of Minas Gerais, using documents provided by D. Luis da Cunha; Amazonia and Peru, where d'Anville had access to sources and information brought by members of the Geodesic Expedition sent to Peru to measure the Equator; and Paraguay, for which he used information from Jesuit missionaries, were all areas whose geography was revealed, where the map became more precise. But in the backlands of the northeast of Brazil, geographical information vanished. In comparison with previous versions, the 1748 map shows the *sertão*, or the unknown backlands of Brazil, as a massive empty space.

1 About the void in d'Anville's maps of Africa see Furtado, "Evolving Ideas: J. B. D'Anville's Maps of Southern Africa, 1725/1749"; Isabelle Surun, *Dévoiler l'Afrique? Lieux et pratiques de l'exploration* [*Afrique Occidentale, 1780–1880*] (Paris: Éditions de La Sorbonne, 2018); Haguet, "Specifying Ignorance in Eighteenth-Century Cartography"; Furtado, *Quebra Cabeça-Africano*.

2 Numa Broc, *La géographie des philosophes*, pp. 28, 35.

3 John Brian Harley, *The New Nature of Maps: Essays in the History of Cartography* (Baltimore/London: The Johns Hopkins University Press, 2001), p. 86.

 | DOI:10.1163/9789004710764_008

If the lack of geographical information about the backlands of the Brazilian northeast could signify that this was a space outside the civilization the Portuguese were building in America, the act of incorporating the region within Portuguese borders—clearly laid out on the map—also lent it a sense of belonging, and the status of a territory waiting to be integrated by the Luso-Brazilian colonizers. In this sense, though the Indians' name is blazoned across this space, these lands did not belong to them. More so than any other form of representation, this void reflected the negation of the Natives' right to the vast territories they had occupied when the Europeans arrived—spreading from the coast of Rio de Janeiro to Peru—and from which they were gradually being separated.

It is important to note that the first sheet of the *Amérique méridionale* (Fig. 4), which shows north and northeast of Brazil, is bordered to the south by a part of the lower region of the São Francisco River, which runs west to east; the river's course in full is drawn on the second sheet and can be seen also on *Suite du Brésil, depuis la Baye de tous les Saints jusqu'a St. Paul* (Fig. 15). Just a tiny stretch, around the town of Cachoeira, penetrates discreetly into the first sheet. The São Francisco River, as already noted, serves as a clear divisor between the coastal and mining territory, one side heavily marked by the signs of colonization, and the other comprised of the vast interior backlands: far from the coast, vague, unknown, isolated, and rough, and hence represented by the geographer as a void.

This disappearance of geographical information is evidenced in the northeast local hydrography. In the 1737 map (Fig. 1), the upper reaches of several rivers are drawn, spreading from the backlands to the coast, where they empty out. This is the case with the Meari, Taboucourou, Siará (i.e., Ceará), and Omarás Rivers, all flowing from the foothills of the Brazilian mountain range to the northern coast on the Atlantic Ocean. Looking at d'Anville's map of America from *c.*1742 (Fig. 2), we can already see that a few of the hydrographical elements have been omitted, reducing the number of tributaries, and that even a few river names have disappeared.

Despite these omissions, with the 1748 map (Figs. 36 and 4) he was able to achieve a great degree of precision in the hydrography near the coast, outdoing many French maps of the period in this respect. For example, his collection contained three versions of Sanson d'Abbeville's map *Amérique méridionale*, from 1650, 1669, and 1709. In all three, the upper parts of the Meari, Taboucourou, and Siará Rivers that flow to the northeast before emptying out at the coast, in the province of Maranhão, are represented (Fig. 37). When we compare Sanson's maps with d'Anville's 1748 *Amérique méridionale*, it becomes clear that although the representation of these rivers is limited to their mouths and the

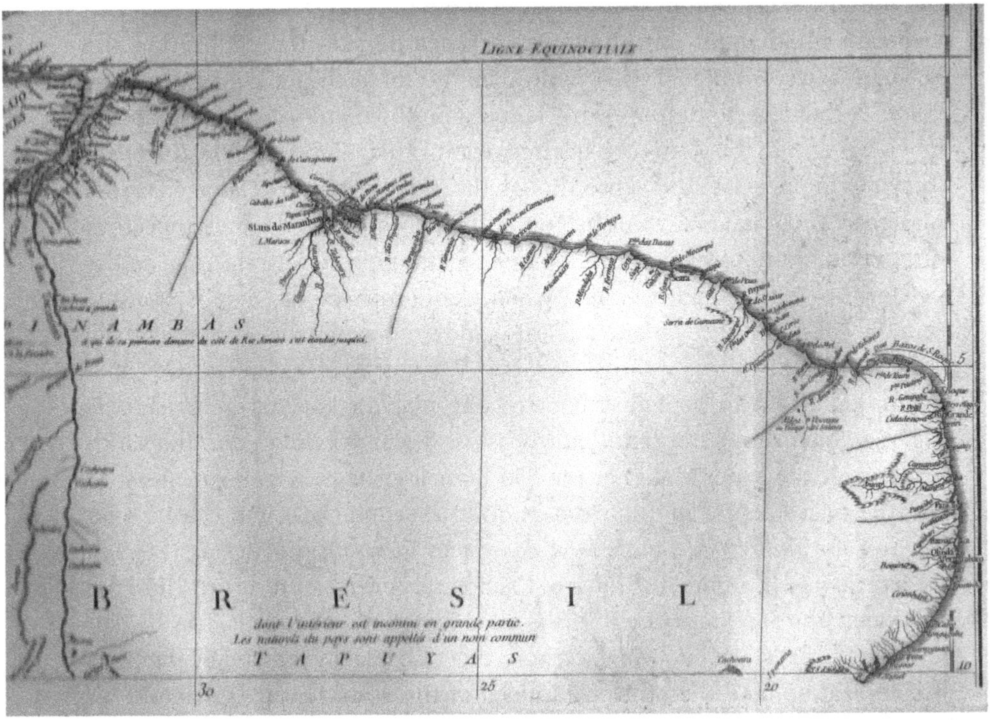

FIGURE 36 Detail of the northeast coast of Brazil in *Amérique méridionale*, J.B.B. d'Anville (Paris, printed by Gravelot, 1779). Furtado Lins Collection
REPRODUCED COURTESY OF FURTADO LINS COLLECTION

stretch near the coast, it is far more precise. For example, the number of rivers displayed is greater, their positions are more accurate, and their nomenclature is better established. This becomes clear in the river system which empties out near São Luis do Maranhão, an urban nucleus situated quite accurately (with latitude and longitude). The Pinaré, Uayou, Meari, Taboucourou, and Moniy Rivers, with no connection to the midwest river system or Lake Xarais, correspond to the modern-day rivers of Pindaré, Grajaú, Mearim, Itapecuru, and Munim. Until the eighteenth century, part of the Paraguay riverbed in the mid center-west of Brazil was represented as Lake Xarais. Although by then the Luso-Brazilian explores and cartographers started to represent it as the Pantanal, d'Anville's map still represents the mythical Lake Xarais.[4] This process reveals that, while the coastal geography of the northeast became better known and better represented during the period these maps were being

4 Maria de Fátima Costa, *História de um país inexistente: o Pantanal entre os séculos XVI e XVIII* (São Paulo: Estação Liberdade/Kosmos, 1999).

FIGURE 37 *Amérique méridionale*, Nicolas Sanson (Paris, by the author and P. Mariette, Printed, 1650). BNF, DCP, Ge DD-2987 (9154)
REPRODUCED COURTESY OF FBN

produced, the interior of the region saw the opposite effect. With this method, the cartographer reveals that, in the 1737 map, much of the geography of the region configured thus far was not based on real knowledge of the territory, but rather pure speculation. In the case of this region, the enlarged blank space in the *Amérique méridionale* presents itself not as a simple absence, but as a cartographical refinement; as the geographer argued, in the absence of sure information, the best strategy was to leave a blank and fill it in later with new, certain information.

The large void that represents the interior of the northeast of Brazil can be explained in part by the fact that d'Anville's collection of maps of the area is massively dominated by Dutch maps from the seventeenth century, produced in the context of the Dutch invasion of the area. The *D'Anville Collection* maps basically represent the coast without going into detail on the geography of the interior, which is covered with pictorial elements. Without information of the northeast interior geography, all that is left is the void. D'Anville had a total of 30 maps of the area, of which six were published after 1748—the date of his printed map—leaving 24 maps and prints which he might have used as sources. Of these, 19 are from the seventeenth century, 17 produced in the context of Dutch Brazil (Blaeu, 7; Post, 7; Marggraf, 1; Visshe r, 1; and Hondius, 1), the majority being prints with city views of Recife, Salvador, or Itamaracá. Just two were not directly produced by the Dutch, but by the Italian Andreas Antonius Horatius—who, nevertheless, reproduced the same Flemish maps in his atlas. This leaves six maps from the eighteenth century. Of these, three are by Amédée François Frézier, but depict only the Bahia de Todos os Santos and Salvador; and three are anonymous, two being prints of Salvador and one of Olinda city.

But what seems to have made a decisive impact on d'Anville's decision to leave the large blank in the backlands was the fact that the main source he used to draw the northeastern coast up to São Luis, capital of Maranhão was the *Carte manuscrite des côtes du Brésil* (Fig. 33). The absence of any recent maps of the local interior in his collection reflects the fact that, around this time, this was not a region in dispute; therefore, the ambassador did not make the effort to put together a cartographical stock to represent it in greater detail. The exception is the São Francisco River, an important means of entry to the mining region—here, D. Luis made multiple maps and itineraries available to the geographer, including the Nunes brothers' report, which allowed d'Anville to represent the river's course with relative accuracy. These aspects are directly reflected in the composition of the *D'Anville Collection* maps.

Within the strategy of leaving a blank in the absence of confirmed geographical information, many other elements of the 1737 map simply disappear in the 1748 version. Another example is the "Brazilian Range," also in the northeast and represented on Sanson's three maps (Fig. 37). In d'Anville's 1737 map this is a mountainous mass which roughly corresponds to the Espinhaço Range. It begins in Serro do Frio county, close to Porto Seguro, and runs perpendicular to the coast until just above the São Francisco River, before turning west toward the interior. Though the map from *c.*1742 has a similar depiction, by 1748 this formation disappears completely in the printed version of *Amérique méridionale* (Fig. 38). Instead, d'Anville depicted it as being divided and unfolding into various other mountains, which corresponds more closely to the mountains' actual configuration in the region. In this latest map he also represents the Serra dos Órgãos (Mountain of the Organs), running parallel to the coast and separating the coastal region from the interior, so that "to arrive there, leaving the edge of the sea, one must cross a chain of mountains called Serra dos Órgãos because their form appears to represent organ pipes."[5] The

5 RBC, 229, n. 539 (1), f. 4.

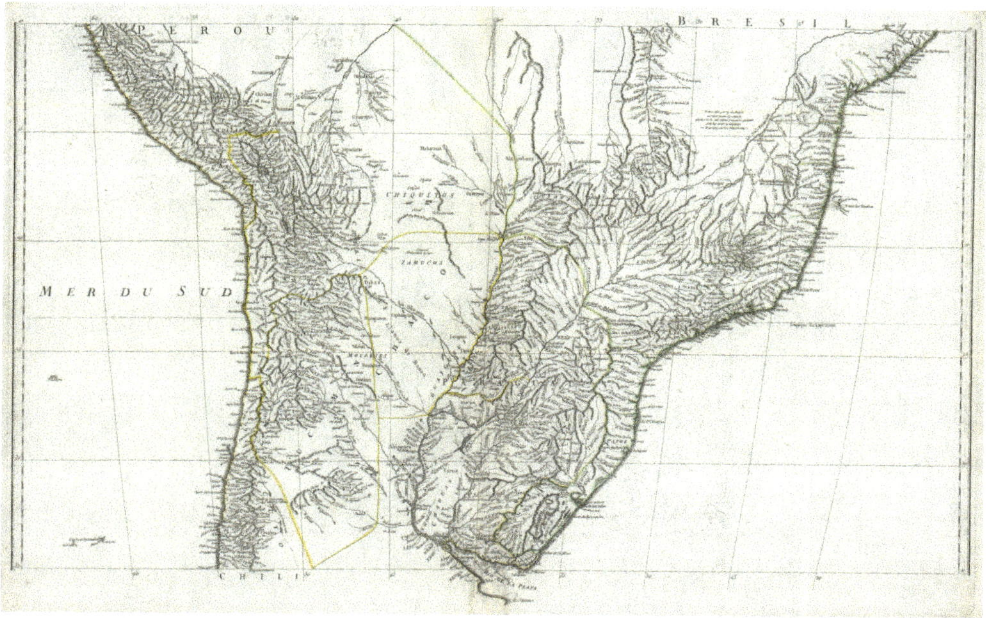

FIGURE 38 Page 2 (center) of *Amérique méridionale*, J.B.B. d'Anville (Paris, printed by Gravelot, 1779). Furtado Lins Collection
REPRODUCED COURTESY OF FURTADO LINS COLLECTION

Mantiqueira is another range he depicts, spreading out southeast of Vila Rica, although left unnamed; and the Emerald Mountain runs perpendicular to the district, heading towards the mining region. As for this range, he writes that the "region of the diamond mines (…) is bordered to the west by a chain of mountains, or a range, separating the maritime stretch from the interior lands."[6] The geographical configuration of all these mountains is due to the maps drawn or informed by Manuel Gonçalo de Lacerda, D. Luis da Cunha's secretary in the embassy (Figs. 30, 31, and 32).

On the 1748 *Amérique méridionale*, west of the São Francisco River—square in the heart of Brazil—d'Anville inserted a cartouche (Figs. 38 and 39) with the following text: "In this part of Brazil can be observed a space of close to 50,000 square French leagues, of which preceding maps had no knowledge." This is essentially an enormous void, represented on the map as a blank space. At first glance this would seem to be a flaw, the result of a lack of knowledge of the geographical elements of this area. But d'Anville affirms this void not as a defect, nor even an absence, but as one of the great additions to the American

6 RBC, 229, n. 539 (1), f. 4.

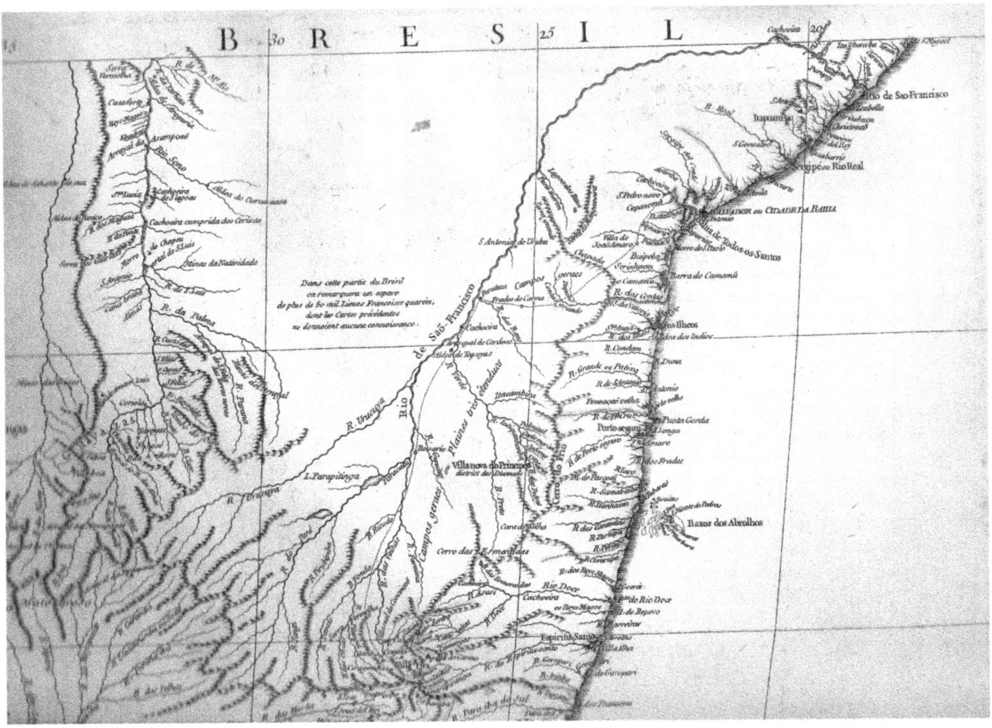

FIGURE 39 Detail of *Amérique méridionale*, J.B.B. d'Anville (Paris, printed by Gravelot, 1779). Furtado Lins Collection
REPRODUCED COURTESY OF FURTADO LINS COLLECTION

territory that he (and only he) had been capable of bringing about. These were 50,000 square French leagues added—as a geographical void—in the midwest of Brazil.

To the east of the center of Brazil, an array of graphic information drawn on the map—rivers, paths, mountain chains, towns, and resting places—allows the reader to get a sense of the local geography. On the other hand, the insertion of the western part—marked by a blank—is legitimated by the geographer's discourse. These paired languages (one pictorial, one discursive) communicate not only the presence of geographical elements as seen through the map, but also of absences not previously perceived by tradition. These languages mirrored "the real void of our knowledge, in these completely bare spaces,"[7] a not insignificant void, representing as it did "more than twice the extension of France."[8] The mapmaker, however, justifies the omission of geographical infor-

7 Anville, *Journal des sçavans* (April 1750), p. 659.
8 RBC, 229, n. 539 (2), f. 4.

mation in this region—information presented in "preceding maps"—as the "vast extension of this continent" had been depicted as "vague things, based on pure speculation."[9]

First of all, we should try to understand the process by which this not insignificant space is inserted into the *Amérique méridionale*; according to its author, after all, no map had shown it before. Secondly, we ought to comprehend the geopolitical meaning of this addition. The apparition of a hitherto unknown territory is explained by the map's establishment of two separate geographic entities, both in the backlands of the Brazilian interior, in longitudes considered more accurate than before: to the east, the São Francisco River, and to the west, the Tocantins River. These insertions meant a meaningful expansion of the Brazilian territory around the 10° and 15° parallels, significantly extending Portugal's possessions in the center-west of South America (Fig. 38).

To represent the Tocantins River, d'Anville says he had crucial access to "a long manuscript map" about the course of this river, "whose wealth in gold mines in this country led to the discovery of the upper region." He claimed that the map had been brought from Pará by La Condamine and Maldonado.[10] This is the *Carte manuscrite de la navigation de la Rivière des Tocantins*—a long Portuguese map 3.33 m long by 0.65 m wide (Fig. 40).[11] D'Anville explained that first he managed to establish a correspondence between this map and the *Carte manuscrite des côtes du Brésil* (Fig. 33)[12] starting around the headwaters of the Tocantins River, near the settlement of Meia Ponte, using the fact that some "rivers that make up the São Francisco River" are also close by. These were the Urucuia and the Paracatu[13] and he succeeded in this.

As was his custom, he first drew up a sketch of his own, based on the original map. This is a schematic version of this Luso-Brazilian map of Tocantins River, d'Anville using more universal and schematic geographical symbols and reducing the size of the features represented to a scale proportional to that of *Amérique méridionale*. He titled this manuscript map *Amérique: Rivière des Tocantins* (Fig. 41), which was then transposed, on a smaller scale onto the printed *Amérique méridionale*, 1748 (Fig. 39). A comparison between the three maps when it comes to the depiction of the Tocantins River shows that d'Anville copied practically the whole run of the river from the Luso-Brazilian

9 Anville, *Journal des sçavans* (April 1750), pp. 658–59.
10 Anville, *Journal des sçavans* (April 1750), p. 659.
11 For detailed analyze of this map see Furtado, "Metamorphoses of Colonization," pp. 367–99; Furtado, "A expansão para o oeste," pp. 119–33.
12 Anville, *Journal des sçavans* (April 1750), p. 659.
13 RBC, 229, n. 539 (1).

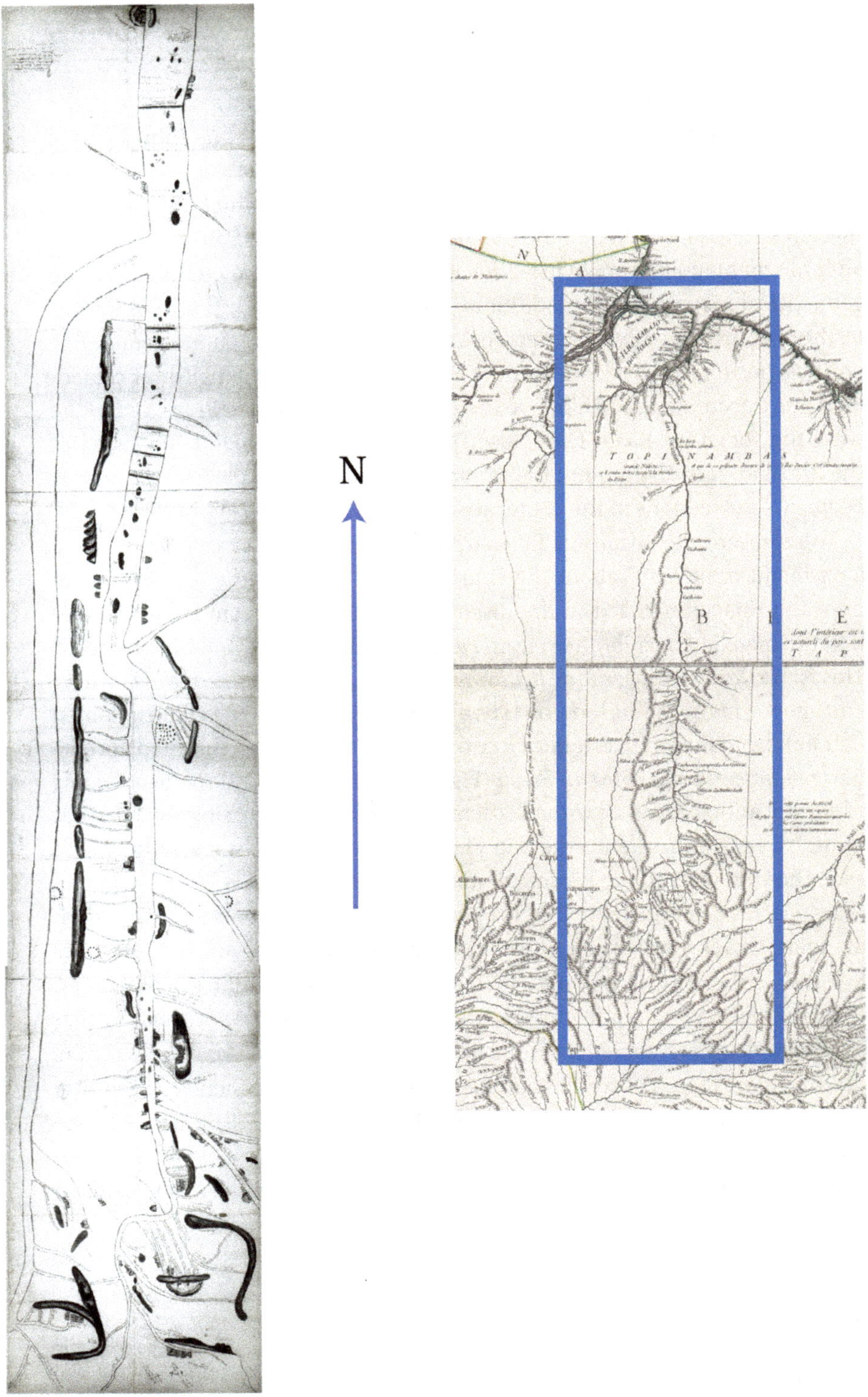

FIGURE 40 *Carte manuscrite de la navigation de la Rivière des Tocantins*, unknown author (Brazil/Belém, manuscript, blue ink on paper, *c.*1740). BNF, DCP, GE DD-2987 (9554 B). On the right the Tocantins highlighted in d'Anville's *Amérique méridionale*
REPRODUCED COURTESY OF FBN

map, from the settlement of Meia Ponte in Goiás captaincy, near its headwaters, to Vila de Cameta, close to its mouth in the Amazonas River (Fig. 42). The Luso-Brazilian manuscript map's main objective was to guide navigation down the Tocantins—all its waterfalls, islands, and sandbanks are duly signaled—and d'Anville reproduced those river markers that helped him to calculate the distances between them, allowing him to draw the river course with accuracy. One can see in the three maps the hills and mountain ranges sighted, as well as the endless waterfalls the intrepid navigators encountered along their way (the quantity and length of which they repeatedly emphasize), which made the trip extremely dangerous. These geographical landmarks, with their positions carefully set down by members of the expedition, were fundamental for allowing d'Anville to establish the measurements of the river's length on his map. And so, he set down a distance of 367 *lieues* between Itaboca Falls, near the mouth of the river, and São José in Goiás, not far from its headwaters.[14]

The text of the *Roteiro da Derrota do Rio Tocantins até Belém do Pará* has a close connection with the Luso-Brazilian manuscript map.[15] It describes a *derrota*, or river expedition, which descended the Tocantins River, from its source in the midwest until Cametá and its mouth on Amazonas River in the north. The expedition resulted in an arduous journey of almost three months, from November 2, 1734 to January 23 of the next year, when travelers reached Belém do Pará. José da Costa Diogo was leader of the expedition, which was part of the movement to explore the river in search of gold, which intensified at this time. The text of the expedition provides a mental map describing precisely most of the same landmarks arranged along the river both in d'Anville's *Tocantins* map, and, consequently, in the *Amérique méridionale*.

Marco António de Azevedo Coutinho, then serving as a diplomat in England, first caught wind of this adventure, "having heard it at this court." But the news was slightly garbled on the way. They said that in the "State of Maranhão, two mines have lately been discovered." Azevedo Coutinho quickly wrote to Portugal asking for information on the matter, fearing that this gateway might pose a threat to Minas Gerais. He was informed that there was a great distance from Minas Gerais and "there were three men who, going through the backlands and using at some points a few of those abundant rivers, came to arrive in Grão-Pará." These travelers "found so many hindrances that the experience

14 Anville, *Journal des sçavans* (April 1750), p. 660.
15 AHU, MAG, caixa 1, doc. 12.

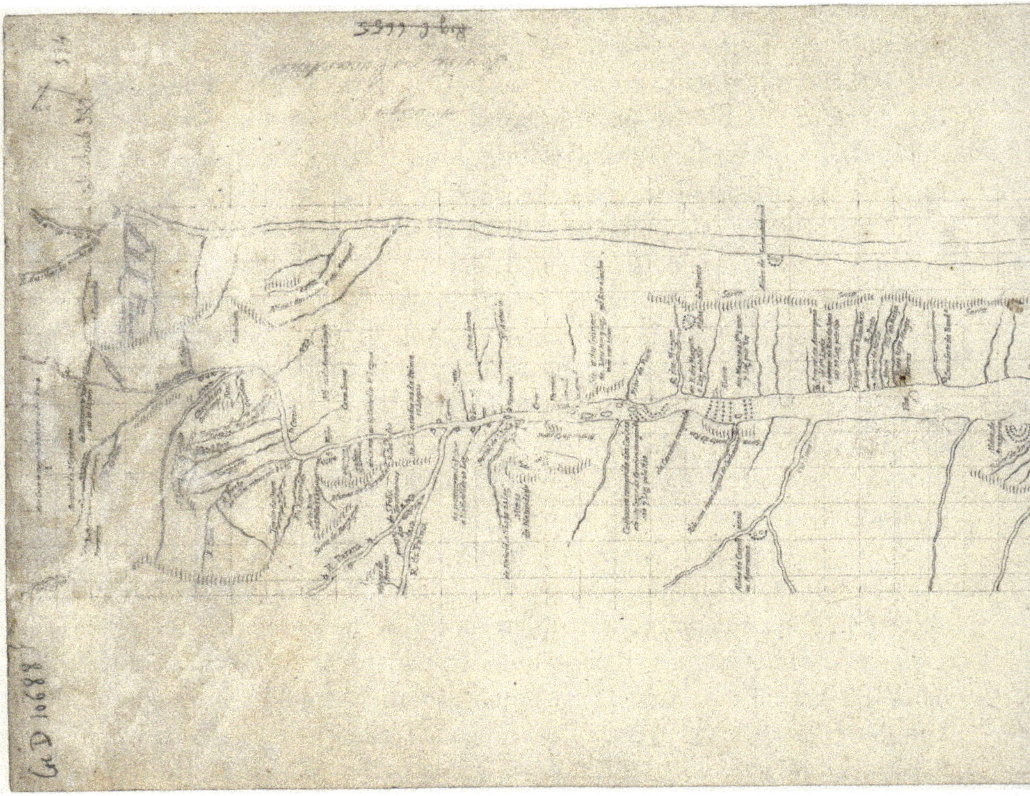

FIGURE 41 *Amérique: Rivière des Tocantins*, J.B.B. d'Anville (Paris, by the author at the Louvre galleries, manuscript in pencil, *c.*1746). BNF, DCP, GE D-10688
REPRODUCED COURTESY OF FBN

dissuaded them and they saw it would not be possible to open an entry from said city to Minas [Gerais]."[16]

A copy of this letter was sent at the time to D. Luis da Cunha, then in Paris, and so the ambassador heard of what was going on.[17] As he always warned, discoveries of gold mines in Brazil would make immediate waves in Europe, especially in England, always jealous of that territory's riches. In fact, the goal of the expedition had been to find gold, and the mines of Tocantins River soon became famous. The travelers, upon encountering the first locals they met living on the banks of the river, declared that their "aim was not to arrive in Pará,

16 ANTT, MNE, livro 16, ff. 23v–24, February 22, 1738.
17 ANTT, MNE, livro 16, f. 24.

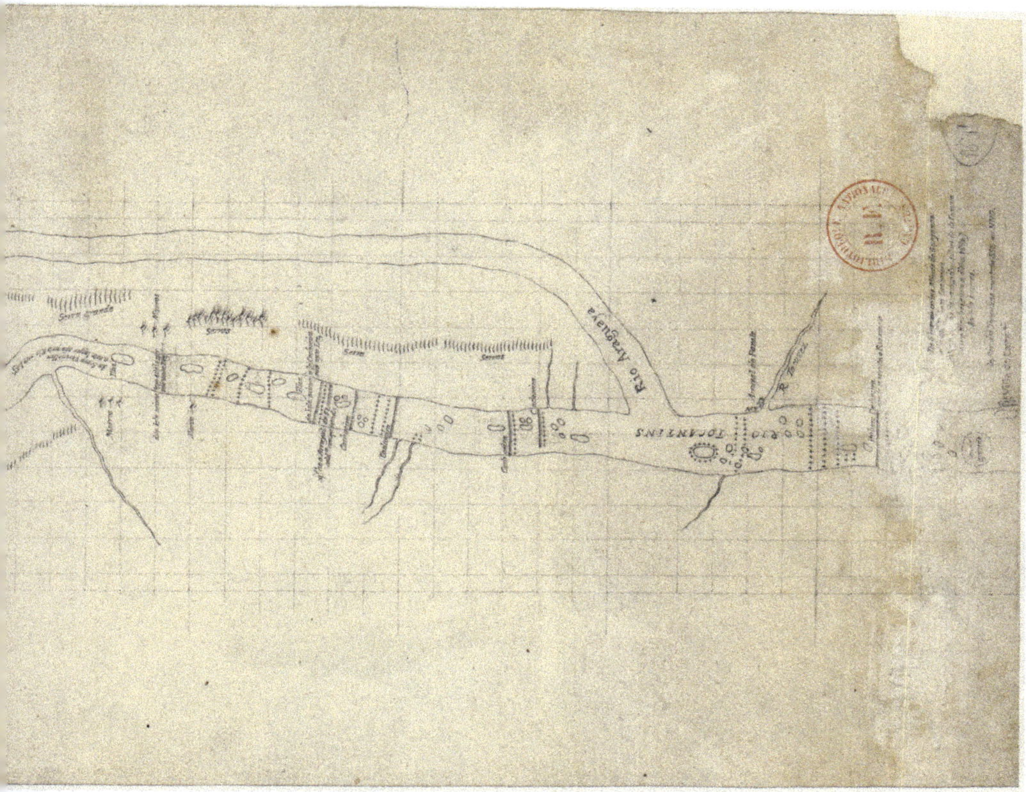

but to seek gold."[18] Similarly, d'Anville referred to Tocantins area as "a country rich in gold."[19] Over the course of the expedition, its members made several attempts to find the metal, some of them successful—such as in the case of what they called "the River of Gold," having found some there. They brought bowls, shovels, and handspikes, tools of the trade for mining. Finally, when they arrived in Belém city, they would affirm yet again the goal of the expedition; there they received license to travel to Portugal and "bear witness to the discovery my comrades and I have made of the gold mines of the Tocantins River."[20] Indicative of the mineral riches around the river, the manuscript map

18 AHU, MAG, caixa 1, doc. 12, f. 6.
19 Anville, *Journal des sçavans* (April 1750), p. 659.
20 AHU, MAG, caixa 1, doc. 12, ff. 14 and 9.

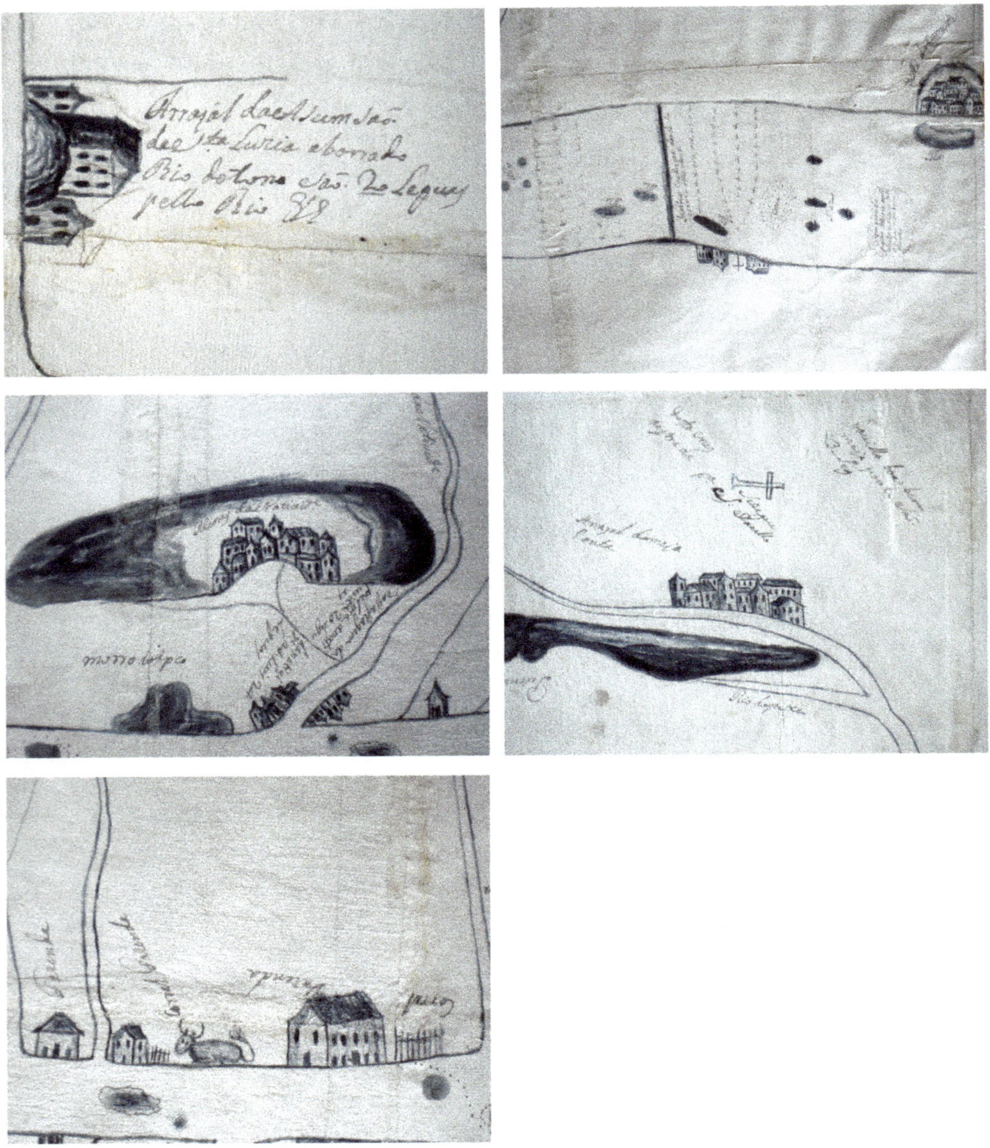

FIGURE 42 Details of *Carte manuscrite de la navigation de la Rivière des Tocantins*, unknown author (Brazil/Belém, manuscript, blue ink on paper, *c.*1742). BNF, DCP, GE DD-2987 (9554 B) REPRODUCED COURTESY OF FBN

also depicts the location of the Natividade Mines (Fig. 42) beyond the river, already explored, which would lead d'Anville to note them on his map as well.

A juxtaposition of this account, the Luso-Brazilian manuscript map, and the configuration of the Tocantins in d'Anville's maps reveals a direct relationship among the four. However, the direction of the texts on the manuscript map is the opposite of the journey described in the manuscript. This inversion has a possible explanation as the map was not the direct result of this particular account, but had been produced in Pará, at the request of local authorities, between 1735–1742, using information from José da Costa Diogo's downriver expedition as well as others. This is made clear as the map included the Araguaia River, which was not followed by the expedition, but is nonetheless represented on the *Tocantins* manuscript map.

The account and the Luso-Brazilian manuscript map are a record of this continual discovery of the local geography. We see that some places were already named, such as the settlement of São Felix das Terras Novas, with a church run by Father José Pires; or the farm of São José, where the adventurers stayed for a few days while ill, near the Preto River, or the fortress where locals hid once in a while when threatened by Indian attacks. Other places are baptized by the adventurers, such as "the waterfall we call Maquines Falls" or "a river we call Mangues River and (…) 7 *lieues* below is a river we call Luzia River, which is within sight of a large, long mountain range, near the Tocantins." As discussed, giving names to geographical landmarks was a symbolic act of possession, continually exercised by those who considered themselves discoverers of these new worlds. This would be an argument in favor of a direct relationship between the account and the map: the names which the voyagers gave to the landmarks along the way were registered in the account and transposed, one by one, to the Luso-Brazilian Tocantins manuscript map—such as the Canabrava River, the long waterfall of Corixás, Santa Luzia, etc.[21] They also appear on d'Anville's Tocantins maps. At some points the spelling slips up, as in the case of the Mangues River—the name d'Anville uses on his map, but which was spelled "Magues" on the manuscript. This map allowed d'Anville to determine a very precise configuration for the course of the Tocantins and Araguaia Rivers (Fig. 42).

The map of the Tocantins and the one by António Álvares da Cunha (Fig. 33) also allowed d'Anville to situate the Tocantins River in relation to the meridians. This measurement was only managed thanks to the position of the settlement of Meia Ponte village, near the source, far upriver.[22] Following the directions of these auxiliary maps, he placed two hydrographic entities in the middle of the Brazilian backlands—the Tocantins and the São Francisco. The first, to the west, was at approximately 32°, with the second, to the east, positioned on average around 25°—revealing a hitherto unknown territory.

21 AHU, MAG, caixa 1, ff. 5, 6, 8, 9, 10, 11, 12.

22 AHU, MAG, caixa 1, ff. 5, 6, 8, 9, 10, 11, 12.

Interestingly enough, in the 1748 map, d'Anville drew two settlements by the name of Meyaponte (*sic*), both positioned north of the point where he wrote the name of Mato Grosso, and positioned on two different rivers with the same name, with one of the settlements just to the south of the other on opposite sides of a mountain range (Fig. 39). This configuration remained through successive editions of the map. The one farther south was placed on the banks of a Meia Ponte River, a tributary of the fluvial system that runs to the southeast of Brazil, connected to the river basins of the das Mortes, Grande, and Paraná, represented as interconnected. The one farther north is placed near the basin of the Tocantins River, of which another Meia Ponte River is a tributary. D'Anville explained this doubling, affirming that he made use of the two maps for this region: the Luso-Brazilian manuscript map of Tocantins (Fig. 40), and the map of the vast territory between Minas and Paraguay, drawn by António Álvares da Cunha, titled *Carte manuscrite des côtes du Brésil* (Fig. 33). Upon comparing the two, he declared himself to have "been very happy, because the position of the settlement of Meia-Ponte, which can be seen high on the Tocantins River, was included in this [first] map," situated "in a land rich in gold mines, occupied by an Indian Nation known as Guayazas." This inclusion allowed him to understand "not only the place near to the river's source, in the Brazilian interior, but also its correspondence or connection with the part represented on the map by M. da Cunha." The latter also exposed him to "a district called Cuiabá, rich in gold mines," situated "above the Mbotetei River, near the Paraguay River."[23] The differences between the two maps concerning Meia Ponte seem to be the reason he carries over both positions for the settlement.

The Genovese geographer Francisco Tosi Colombina (1701–17?), who conducted cartographical surveys of Brazil in the service of the Portuguese, carried out an evaluation of the *Amérique méridionale* in August 1750 at the behest of the governor of Goiás captaincy. He concluded that this stretch of the river was riddled with errors, and in his report concluded that "one cannot base oneself on this map of d'Anville's." As for Meia Ponte, Colombina does not mention the two settlements with the same name on the map, only concluding that the latter was off by 2°55′. The settlement of Meia Ponte corresponds to the modern city of Pirenópolis; and if we compare d'Anville's map to a current IBGE map, the place labeled Meia Ponte (farther to the north) is the closest to the current location of the city. Colombina certainly overstates the map's imprecision! The differences in position of places in the midwest on the *Amérique méridionale*, compared to current measurements, varied in longitudes, and in latitudes,

23 RBC, 229, n. 539 (1), ff. 2, 3.

which is almost insignificant, principally when one considers the dimensions of the map and the lack of precise figures available for these locations when d'Anville was drawing up his map. On the contrary, the map appears fairly precise; and if there is a difference of 2°14′ in latitude between Meia Ponte and Pirenópolis, in longitude the average difference is just 1°01′.[24] Colombina also notes that Meia Ponte was in fact on the banks of the Rio das Almas, which empties into the Maranhão River, which then reaches the Tocantins, and that "on this map one does not find the Maranhão River, which seems to have been taken for what is called [the river] of the settlement of das Almas."[25] In this case he was right, as d'Anville put the settlement not on the banks, but to the south of the Rio das Almas.

The other Meia Ponte, farther to the south from the first one, seems to correspond to modern-day Rochedo, lying on the banks of a river in the basin of the fluvial system of the Grande and Paraná Rivers—a system running towards the south. If Rochedo is the urban nucleus represented, d'Anville is quite accurate; there is a difference of just 1°40′ in latitude and 0.86′ in longitude south in relation to current measurements. The settlement he labels Santa Ana seems to be the modern location of Dourada Mountain, on the banks of the Pilões River (feeding into the Claro River). In this case, he has also achieved a fair degree of accuracy. Cuiabá and Vila Boa are also fairly close to their actual positions. The five locations in the midwest display a larger variation in latitudes (1°87′ on average) than in longitudes (an average of 1°01′)—and so in Colombina's critique of the map, he only makes reference to the former. Interestingly enough, the latitudes—which were easier to measure, requiring only a measurement of the declination of the Sun with the proper instruments (such as a quadrant)—are less accurate than the longitudes, which were more complicated and difficult to achieve, demanding as they did the observation of eclipses (generally of the satellites of Jupiter) and instruments of great accuracy and range, such as pendulum clocks and telescopes.

One curious fact is that his depiction of Rio de Janeiro—whose meridian had been measured by the Priest Mathematicians, the Portuguese Diogo Soares (1684–1748), and the Italian Capassi, in 1730, of which d'Anville seems to have been aware—is stunningly accurate. In fact, as he himself argued, armchair geography was wholly capable of representing the planet accurately, simply through a critical reading of sources. For success in this respect, good maps and accounts were essential. These were the lenses that could magnify the geography of South America so as to be legible from a study in Paris.

24 Calculations by Jorge Pimentel Cintra using a digital cartography program (Map Info).
25 Cortesão, *Alexandre de Gusmão e o Tratado de Madrid*, vol. 2, pp. 139–44.

From another point of view, the presence or absence of certain pieces of information on a map can also be important for our understanding of how people imagined the region. Mapmaking was not merely the product of direct and empirical observation of nature—despite the work of some individuals in surveying circumscribed regions, geographers like d'Anville, and ambassadors like D. Luis da Cunha, who believed that maps were pure mirrors of the real world. In fact, despite the banishment of dreamlike or mythological images from this enlightened cartography, several such visions—echoes of the belief in the existence of Paradise on Earth[26]—still stubbornly and surreptitiously made their way into maps. Though no longer represented in pictorial form, oneiric spaces found expression in cartographic symbols which Enlightenment-era geography sought to universalize.

2 Simulacra

If the blank spaces in the northeast and the midwest are left on purpose, there are other areas in the *Amérique méridionale* where the geographer plays a sort of intentional sleight of hand with the use or absence of certain pieces of information. This becomes more evident around the region of the Amazonas basin. To understand the process of establishing the positioning of the river and its tributaries on the map—and the way in which he ranks, selects, or discards material at hand—we must first undertake a critical reading of the memoirs in which d'Anville describes the process of producing the map. Secondly, using modern systems of information, we will compare the coordinates of various geographic points along the river in the *Amérique méridionale* with the auxiliary maps on which he based his work, as well as with current maps. These digital tools are not meant to evaluate the quality of d'Anville's data, but rather to help shed light on the process of critiquing the geographer's sources.[27]

As one sees on the cartouche of the *Amérique méridionale*, La Condamine played an important role in the spatial configuration d'Anville lent to the region of the Amazonas basin—even along stretches, such as the Negro River, where he never traveled. Maldonado had accompanied La Condamine on his expedition to the Amazonas River from 1743 to 1744. The two set out from around Quito and took different routes at the start. D'Anville considered this strategy

26 Sérgio B. Holanda, *Visão do Paraíso* (São Paulo: Companhia das Letras, 1994).

27 Cintra and Furtado, "A *Carte de l'Amérique méridionale* de Bourguignon D'Anville," pp. 273–316.

"worthy of fine praise," as this meant that they "multiplied the knowledge"[28] about the geography of the region. La Condamine headed for the river by way of Jaén de Bracamoros in Spanish America, the first leg of the navigable stretch. Maldonado went down a tributary called Pastaça; the two met up in Laguna, in the province of Mainas on the banks of the Amazonas, still in Spanish territory.[29] Maldonado took a shorter route and La Condamine chose the longer, better-known route, starting with a long overland trek, hitting the Amazonas at Jaén de Bracamoros. By the time La Condamine arrived at La Laguna, at the source of the Guallagua where they met, Maldonado had been there for six weeks already.[30] From there they went down the river together to Belém do Pará city.[31]

In Belém, Maldonado and La Condamine would part ways again. The latter writes in his diary: "During my stay in Pará, I went on a few little trips around the area by canoe, taking advantage of the occasion for details of my map." Curious about the phenomenon of the *pororoca*, the encounter of the river and sea waters that causes a succession of waves upstream, he went by canoe from Pará to Caiena, where he caught a ship bound for Europe, while Maldonado set off for Lisbon directly from Belém.[32] Throughout his excursions in the Amazonas River delta, La Condamine not only established multiple distance measurements but also measured the latitudes of various points, producing a manuscript cartography with multiple numbered intervals, complete with the names of the observation points. D'Anville had one of these manuscript originals, depicting the northern part of Marajó Island, titled *Route de Mr. La Condamine le long le la Côte Nord de l'Ile Marajó ou Joannès*, attributed to La Condamine, as well as a copy in color drawn up by the geographer himself, of the southern part of the Island, between Yaraoubi and Anajaheba, *Carte de la côte septentrionale de l'Ile de Marajo depuis Yaraoubi jusqu'à Anajaheba* (Figs. 43 and 44). D'Anville also owned fragments of chapters from a travel diary, probably belonging to La Condamine, with observations and geographical notations on the mouth of the Amazonas, entitled *De l'entré du Pará*, copied out in his own hand, with no clear attribution.[33] La Condamine's travels at and descriptions of the mouth of the Amazonas River were fundamental to d'Anville being

---

28 Anville, *Journal des sçavans* (March 1750), p. 532.

29 Neil Safier, *Measuring the New World*, pp. 103–04.

30 Charles-Marie de La Condamine, *Viagem na América meridional descendo o Rio Amazonas*, pp. 45, 47, 64.

31 J.P. Cintra, J.C. Freitas, "Sailing down the Amazonas River: La Condamine's Map," *Survey Review*, 43, 323 (2011), pp. 550–66.

32 La Condamine, *Viagem na América meridional descendo o Rio Amazonas*, p. 113.

33 RBC, 229, n. 558.

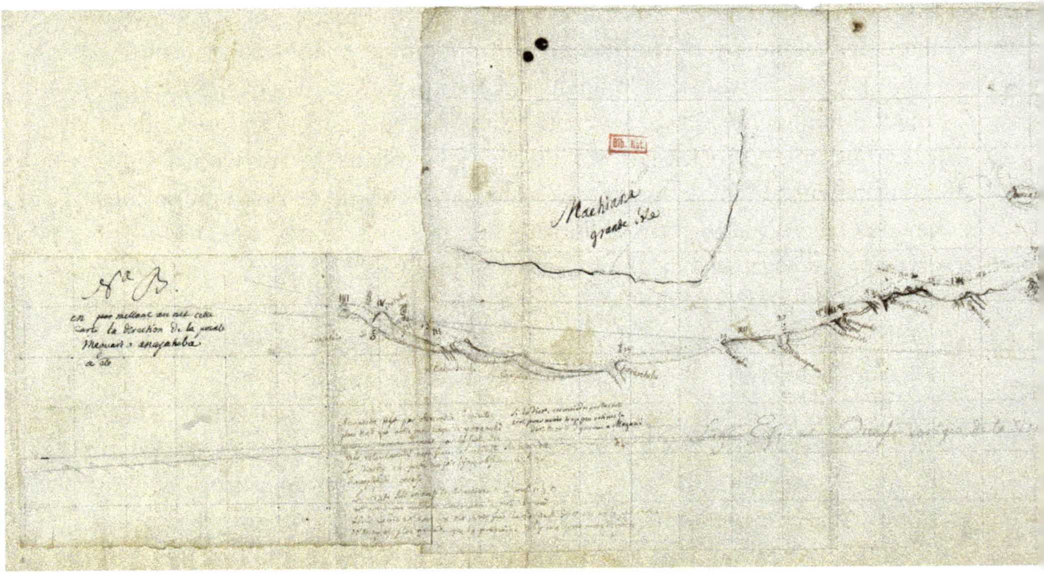

FIGURE 43 *Route de Mr. La Condamine le long de la Côte Nord de l'Ile Marajó ou Joannès*, La Condamine, 1736 (Brazil, manuscript, pencil on brown paper, *c.*1736). BNF, DCP, Sh Pf 166 Div 1 pièce 8 B
REPRODUCED COURTESY OF FBN

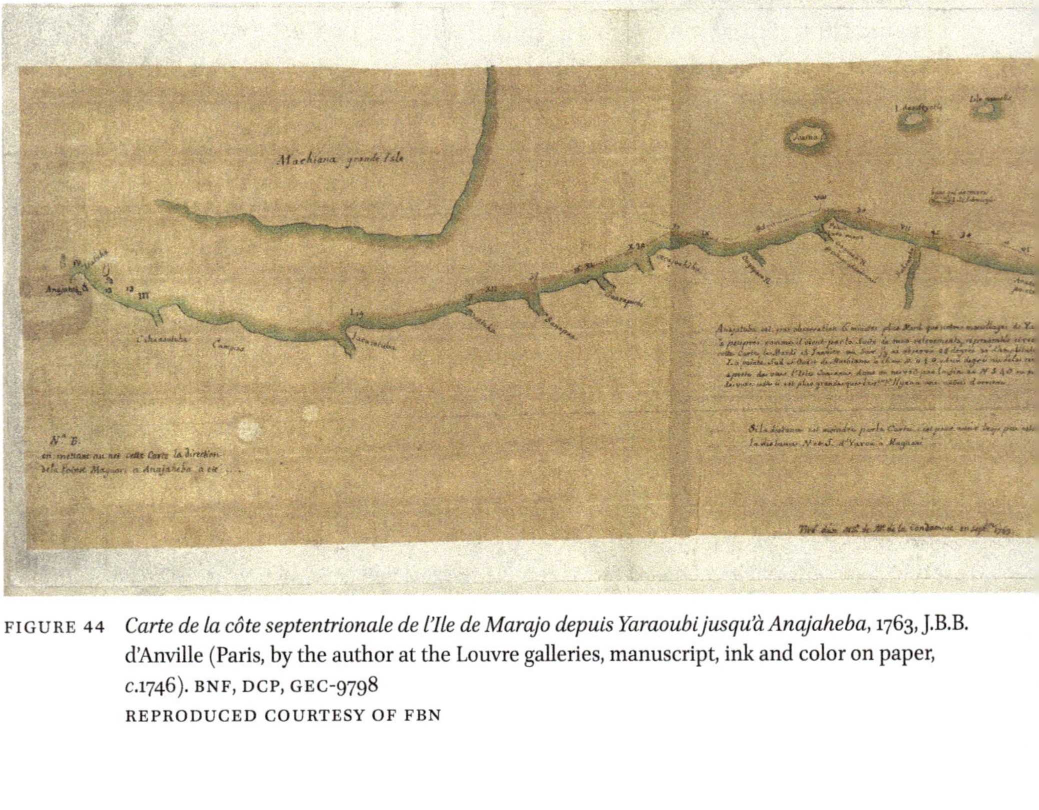

FIGURE 44 *Carte de la côte septentrionale de l'Ile de Marajo depuis Yaraoubi jusqu'à Anajaheba*, 1763, J.B.B. d'Anville (Paris, by the author at the Louvre galleries, manuscript, ink and color on paper, *c.*1746). BNF, DCP, GEC-9798
REPRODUCED COURTESY OF FBN

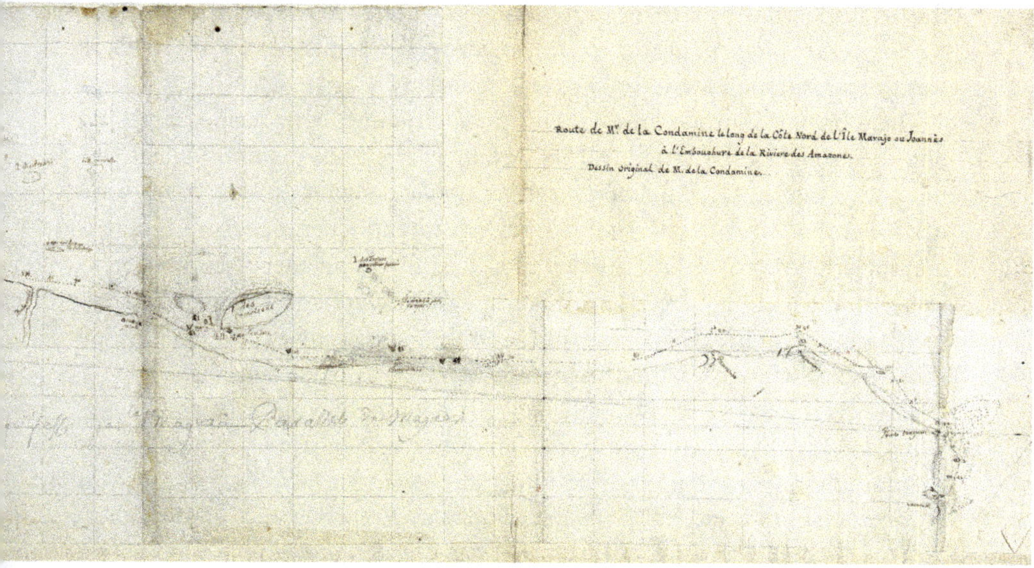
Route de Mr. de la Condamine le long de la Côte Nord de l'Île Marajo ou Joannès
à l'Embouchure de la Riviere des Amazones.
Dessin Original de M. de la Condamine.

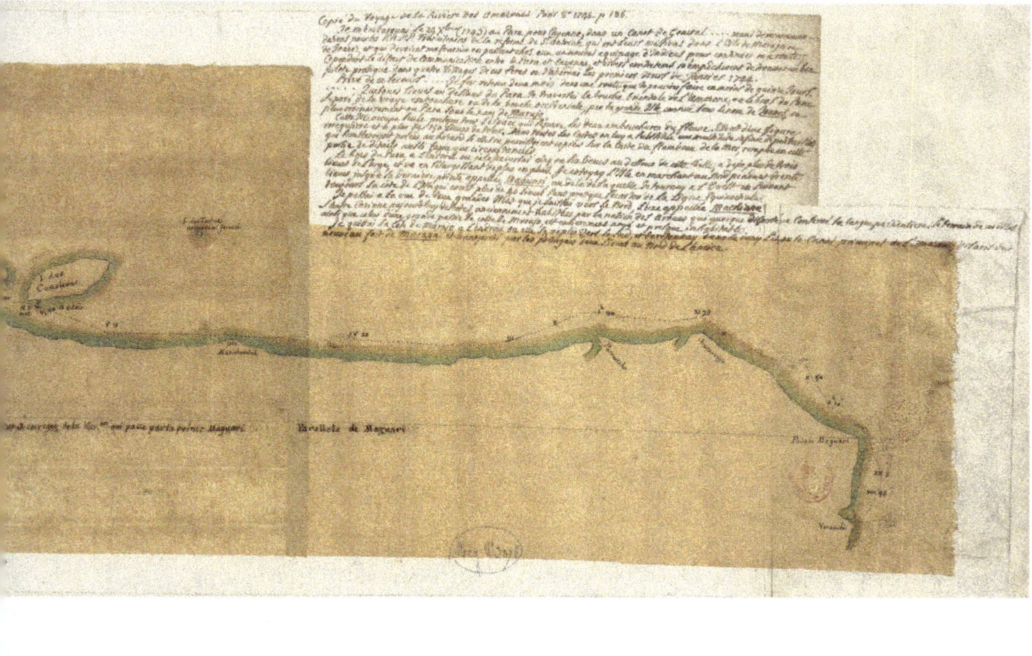
Parallele de Maguari

the first to portray the large island of Marajó on his maps, as it was portrayed in *Amérique méridionale* (Figs. 4 and 12), instead of a fragmented set of islands as was the cartographic tradition (Figs. 16, 17, and 18).

Once back in Europe, in 1745, La Condamine and Bouguer began falling out once they started to publish the results of their observations, which did not always agree.[34] In the *Amérique méridionale*, Bouguer's observations were d'Anville's basis for delineating the region around Cuenca, where Bouguer and the other members of the expedition took the principal measurements to establish the meridian arc.[35] La Condamine and Maldonado, however, had collected important information about the Amazonas River basin; and in Europe they both collaborated with d'Anville, although for different reasons.

La Condamine focused on publicizing his feats on the Amazonian odyssey,[36] worked with d'Anville to produce a number of maps accompanying his account, narrating the journey and illustrating various aspects. The *Carte du cours du Maragnon ou de la grande route de Amazones* (Fig. 24) came about in 1745 to be included in that edition of his *Voyage sur l'Amazone*, and was the first map of this river to be based on La Condamine's longitude measurements. He drew an initial sketch, gave it to d'Anville, "whose ability," the traveler writes, "is well known, [and] was of great advantage to me in reducing into order and putting together these scattered materials, to embellish my map wherewith."[37] In the natural division of labor of that period's armchair geography,[38] La Condamine took on the surveying (*lever*, in French), while d'Anville would draw up the map (*dresser*), which also meant joining up the information available to him. We can see, however, that many of the geographical conformations around the Amazonas basin, which La Condamine presented to the *Académie* as his own, were in fact based on information provided by third parties.[39] Though the *Carte du cours du Maragnon* cartouche reads, "Drawn in 1743 and 1744 and regulated by Astronomical Observations, by Monsieur de La Condamine," in the preface to his book the French traveler confesses that these documents were "communicated to me in the country by several intelligent missionaries and travelers," and that d'Anville, "whose ability is well known," used his method of cartographical criticism in "reducing into order and putting together these scattered materials."[40]

34 Safier, *Measuring the New World*, pp. 123–65.
35 La Condamine, *Viagem na América meridional descendo o Rio Amazonas*, pp. 37, 47.
36 Pratt, *Os olhos do império*, pp. 47–48.
37 La Condamine, *Viagem na América meridional descendo o Rio Amazonas*, p. 34.
38 Pedley, *The Commerce of Cartography*.
39 Safier, "Como era ardiloso o meu francês," pp. 91–114.
40 La Condamine, *Viagem na América meridional descendo o Rio Amazonas*, p. 34.

Maldonado on the other hand was collaborating on the *Carta de la provincia de Quito au Pérou dressée sur les observations astronomiques, mesures géographiques, journaux de route et mémoires de Mr de La Condamine et sur ceux de Don Pedro Maldonado* (1751), produced at the request of the Spanish crown. Around this time, as Portugal and Spain prepared to negotiate borders in America, the Spanish crown sought to fund scientific institutions with an Enlightenment bent and simultaneously map its possessions in America. Given the urgent nature of the task, it used French printers (as in d'Anville's case) and sent specialists to France to study the latest in the art of cartography (as was the case with Maldonado). It was through this Franco-Spanish collaboration that d'Anville had access to Spanish maps of the region.

To establish the shape of the Amazonas basin on the *Amérique méridionale*, d'Anville certainly used the *Carte du cours du Maragnon* by La Condamine (Fig. 24) as an important source. Four of this map's significant geographical contributions are present in the his first edition of the 1748 map (Fig. 11) the course of the Negro River; 2) the disappearance of Lake Parima and its substitution by the Amucu, through the connection of the mouths of the Rupunuwini, Essequebé, and Negro Rivers; 3) the connection between the Amazonas and the Orinoco via the Negro River; 4) the shape of Marajó Island, seen for the first time as a single stretch of land instead of an archipelago of smaller islands, as had been common in maps up until then.

Despite what one might expect, however, the two maps diverge on multiple points, even in areas around the mouth of the Amazonas where La Condamine had conducted several reliable astronomical observations. Along the way, La Condamine carried out several other longitude and latitude measurements, according to the record of his journey. However, there are discrepancies between what he observed and what he wrote.[41] When referring to the astronomical measurements La Condamine took, he seeks to support a new trajectory for the river's course (going against accepted knowledge). D'Anville, a master of the art of cartographical criticism, would at times disregard these measurements when they come into conflict with other documentation he possessed. The differences between these two maps, especially in terms of longitudes, suggest this. Let us examine the main auxiliary documents he made use of.

Firstly, we might highlight the map by the priest Samuel Fritz (1654–*c.*1725), a missionary in the service of the Spanish to the Omágua and Maina Indians. On one of his many voyages through Amazonas between 1686 and 1725, the

41 Safier, "Como era ardiloso o meu francês."

missionary went down the river to the city of Belém do Pará, close to its mouth. Fritz wrote a diary or a manuscript account of his experiences and, in Quito, the 2nd Marquis of Valleumbroso, Diego de Esquivel y Navia (1672–1730) gave a copy of this diary to La Condamine, who then took it to Paris.[42] Fritz—with the aim of defending Spanish control of the region—drew up several maps of the area, but only a few have survived. One of them, dated 1691 and entitled *Mapa geographica del Rio Marañon o Amazonas*, La Condamine also took to Europe.[43]

Going down the Amazonas River, La Condamine used Fritz's diary and map—considered the most accurate representation yet produced and was very good indeed.[44] But as soon as La Condamine arrived in Europe, he launched a campaign to discredit Fritz's findings[45] and guarantee the hegemony of his own representation, with an eye to immortalizing his own contribution to European knowledge of the geography of the New World. The *Carte du cours du Maragnon* (Fig. 24), which he created with d'Anville, was an important weapon in this campaign. The two included a faint traced line on this map, representing the Jesuit's depiction of the river, and overlaid the course they considered more accurate. This tool—fairly visible around the lower region of the Amazonas, around its mouth, and Marajó Island, presenting accentuated shifts to the east from Fritz's map—allowed them to call attention to the corrections they were making to the river's course.

La Condamine used information from the Fritz's own diary to underscore the map's lack of accuracy. He wrote that "Father Fritz, without a pendulum or telescope, could not possibly determine any point of longitude." As for instruments, he reported that the priest had "only a wooden semicircle of three inches radius for the latitudes." He lumped this together with Fritz's poor health, adding that as he was "indisposed when he went down that river to Pará," his reports could not be trusted.[46] However, d'Anville (despite not mentioning Father Fritz on the cartouche) admits in one of the memoirs that he did in fact use the Jesuit's map to fix "the detail of the upper region of the Maranhão,"[47] one of the denominations for the Amazonas around that time

42 Safier, "Como era ardiloso o meu francês," p. 95.

43 André Ferrand de Almeida, "Samuel Fritz and the Mapping of the Amazon," *Imago mundi*, 55 (2003), pp. 113–19.

44 See Furtado and Cintra, "'Bourguignon D'Anville's *Carte de l'Amérique méridionale*," pp. 301–10.

45 Safier, *Measuring the New World*, p. 77.

46 La Condamine, *Viagem na América meridional descendo o Rio Amazonas*, p. 44.

47 Anville, *Journal des sçavans* (April 1750), p. 627.

and for different regions of the river, since La Condamine had not been down those parts.

Another source in fixing the positions of these areas was a large manuscript map drawn by Father Joannes/Juan Magnin (1710–1701)—who wrote *Descripción de la provincia y misiones de Mainas en el Reino de Quito*, 1740—and given to La Condamine in Borja. Though the geographer found fault with this map, for a lack of "geometric rigor" that led it to "exaggerate the extension of spaces," he would incorporate various aspects of it, checking them alongside memoirs given to him by Maldonado when the two had worked together on the *Carta de la provincia de Quito*. As for the mouth of the Amazonas, d'Anville could use measurements taken by La Condamine as well as those from a manuscript map that he drew according to information from the Mercedarian Father Ignácio dos Reis, on September 10 and 11, 1729, in Pedro Nolasco Couvay's house in Paris.[48] D'Anville would produce the three sketches of upper parts of the Amazonas and its tributaries (Fig. 35) on the first day, and the map entitled *Carte particulière de l'Entrée du Pará* (Fig. 45) over the two days. This map, compared with the Luso-Brazilian manuscript map *Carte manuscrite de la navigation de la Rivière des Tocantins* (Fig. 40),[49] as well as the map provided by D. Luis's nephew António Álvares da Cunha, were important in fixing where the mouth of the Tocantins flows into the Amazonas. It allowed him to locate Cametá settlement in relation to the position of Belém do Pará. This manuscript map was used to draw the south banks of the Amazonas River and Marajó Island in the *Carte particulière du cours de la Rivière des Amazones ou de Maragnon depuis l'entrée du Para en remontant jusqu'au confluent de la Rivière Noire* (Fig. 17), published by d'Anville in 1729 according to Father Reis's information. The three sketches were used to draw the confluence of the Amazonas, Madeira, Abacaxi, Guatuma, and Urubu Rivers. The configuration of all these rivers was carried over into the *Amérique méridionale* (Figs. 4 and 11).

Let us see how d'Anville compiled all the information he had about the Amazonas, from headwaters to mouth. For the area around Cuenca and Quito, he used the data from the geodesic expedition. In his memoirs he writes that La Condamine considered the correct measurement for the meridian at Quito. For the *Amérique méridionale*, he positioned it at 60°20′ and a few minutes to the west of the Ferro Meridian, which meant a divergence of seven or eight minutes from La Condamine's findings. D'Anville was fully aware of this, but believed that, given the size of the map, this would not lead to significant

48 Anville, *Journal des sçavans* (March 1750), pp. 438–39, 554–55.
49 Anville, *Journal des sçavans* (April 1750), p. 659.

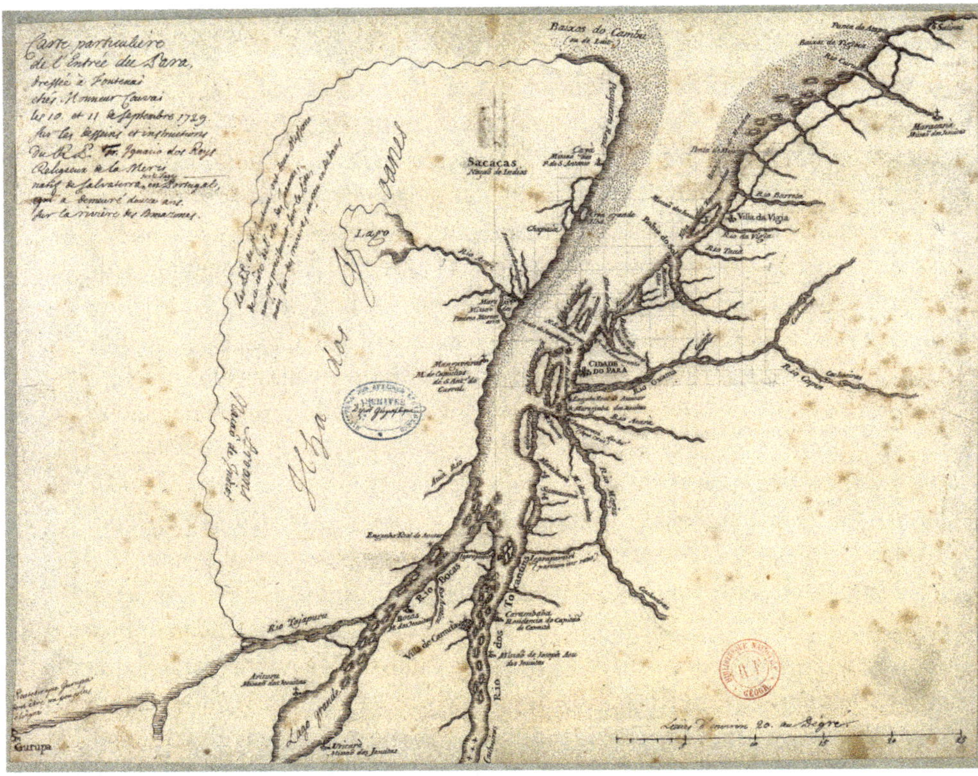

FIGURE 45 *Carte particulière de l'entrée du Pará dressée (à Fontenai) chez Monsieur Couvay, (les 10 et 11 Septembre 1722) sur les desseins et instructions du* R.P. *Ignacio Dos Reys*, J.B.B. d'Anville (Paris, manuscript, brown ink on paper, September 1729). BNF, DCP, GE DD-2987 (9553)
REPRODUCED COURTESY OF FBN

modifications.[50] He accepted La Condamine's measurements between Quito and Jaén, despite all the difficulties the latter had in taking them, and fixed the longitude of Jaén at a difference of just 0.25° from La Condamine's *Carte du cours du Maragnon*. But this difference grows as we move downriver, peaking by Belém do Pará, a difference suggesting that the two maps are not coincident, and that d'Anville did not use the French scholar's measurements for the middle and lower regions of the river.

D'Anville relates that, in order to draw the stretch between Quito and Laguna, he compared La Condamine and Maldonado's observations on the two routes they took to the Amazonas,[51] using the latter to draw the course of

50 Anville, *Journal des sçavans* (March 1750), pp. 524–25.
51 Anville, *Journal des sçavans* (March 1750), pp. 535–38.

the Pastaça River. La Condamine arrived at the mouth of the Napo River on the 31st in time to observe an eclipse of the satellites of Jupiter, which was not possible, despite his efforts, due to vapors on the horizon blocking his view of the phenomenon.[52] As they had not measured longitude until this point, d'Anville was obliged to calculate longitude using the measurements of the distances covered between several points noted by the two travelers between Quito and Laguna, passing the mouth of the Pastaça, as well as using auxiliary documentation. The fact that they chose two different itineraries allowed d'Anville to compare the measurements each took separately and thus establish the difference in longitude between the two points which he considered most accurate. For several reasons, in fact, including the technology being used, these measurements were far from perfect, but d'Anville was a specialist in not only the equivalence but the transformation of itinerary measurements into geographical measures of distance.

From Jaén, still using La Condamine's data, d'Anville positioned the cities of S. Tiago and Borja—the latter still in the river channel—which allowed him to establish the first point of the course of the Amazonas on the map. Between Jaén and the mouth of the Pastaça, d'Anville had two main options: La Condamine and Fritz. He set the longitude of Jaén based on La Condamine and used the differences in distance (converted into longitude) measured by Father Fritz, a source likely to be more trustworthy, since the Jesuit had been along this path many times when visiting the villages that he served.

For the region between the mouth of the Pastaça and Laguna Rivers, d'Anville had a third source available to him: the data Maldonado had collected on-site with an eye to positioning the course and mouth of this river. With these in hand, he modified the longitude measurement of the mouth of the Pastaça, changing it from the *Carte du cours du Maragnon*, as he considered Maldonado's measurements more accurate. He forgave La Condamine for the inaccuracy of his measurements, however, as they must have been caused by the rough terrain of the mountainous area at the beginning of the route, before reaching the river, or by La Condamine's having minimized the strength of the wind, which would have influenced his calculations of the distance covered by canoe.[53] As for the mouth of the Napo River, though La Condamine and Maldonado measured the longitude of this point, the *Carte du cours du Maragnon* is 3°79′ off, a difference contaminating the whole of that map. D'Anville, in the critical elements of the construction of the *Amérique*

52 La Condamine, *Viagem na América meridional descendo o Rio Amazonas*, pp. 64–65.

53 Anville, *Journal des sçavans* (March 1750), pp. 537–38, 540–41.

méridionale, noted this difference and sought to eliminate it, mainly using information collected by Father Fritz.[54]

Access to Portuguese sources provided by D. Luis da Cunha allowed d'Anville to alter the position of a number of points situated around the mouth of the Amazonas in relation to La Condamine's mapping, which produces a series of divergences from current measurements in both latitude and longitude around this area. La Condamine, using astronomical methods, had taken the longitude of Belém do Pará city on three separate occasions and his measurements were precise.[55] In terms of the position of Belém, near the mouth of the Amazonas, d'Anville differs from La Condamine's figures, but neither does he use the position Father Fritz gives for the city. Suggestively, and unlike his method for other cities and important points in South America (such as Caiena, Buenos Aires, Valparaíso, Santos, São Vicente, Olinda, and many more), across all the versions of his memoirs, d'Anville does not mention La Condamine's longitude measurements in Belém, much less referring to the sources, data, or calculations used to arrive at the position of this site. In the document *De l'entré du Pará*[56] there is mention of the city of Belém do Pará—but this is only in terms of latitude, which fits with the figure presented by La Condamine (about 1°20′ to the south of the Equator), with no mention of longitude, and referring only to distances in relation to other points around the river mouth. In the *Amérique méridionale*, the repositioning of these nearby sites around the mouth of the Amazonas, such as Guanapu and Jacunda—or in the stretch just above it, such as the mouth of the Tocantins—may have contributed to the repositioning of Belém. In the case of the *Amérique méridionale*, the adjustments in the area around Belém were significant; and, given that d'Anville had discarded La Condamine's measurements, it displays a divergence from current measurements, which evidently could have been avoided.

Generally speaking, in the *Amérique méridionale*, d'Anville traced the hydrography of the region on the following basis: establishment of the base longitude in Quito using data from the geodesic expedition; use of Maldonado's figures for the Pastaça region; use of Father Fritz and other missionaries' maps for the areas between Jaén and Borja, and from Borja to the Napo and the Javari. From there, he bases himself on measurements taken by La Condamine and Maldonado, and, finally, from the Xingu to the mouth of the river, he extracts coordinates from Portuguese maps. In the upper region of the basin, the

---

54 All the data can be seen in Furtado, and Cintra, "Bourguignon D'Anville's *Carte de l'Amérique méridionale.*"

55 La Condamine, *Viagem na América meridional descendo o Rio Amazonas*, p. 112.

56 RBC, 229, n. 558.

elimination of Lake Parima and the establishment of a connection between the Amazonas and Orinoco basins via the Negro River were both based on La Condamine—considering that he, for his part, relied on information provided by third parties, not having traveled through these areas himself. The substitution of Lake Amucu for Lake Parima was brought about using a map and a journal which La Condamine brought with him, from a Prussian, Nicolas Horstman, who had traveled from Caiena to Belém, as discussed below (Fig. 47).

As for the connection between the Amazonas and the Orinoco, d'Anville draws not one, but three connections branching out from the confluence of the Caquetá (high up the Orinoco) with the Iça, Yupura, and Negro Rivers, respectively (Figs. 4 and 11). This triple connection was first established by Father Cristóbal de Acuña. According to d'Anville, "the Caquetá, which forms this junction, according to Father Acuña, after having lent its name to a small place situated on its banks, located at one degree of latitude north, spreads its waters out into different branches which will form the Negro River, Yupurá River, and Iça River." From the headwaters to the mouth of the Caquetá, d'Anville follows these three connections. The first is at the confluence of the Joya and Iça Rivers; the second, directly with the Yupura, in the country of the Acanejos; and the third, with the Negro, in the territory of the Orelhudos. The latter was affirmed to La Condamine by a group of Portuguese missionaries who traveled down the Negro River in 1744 and then, "by going from river to river, found themselves on the Orinoco."[57] This connection, which does exist, was "always known to the Natives." At the dawn of the nineteenth century, Humboldt would navigate 12 days in a canoe to find it, managing to take the measurements of the place and verify it scientifically, via the Casiquiare.[58]

The apparent lack of correspondence between the text on the cartouche, the memoirs, and the data we can extract through the use of digital tools merits some consideration. These, in some areas of the map, allow us to lay bare what lay behind d'Anville's discourse—that is, the data actually incorporated into the map, those discarded, those not fully revealed to his interlocutors. We thus see that his discourse, like a distorting mirror, transforms itself from audience to audience. Assuring the credibility of the *Amérique méridionale* meant using vague or ambiguous information on some parts of the map to disguise the data sources he actually compiled and those he wound up discarding. In the same way as on the cartouche, although in more detail, the members of the geodesic expedition were praised over less prestigious sources that d'Anville

57 RBC, 229, n. 539 (2), f. 10.

58 Ana Pizarro, *Amazônia: as vozes do Rio, imaginário e modernização* (Belo Horizonte: Ed.UFMG, 2012), p. 108.

also used. But it is possible to clearly make out how he actually positioned himself with respect to the diverging datasets compiled by the members of the expedition. Despite La Condamine's prestige, he opted for Maldonado's figures in some locations, considering them more accurate. Finally, the use of digital tools was able to reveal that at other points d'Anville in fact used sources which on the cartouche are referenced simply as "material," without naming individual authors. Some of them, such as Father Fritz's map, allowed for a refinement of the representation of the channel of the Amazonas River; others, not so much. How furtive was the French!

But Anville included a number of unsettling, persistent presences in the *Amérique méridionale* that can't be explained only by his methodology of criticism: one major mythological image represented as real geographic entities—Lake Parima—as well as the name "River of the Amazonas," which substituted the traditional toponyms of Solimões or Maranhão River (mainly used by the Portuguese up until this point). The representation, and the choice of this name, are evident symptoms of the ways in which ancient myths linger on in this map, almost stealthily.[59] If maps such as the *Amérique méridionale* could legitimate the dominion of an empire over its territory by means of a cartographical representation designed to mirror reality, they could also—even with the same aim, or even if they sought the opposite—mythicize the land[60] by continuing to represent a fantastical geography.

3 The Amazonas Warriors

La Condamine played an important role in the spatial configuration d'Anville lent to the region of the Amazonas basin—even along stretches, such as the Negro River, where he never traveled. The Frenchman's impressions were also crucial in selecting the name of the river, since by the mid-eighteenth century several different toponyms were still being used to designate this vast sea of fresh water. D'Anville paid attention to the nomenclature he used on his maps, trying to respect the local language and usage. The name *Riv. des Amazones* is printed on the 1748 *Amérique méridionale* (Fig. 3), set apart from the other rivers in its system and in the upper-middle third, just after its waters flow into Portuguese territory. For all the other rivers, he uses the Portuguese word *rio* and not *rivière*. In this sense, the chosen name reveals two inconsistencies

59 Safier, "Como era ardiloso o meu francês," pp. 91–114.

60 J.B. Harley, "Maps, Knowledge and Power," in Denis Cosgrove and Stephen Daniels, eds., *The Iconography of Landscape* (New York: Cambridge University Press, 1988), pp. 277–312.

in terms of his own method for choosing toponyms: *Amazones* is written in French, not in the local language—*Amazonas*—and the name was barely used in the Portuguese portion of the territory. For this last reason, he inserts a text near the confluence with the Negro River, saying that the Amazonas, "below the Rio Negro, is called Solimoens or Rivière des Poisons by the Portuguese."

In fact, the river's nomenclature at that period still varied widely: Orellana, Maranhão, Grão-Pará, and Peixes (fish), from the colonizers; or Paranguazú, Guyerme, and Solimões, in the languages of the Natives who lived on its banks. Well before, back in the sixteenth century, "the first European who sighted it from the Atlantic, where it flows into the sea, gave it the name of 'Santa Maria do Mar Doce' (Saint Mary of the Sweet Sea), thinking it was a sea of fresh water."[61] This denomination was related to the context of the European maritime expansion, where the designations chosen by the Portuguese usually made reference to Catholicism, with its festivities, saints, and divinities: Monte Pascoal (Easter Mount), Bahia de São Salvador (Bay of San Salvador), or Santa Maria (Holy Mary) de Belém do Grão Pará, which recalls the first name given to the river by its discoverer.

D'Anville does not shy away from this discussion, knowing that he will have to justify his choice of the toponym *Amazones*. And so, in one of his memoirs on the *Amérique méridionale*, he writes that, as for the "Maranhão River, one must say something about the origin of the name." He wrote that the Jesuits Pedro Simão (1510–1579) and Manuel Rodrigues (?–1570) credited its etymology to the expedition of the Spanish conqueror Lope de Aguirre (*c.*1510–1561), who accompanied Pedro de Ursúa (1526–1561), heading up an expedition which left Peru in 1559 in search of El Dorado. As they went downriver, Aguirre transformed the mission into a bloodbath, killing Ursúa and others, including Fernando de Guzmán (1568–1636), the former's successor. And so those who remained with Aguirre after these violent clashes took to calling themselves *maranhões*, or "***marañones***, which they invented to call one another, or to refer to their troupe in general, and *Moragnon* (or Maranhão) (…) became (the name) of the river where these bloody episodes took place."[62]

However, d'Anville does not choose this toponym to designate the river on his map. The reason, he says, is "the women who gave the Maranhão the name of the River of the Amazonas." The myth of the Amazonas hearkens back to Antiquity, one of its most well-known versions stemming from the Greek philosopher and historian Strabo. D'Anville was well aware of this and mentions the philosopher's Book XI, where he describes these women, "who

61 Pizarro, *Amazônia*, p. 39.
62 RBC, 229, n. 539 (2), f. 8 (italics are the author's).

live around Terme," at the edge of the world known to the Greeks. In characterizing the Amazonas, authors of Antiquity display some divergence: some say that they cut off one of their breasts to better aim their bows, others do not; some, that they killed their male children, others that they gave them over to their fathers to be raised and kept only the girls. D'Anville sees that, given so many possible variations, the American Amazonas have assimilated the characteristics of the classical standard, of those who lived along the Terme River.[63]

The transposition of this myth to America is initially set about by Christopher Columbus (1451–1506), who places these warrior women on some islands of the Caribbean, but sticks definitively to the space between Portuguese America and the Guianas. This was at first the result of the expedition led by Francisco de Orellana (1511–1546), who went down the river in search of the Land of Cinnamon between 1541 and 1542. Shortly after they set off, in the region of the Omaguas Natives, the travelers heard from Aparia—the chief, who had traveled through those lands—that "downwater, on the great river, there are Amazonas, and that far from the river, inland, were the lands of the chief Ica, which were abundantly rich in yellow metal."[64] They were warned, however, that if they "saw the Amazonas, which they call in their language *coniupuiara*, which means large ladies, we should watch what we did, because we were few and they were many, and they would kill us." Father Gaspar de Carvajal (1500–1584), the chronicler of this trip, narrated the violent and bellicose encounter that finally came about when they arrived in the lands they called "de São João," near the Madeira River; there they stopped to celebrate that saint's feast, on June 24 Carvajal also left a rare record of the Amazonas' appearance: "These women are very pale and tall, with very long hair, braided and wound around their heads. They are very strong-limbed and go about naked, with their private parts covered, carrying their bows and arrows."[65] The English traveler Sir Walter Raleigh (1552–1618), writing in 1591, also bears witness to their warrior spirit: "Amazonas, those women shall hereby hear the name of a virgin, who is not only able to defend her own territory and her neighbors', but also able to invade and conquer so great empires and so far removed."[66] Despite having been the first to find them, Orellana did not

63 RBC, 229, n. 539 (2), ff. 8, 13.

64 RBC, 229, n. 539 (2), ff. 24–25, 27.

65 Gaspar de Carvajal, *Descobrimento do Rio Orellana* (São Paulo: Companhia Editora Nacional, 1941), pp. 30, 60–61.

66 Walter Raleigh, *The Experience of the Large, Rich and Beautiful Empire of Guiana, with a Relation of the Great and Golden Citie of Manoa (Which the Spaniards Call Eldorado), 1596* (Manchester: Manchester University Press, 1997), p. 199.

associate the river with them, but with himself as Carvajal's account constituted a public baptism of the river with the name of its explorer, contributing to the polyvalence of existing denominations.

The journey up the river by the Portuguese officer Pedro Teixeira in 1639 was the first time the river was given the name "Amazonas," even though he tried to instill that of "São Francisco de Quito," which he said was more widely used being "given to the river by the Portuguese, believing that it had been discovered and navigated by the Franciscans."[67] Pedro Teixeira's expedition, under the União Ibérica, sought to establish Portuguese sovereignty over the river's navigation—thus the move to abandon the name "Maranhão," given by Aguirre, or "Orellana," both impregnated with references to Spanish hegemony over the region. Father Cristóbal de Acuña, who was one of the expedition's chroniclers, makes his preference for Amazonas clear, refuting the rest, and starts with a description of "the discovery of that great river of the Amazonas, called, by a vulgar error among those little versed in geography, by the river of Marañon," which would kick off the spread of this toponym. In describing the river, he affirms that "it is not known by any other [name], but only this, to all cartographers who have studied it," commenting that it would be remarkable if the Amazonas had usurped the name of the river without cause—that is, if they did not, in fact, exist,[68] as Pedro Teixeira never spotted them.

Father Alonso de Rojas (1588–1653), director of the Jesuit College of Quito, another chronicler of the expedition (1636), relates that

> the Indians said to the soldier who understood them, that in the northern lands, where they went once a year, there were women, and they stayed with them two months and if that union had produced sons, they brought them back and the daughters stayed with their mothers.[69]

Acuña swears as for the famous Amazonas, "of all the long stories, along the river," which the travelers gathered, the most concrete information came from the Tupinambás.[70] This tribe could communicate in the *Língua Geral* that the Jesuits created to talk to the Tupi-speaking Natives, which meant that the Portuguese needed no interpreters to speak with them.

---

67 Rafael Díaz Maderuelo, ed., *La aventura del Amazonas* (Madrid: Historia 16, 2012), pp. 214, 10.

68 Acuña, *Nuevo descubrimento del Gran Rio de las Amazonas*, pp. 49, 152.

69 Alonso de Rojas, *Relação do descobrimento do Rio das Amazonas* (São Paulo: Companhia Editora Nacional, 1941), p. 111.

70 Acuña, *Nuevo descubrimento del Gran Rio de las Amazonas*, pp. 152–53.

Acuña is the great authority that both d'Anville and La Condamine rely on for the existence of the Amazonas and in naming the river after them, even though this was not the name most used by the Portuguese at the time.[71] As copies of his book were destroyed after publication by orders of the King of Spain, d'Anville had access to his writing through D. Luis da Cunha. The latter received a copy of Acuña's writings from the Portuguese ambassador José da Cunha Brochado when in Utrecht negotiating with the French over Brazil's northern border with the French Guiana.[72] La Condamine wanted to be recognized as the new Acuña, whose account he considered of historical value[73] and d'Anville effectively concedes this credit to him, affirming that La Condamine "revived the writings of Acuña."[74]

D'Anville and La Condamine demonstrated knowledge of the diversity of names the river had been designated with. Why, then, choose one name over the rest, and one of the less-used names at that? La Condamine reflects on the river's name and affirms that historical Spanish authors called the river Marañon, but that in modern times, after Orellana's sighting of the Amazonas, the latter name had become the only one in use, as it "has always retained this name, without interruption, amongst the Spaniards, for over 200 years, for its whole extent." The Portuguese, meanwhile, displayed a wide range of names. Near the headwaters they "only know it (…) by the name of the River of Amazonas; and up higher, by that of Solimões (River of Poisons); and they have transferred the name of Marañon, or Maranhão, in their idiom, to a city, and a whole province." After these considerations, he concludes: "As for my part, I shall call it indifferently by the name of Maranhão, or the River of the Amazonas."[75]

The authority d'Anville confers on La Condamine is, beyond a shadow of a doubt, fundamental in the geographer's decision to opt for the name "Amazonas" for the river. The French traveler devotes an entire chapter to these famous women warriors in the book narrating his adventures down the river. He bears witness to their existence and naturalizes it, which would of course cause a certain unease among European readers.[76]

---

71 Charles-Marie de La Condamine, *A Succinct Abridgment of a Voyage Made within the Inland Parts of South-America* (London: E. Withers, 1747), p. 8.

72 Furtado, "O Tratado de Utrecht e a construção da cartografia como instrumento da diplomacia portuguesa."

73 Safier, "Como era ardiloso o meu francês," p. 95.

74 RBC, 229, n. 539 (2), f. 10.

75 La Condamine, *A Succinct Abridgment of a Voyage*, p. 6.

76 Safier, "Como era ardiloso o meu francês," pp. 106–11.

La Condamine writes that everywhere he went, "with great care," he asked the "Indians of divers Nations (...) if they ever had any knowledge of those warlike women, whom Orellana pretended he had met with, and fought." His main source of information was, once again, Acuña; seeking to confirm the latter's information about the brave women warriors, he asked those he met "whether it was really true [that] they lived separated from all intercourse with mankind, unless it was once a year." Their existence was confirmed everywhere, but nobody had seen them; all affirmed to have heard the story from their parents, each adding a new detail. La Condamine says he got the most concrete information from the Omaguas, as they pointed him to an Indian in the village of Coari, one of the rare living eyewitnesses.[77]

When they arrived at the village, they did not find the Indian—who had already died—but they spoke to his son, around 70 years of age, who confirmed the fact that his grandfather (not his father) had seen the Amazonas and even spoke with four of them. La Condamine entitled the chapter on the question *The Amazonas of America—the Asiatic and the African*—and points to correspondence between the stories he heard from this Coari Indian, the Tapajós Indians, a Mortigura Indian, and an old soldier from the garrison at Cayenne, which seemed to underscore these reports' veracity. All associated the Amazonas with the emeralds of their lands, located in distant mountains in the interior of the Guianas, where neither the French nor the Portuguese had yet penetrated. La Condamine concludes that, though they probably no longer existed, this was not proof that they had never existed.[78] Rather, what we see is a sophism: if one cannot prove the contrary, then it must be true, concluding that while the Amazonas of Asia and Africa were pure legend, the American Amazonas concretely existed.

"In the case of the American Amazonas, La Condamine thus painted a portrait fusing the bitter realities of African slavery and the harsh conditions of Amerindian domestic life," establishing an association between the Amazonas and the *quilombos* (communities of runaway slaves). This was heavily influenced by the presence of a runaway slave by the name of Louis, in whose canoe La Condamine traveled from Belém to Cayenne.[79] Interestingly enough, in a France where Enlightenment thinkers did not tend to construct a positive image of women in their works, La Condamine's Amazonas, located on the fringes of Western civilization, were the very picture of female liberation.

77 La Condamine, *A Succinct Abridgment of a Voyage*, p. 51.
78 La Condamine, *A Succinct Abridgment of a Voyage*, pp. 51–54.
79 Safier, "Como era ardiloso o meu francês," p. 107.

D'Anville's other source was Father Ignácio dos Reis, with whom he met at Pedro Nolasco Couvay's house in 1729. Though his printed map, representing the mouth of the river and based on information from this priest, presents two names for the river—*Carte du cours de la Rivière des Amazones ou de Maragnon* (Fig. 17)—one observation on the northern bank of the river signals the existence of Amazonas: "Higher up the Cunuris River, Father Acuña believes that there is a particular province, situated in the mountains, which is inhabited by the Amazonas." D'Anville, however, does not make explicit reference to the Amazonas on his map. Between the northern margin of the river and the southern border with the Guianas, in an explanatory text, he writes that "around this region reigns a mountain chain." This approach fits with his method of leaving unknown regions blank in the absence of consistent geographical information; this was the case with the land of the Amazonas, whose "common center," according to La Condamine, was in "the mountains of Guiana, a region whither neither the Portuguese of Pará nor the French of Cayenne have ever yet penetrated."[80] Farther to the north, past the border, d'Anville inserted—in an allusion to the association between Amazonas and *quilombos*—a note about the existence of "a community of runaway blacks," high up the Suriname River, which flows to Paramaribo.

If the allusion to the Amazonas can only be made out indirectly on the map itself, in the choice of the river's name and in the memoir written about the map, it becomes clear that this toponym was chosen not only because d'Anville believed in the existence of the Amazonas, but also that this belief was due to La Condamine. D'Anville wrote that, "though one runs the risk of going too far, one cannot avoid speaking of these women who gave the Maranhão the name Amazonas" and affirmed that La Condamine had sought out all information about them, from which he concluded, in a double negative, "one cannot say that they did not exist."[81]

And so, in giving their name to the great freshwater river, the Amazonas subtly made their way onto the map and gave a mythological cast to a region that the geographer was doing his utmost to represent rationally. On the other hand, the choice of the name Amazonas represented Portuguese hegemony over the entire course of the river—d'Anville taking their side—as the toponym was commonly used by the Portuguese explicitly for the region around the river delta, near Belém. Choosing a single name from Belém to the headwaters, then, was an expression of the Lusitanian project first expressed in Pedro Texeira's expedition, to have the Portuguese be the unquestioned lords

80 La Condamine, *A Succinct Abridgment of a Voyage*, p. 54.

81 RBC, 229, n. 539 (2), f. 13.

of the navigation of the entire river's basin. This very principle would guide the negotiations between D. Luis da Cunha and the French at the Congress of Utrecht; this adoption revealed that the *Amérique méridionale* was in tune with the ambassador's geopolitical project for the region, seeking to guarantee Portuguese control over both banks of the river and thus ensure a Lusitanian monopoly over its navigation.

4 El Dorado

In the first manuscript (Fig. 11) or print version of the *Amérique méridionale* in 1748, on sheet 1—depicting the Amazonas region—one can see that Lake Parima is not present, an absence also visible in the two simpler versions from 1737 and *c.*1742 (Figs. 1 and 2). The first reports of the existence of El Dorado in the region (coinciding with Lake Parima) date back to the expedition led by Gonzalo Pizarro in 1541. After conquering Quito, where he supposedly met an Indian who gave him the first descriptions of a kingdom on the banks of a vast inland lake and whose emperor covered himself in gold, Pizarro decided to strike out from Peru in search of El Dorado.[82] But it was the publication of Walter Raleigh's travel account[83] in 1596 that "installed [the] mythical empire at the iconic center of the Guianas"[84] as represented in *Amérique méridionale*, by Mercator in the sixteenth century (Fig. 25) or Sanson and Johannes Janssonius (1588–1664) in the seventeenth century (Figs. 37 and 46). Raleigh would lead two expeditions in search of El Dorado, though he never doubted its existence, he would never reach it. He spent 22 years on this quest, and while he never found the mythical El Dorado, he sought to confer a semblance of veracity on his account as with his travels he "came to understand the situations, the rivers, the kingdoms from the east sea to the borders of Peru, and from Orinoco River southward as far as Amazonas or Maranon."[85] His principal source, the Indian Juan Martinez de Albújar, whom he had met in Trinidad in 1595, told him that he had been taken prisoner by a tribe of cannibals in the south of (British) Guiana, in a region surrounded by mountains, where there was a city covered in gold, just like its chief, and all was built around a great

82 RBC, 229, n. 539 (2), f. 11.

83 Raleigh, *The Experience of the Large, Rich and Beautiful Empire of Guiana.*

84 D. Graham Burnet, *Masters of All They Surveyed* (Chicago: Chicago University Press, 2001), p. 25.

85 Raleigh, *The Experience of the Large, Rich and Beautiful Empire of Guiana*, p. 149.

FIGURE 46 *Guiana sive Amazonum regio*, Johannes Janssonius (Amsterdam, printed by Johannes Janssonius, 16??)
REPRODUCED COURTESY OF FBN

inland ocean.[86] This kingdom would be the Inga and the capital city of Manoa, he assured, "hath more abundance of gold within [its] territory than all Peru and the West Indies."[87] Raleigh had never seen the kingdom, nor the city or its emperor, but he trusted his sources, such as other Indigenous nations living on the borders of the empire of Inga, whom he spoke with as he traveled through the interior of Guiana.

Raleigh's account was crucial, not only in the construction and circulation of the myth of El Dorado, but also in establishing where it stood.[88] The empire was located on a vast interior plain, halfway between the Guianas and the

86 Burnet, *Masters of All They Surveyed*, p. 27.
87 Raleigh, *The Experience of the Large, Rich and Beautiful Empire of Guiana*, pp. 193–94.
88 Burnet, *Masters of All They Surveyed*, p. 26.

Amazonas River, surrounded by a dense forest and tall mountains, beside an enormous inland lake, the Parima. For centuries to come, innumerable traveler/scientists and adventurers, including Alexander von Humboldt, tried to locate Lake Parima; and geographers, following these descriptions, tried to include it in (or leave it off) their maps.

However, in the first version of the 1748 *Amérique méridionale* (Fig. 11), following the information La Condamine had gathered on his expedition traveling down the Amazonas and scorning tradition, d'Anville did not include Lake Parima but Lake Amucu, placing it not in Guiana, as Raleigh's report indicated, but in the Portuguese portion of the territory, a bit farther to the south, below the Orinoco River and to the north of the Amazonas. This was due to a manuscript map, titled *Carte huilée de la route de Nicolas Horstman natif de Hidelsheim en Westphalie depuis Rio Esquibé jusqu'à Rio Negro* (Fig. 47), drawn up by Nicolas Horstman, a Prussian whom La Condamine had met during his trip down the Amazonas.[89] Nicholas Horstman was tasked in 1739 by the Dutch Guiana governor Laurens Storm van's Gravesande (1704–1775) to explore the southern part of Guiana. He traveled up the Essequibo River accompanied by two Dutch soldiers and four Indian guides.[90] Instead of returning to his Dutch superiors to report on his discoveries, Horstman pressed even farther south, pushing into Portuguese territory. Rumor said he had become an Indian, or that he had organized an Indigenous revolt, or that he had betrayed the Dutch. Imprisoned in the city of Belém he finally passed state secrets to the Portuguese whom he started serving.[91] La Condamine met him in Belém where he was in jail suspected of being a spy and returned to Europe with a copy of his diary written in Portuguese and map of Lake Amucu, with the rivers that flowed into it,[92] all of which are now in the BNF. Horstman had made use of the network of rivers in the region, going up the Essequebé (whose source was in Guiana), going down the Rio Negro to reach the Amazonas, and from there going to

89 La Condamine, *Relation abrégée d'un voyage fait dans l'interieur de l'Amérique méridionale, depuis la côte de la Mer du Sud, jusqu'aux côtes du Brésil et de la Guiane* (Paris: Vve Pisot, 1745), pp. 127–29.

90 According to his diary at BNF, which belongs to the Collection Heinrich-Julius Klaproth, accompanied by his map. His diary and map belonged to d'Anville and were used to draw this part of the map. See also C.A. Harris and J.A.J. de Villiers, *Storm van's Gravesande: The Rise of British Guiana*, vol. 1 (London: The Hakluyt Society, 1911), pp. 168–74; Burnet, *Masters of All They Surveyed*, pp. 30–31. Graham Burnet, based on information from Humboldt, declares that Horstman was of Dutch origin and was a merchant. However, in his papers, we see that he was of Germanic origin and a surgeon. La Condamine informs us that he was born in Hildesheim.

91 Burnet, *Masters of All They Surveyed*, pp. 30–31.

92 Harris and J.A.J. de Villiers, *Storm van's Gravesande*, pp. 168, 174.

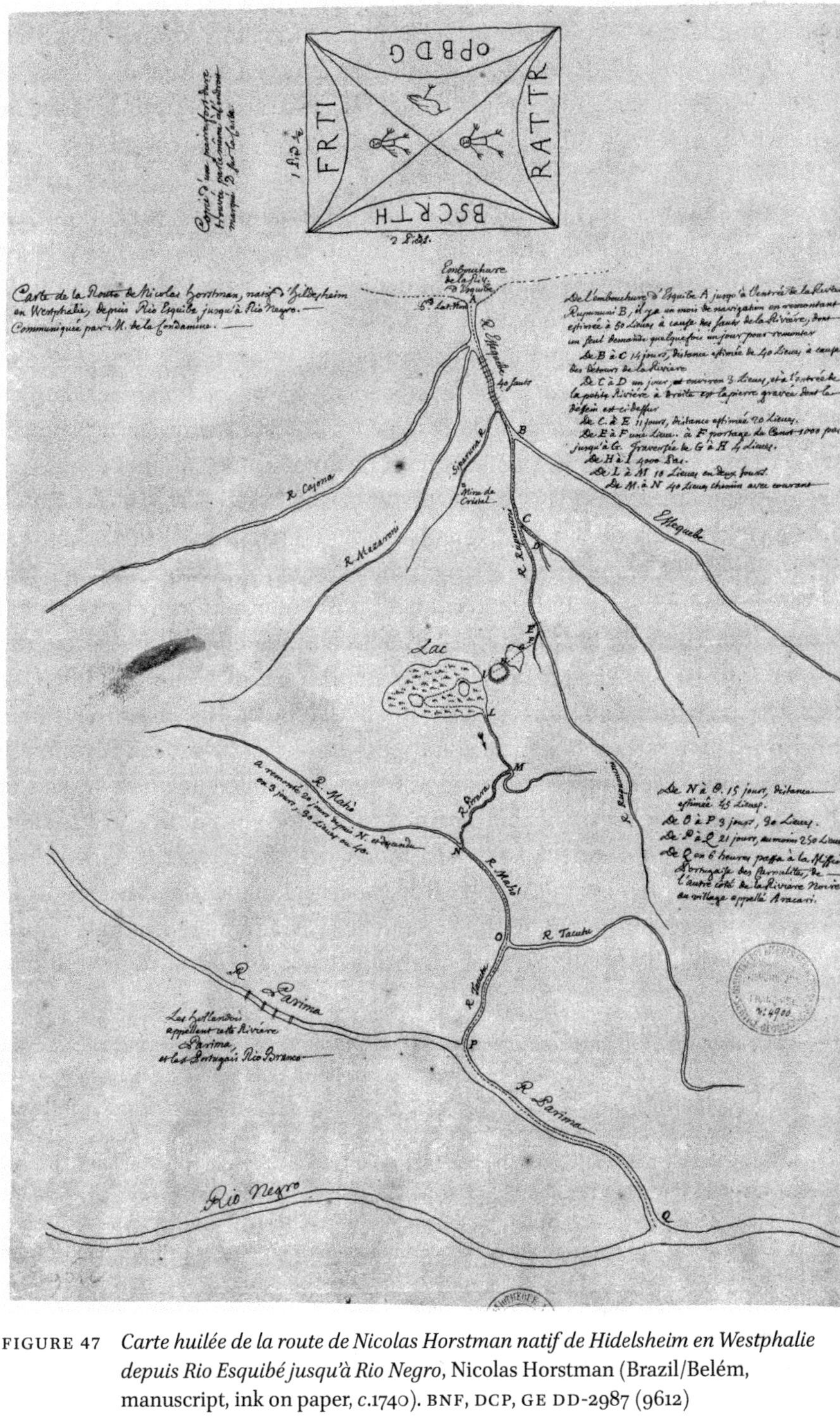

FIGURE 47 *Carte huilée de la route de Nicolas Horstman natif de Hidelsheim en Westphalie depuis Rio Esquibé jusqu'à Rio Negro*, Nicolas Horstman (Brazil/Belém, manuscript, ink on paper, *c.*1740). BNF, DCP, GE DD-2987 (9612)
REPRODUCED COURTESY OF FBN

Belém, where La Condamine came across him.[93] This was the configuration d'Anville lent to the rivers of this area on his map, identifying Lake Amucu with the mythical Parima, which would feed future discussions and journeys by discoverers to the region in search of the "true" local geography. This debate would go on until Humboldt, who used it as a basis for his expedition to the area, having confirmed d'Anville's geographical configuration.[94]

The relationship between Horstman's map and the geography of Lake Amucu and the surrounding river network in the *Amérique méridionale* is direct and obvious, striking the observer immediately. In his first memoir on the map, d'Anville confirms that for this stretch he used the map and an account written by the Prussian and that when La Condamine was in Belém, he made copies of "the drawing in the form of a map" by Horstman and the memoir in which the Prussian described his journey. As was his custom, d'Anville compared the two documents, and the memoir allowed him to "note the flaws in the drawing and make the [necessary] corrections."[95]

The geographer was struck by the mention of the "name of Rupununi, given to one of these rivers which served to make this crossing, and the proximity of a lake," as the old maps often affirmed "that Parima Lake is called Rupunuwini by the Indians." He then concluded, based on the existence of a Rupununi River flowing into this lake where Horstman had traveled, that "as regards this famous lake, (…) the name of Parima is of another river in this region, which flows into the Rio Negro," not being the name of the lake labeled Amucu.[96]

This part of the map provides a good illustration of his method of cartographical criticism. He starts out with side-by-side comparisons of a manuscript map, product of a recent empirical observation, with a memoir drawn up on the occasion; and then juxtaposes these sources with old maps from his collection. In this way, older knowledge can be renewed in the light of newer information, as we see when he refers to a map in his collection where Lake Parima had been referred to as Rupunuwini by the Indians. He had seventeenth-century maps with this reference such as *La Guaiane ou Coste Sauvage, autrement El Dorado, et Pais des Amazones*, from 1654, by Pierre Duval d'Abbeville (1618–1683), royal geographer, and several which depicted Lake Parima, notably one by Johannes Janssonius, titled *Guiana sive Amazonum regio* (Fig. 46), one by Willem Blaeu (1571–1630), and one by Vincenzo Maria Coronelli.

Horstman's manuscript map was the basis for d'Anville to set down the geographical configuration of the area. Once again trusting in La Condamine, who

93 RBC, 229, n. 539 (2), f. 11.

94 Burnet, *Masters of All They Surveyed*, p. 31.

95 RBC, 229, n. 539 (2), f. 11.

96 Anville, *Journal des sçavans* (March 1750), pp. 552–53.

had brought information from the Prussian and sworn to its veracity, he stuck to this adventurer's narrative, drawing a nearly perfect link between the basins of the Negro/Amazonas and Rupunuwini/Essequebé Rivers, separated by two small isthmuses and Amucu Lake, respectively. We see on both Horstman's and d'Anville's maps that from its headwaters, the Essequebé River takes in two tributaries from the right, the Cajona and the Mazaroni, with two stretches of waterfalls shortly afterwards. In the second stretch, between the tributaries of the Arussari and the Rupunuwini, d'Anville writes that there are 39 consecutive falls, a piece of information surely taken from Horstman's memoir, as it does not appear on the map. After this second leg of waterfalls, the Rupunuwini flows into the Essequebé. Back at the source of the Rupunuwini, on Horstman's map, one finds three lakes separated by small isthmuses. The third and largest is Lake Amucu. D'Anville, however, makes these first two lakes into one on his map, the second being the Amucu. From this lake flows the Pirara River, which runs into the Maho, which empties into the Tacutu, all these being tributaries of the Parima River, "which the Portuguese call Rio Branco." The expression, "couvert de joncs" (covered with rushes) which he writes just to the north of the lake and below the toponym, indicates the seasonal nature of these waters, which spread outward in the rainy season.

Lake Parima, on d'Anville's map, following the fluvial geography suggested by Horstman, is transformed into a river; but another lake appears in its place—the Amucu. At first, then, the *Amérique méridionale* breaks with a mythological tradition that had identified the golden lake with Parima. The only reference to Parima appears in the writings that identify the Branco River, a tributary of the Rio Negro, which connects to the Amucu. In another distinction from preceding maps, there is no reference to El Dorado or mineral bounties. In line with Enlightenment tastes, everything pointed to the triumph of reason over mythical geography. Indeed, d'Anville had carried out his method once again and with apparent success, working from his study in the galleries of the Louvre Palace and critiquing the cartographical and textual sources he received, produced by explorers who wandered the vast interior of the New World.

5 The Return of Parima

Locating Lake Amucu in Luso-Brazilian lands did not reflect the interests of the French, nor the Spanish or the Dutch—all with territorial holdings in the region, all still pushing against the Portuguese at their borders. In the 1748 *Amérique méridionale*, d'Anville places the dividing line to the northeast of the lake, almost parallel to a mountain range (the habitat of the mythical

Amazonas?), which fit with the custom of using geographical features as landmarks to establish territorial boundaries. However, the line drawn to the northeast does not correspond to any natural barrier in the terrain; it has simply been placed there by the mapmaker. D'Anville's map thus allowed the Portuguese to claim possession of a territory that, according to mythological tradition, oozed mineral riches and was in accordance with D. Luis's geopolitical view of stretching the border as far northwest as possible.

Though he sought to demystify the geography of the area in the first edition of the map, when one looks at d'Anville's alterations to *Amérique méridionales*'s sheet 1—corresponding to the Amazonas region, after the 1760 version (Fig. 4)—we see to our surprise that the geographer has included Lake Parima. The process of reconfiguring this area may be observed in a fragment from his archive (Fig. 48), meant to substitute the original area, and in the manuscript version of the map where in crayon he tested out a few possibilities for inserting the lake (Fig. 11). From then on, until d'Anville's last revision of the *Amérique méridionale* in 1777, the Amucu appears in the configuration suggested by Horstman and Lake Parima appears to its northeast, outside Portuguese territory. We also see that the network of rivers, inserted in the area around 1760, establishes a connection between Amucu and Parima through tributaries of the Mahos, Parima, and Essequebé Rivers, as well as a connection between Lake Parima and the Orinoco River. A traveler leaving the Amazonas via the Negro and Maho Rivers, then, would arrive at Lake Parima and from there travel by river between Portuguese and Spanish territory, all the way to the Orinoco.

In his 1779 memoir, d'Anville justified the introduction of Lake Parima in all editions of the map post-1760. "In the eastern part in relation to the Orinoco," he wrote,

> in the direction of the upper part of the river, is the location previously given to a country judged to be very rich. This province, so eagerly sought by the Spanish, which they aimed to reach in a number of undertakings, all failed, is the famous [El] Dorado.

This lake, with its golden king, "is none other than the [lake] discovered and named Parima, which occupies a vast space on the map of Guiana by Jean de Laet." He then adds that he quickly "had this area, with all its particular details, added to my *Amérique méridionale*."[97] We see that he not only includes Lake

97 RBC, 229, n. 539 (2), f. 11.

FIGURE 48 *Fragments d'une carte de l'Amérique du Sud montrant partie de la nouvelle*, J.B.B. d'Anville (Paris, by the author at the Louvre galleries, manuscript in ink, 1760). BNF, DCP, GE D-10658
REPRODUCED COURTESY OF FBN

Parima, but in his justification also relates this geographical feature to the old myth of the empire of El Dorado, revealing that the lake had been discovered by the Spanish. As is his wont, he also references older maps—in this case, one by Laet—to underscore the authority of the cartographical elders in confirming the geographical landmark now included in the map. However, if the existence of El Dorado is dismissed as a chimera by him, the geographical entity identified as Lake Parima is inserted, without greater scrutiny, into the map.

D'Anville had been involved with the production of maps of the area since 1742, and though he claimed to have all the documents used to configure the region in hand since 1745, his maps of the region before 1754 did not include Lake Parima; neither did this first revision of the map of South America, also released in 1754. As for the documents available to him at the time, he contradicts himself in the different versions of the memoirs he wrote about the *Amérique méridionale*; but these contradictions are not random. In the one of April 1750, d'Anville writes that he had on hand a map, among others, bound "in a compilation of many folio volumes of divers memoirs, written by a Spanish official for his king shortly after the ascension of Filipe V (1683–1746) to the throne."[98] In the memoir he drew up in 1779, however, he affirmed that the inclusion of Lake Parima came out because, "in 1749, M. de la Condamine, placing before my eyes a map of the region of this lake, told me that this document had come from the study of the Marquis of La Encenada"[99] (Zenón de Somodevilla y Bengoechea, 1st Marquis of Ensenada, 1702–1781). This was the then-powerful minister of the Spanish king Fernando VI (1713–1759). All signs seem to indicate that this was the same map—and if it were, when would he have gotten it? In 1745, as he says in the first text, or in 1749, as he says in the second? The first date is more probable; after 1744, when he started working more closely with La Condamine, and with the arrival of Maldonado in 1746, d'Anville was given access to an assortment of Spanish documents. Referencing 1749 would only justify the absence of Lake Parima in the first version of the *Amérique méridionale* in 1748—he would not have had access to said Spanish map by that time. If Spanish documentation confirming the existence of Parima had been in his possession since at least 1749, however, this does not explain why in the first new edition, in 1754, he merely added "a few details to the mountains above Lima."[100] The fact that he did not incorporate Lake Parima in the first edition or this revision of the map merits analysis. Why would someone so bent on achieving "cartographical perfection" have neglected to refine his map if he believed this information about the existence of the lake? And why does he then add it in 1760?

---

98 Anville, *Journal des sçavans* (April 1750), p. 664.

99 RBC, 229, n. 539 (2), f. 11.

100 Manne, Bocage, and Dacier, *Notice des ouvrages de M. D'Anville*, p. 98.

Before seeking an answer to these two questions, we must make a few observations. Neither Laet's map—which he references in the memoir—nor the Spanish map can be found in the *D'Anville Collection*. This does not mean in the slightest that he never consulted the maps, neither does it prove that he did not own them, as the collection has a few lacunas. However, it is at the very least striking that the collection does not include at least a personal copy (when all the others he refers to are there), and that he leans on Laet to justify his continuity with tradition. Laet has no major map of the region, unlike those by Sanson, Hondius, or Blaeu, which d'Anville owned and which depict Lake Parima. The only reference to Laet in the collection is a map by Guillaume Delisle entitled *Tabula Americae specialis geographica regni Peru, Brasiliae, terrae firmae et reg Amazonum* in which Parima does not appear. This map has a lone textual reference to the lake, near the Essequebé River, noting: "in this region is found Lake Parima, the city Manoa of El Dorado." D'Anville, then, seems to feint so as to conceal what might be understood as a blatant error in the first edition of his map.

How does he reconcile the explanation given in the memoir from 1750, when he affirmed that Lake Amucu was in fact Parima, with the 1760 reconfiguration of the map, when the two lakes appear side by side? First of all, he declares that this development was presented to him by La Condamine in the Spanish map where Parima is seen to be separate from Amucu. Secondly, he points out Horstman's lapse in not noting the existence of Parima. Nevertheless, he writes nothing to explain the fact that in the 1748 version he took Amucu for Lake Parima. He relates that Horstman set out from the coast of Guiana on November 3, 1739, then beginning his journey along the Essequebé. Taking advantage of the rainy season, he arrived at the Rio Negro on July 16 of the next year and followed the current to the city of Belém do Pará. He forgives the traveler his error, saying that

> on his journey, which caused him innumerable hardships, because he traveled by water, taking transports, passing close by Lake Parima without knowing of it. He did enter, however, into a lake all covered with rushes and named Amucu by the Indians, but was wrong in taking it for Parima, which he sought.[101]

It is important to note that by late 1747 d'Anville had started to prepare another map alongside La Condamine and Maldonado. This was the *Carta de la*

101 RBC, 229, n. 539 (2), f. 12.

provincia de Quito. This map, produced at the request of the Spanish crown, was finally finished in early 1750 and sent to Madrid. During this period La Condamine was in contact with Spanish authorities, including José de Carvajal, then Secretary of Foreign Affairs to the Spanish king, Fernando VI. Around this time, as it prepared to negotiate borders in America with the Portuguese, the Spanish crown sought to fund scientific institutions with an Enlightenment bent and simultaneously map its possessions in America. Given the urgent nature of the task, it used French printers (as in d'Anville's case) and sent specialists to France to study the latest in the art of cartography (as was the case with Maldonado).[102] It was through this Franco-Spanish collaboration that La Condamine, and, by extension, d'Anville, had access to Spanish maps of the region, important tools for the production of the new map in terms of complementing information about areas not visited by travelers since at least 1746.

Having tracked the Spanish map's path to d'Anville's hands, we must face the question of the reasons that led him to trust it, so that we may then understand why he did not include Lake Parima in the 1754 revision of the map and then comprehend why he then did include it in the next revision, in 1760.

Throughout the 1740s, the geographer struck up a friendship and symbiotic relationship with La Condamine. Armchair geographers depended on such collaborators to furnish them with accounts and maps of recently explored areas and in La Condamine's accounts and public readings at the *Académie Royale des Sciences de Paris* on his expeditions to Quito and the Amazonas, he tried to guarantee that his descriptions were based principally on direct visual observation, disguising many of his sources as best he could; but most of what he presented to European audiences, however, came from elsewhere,[103] such as Horstman's report. In some places he in fact made astronomical observations, but "in others, he relied on maps, travel logs, and oral accounts by missionaries and Indians."[104] Filtering the information received from informants along the way was a challenge for these travelers, as the length of their journey did not allow for personal visits to all the places of interest. Though empiricism ought to guide the knowledge being produced in keeping with Enlightenment rationalism, what we see is that—as in La Condamine's case—the vastness of the territory meant that many explorers had to make use not only of what they saw, but what they heard tell.

The question was: among so many narratives being collected, which deserved to have the status of truth conferred on them? This information

102 Safier, *Measuring the New World*, pp. 149, 134.
103 Safier, "Como era ardiloso o meu francês," pp. 97–102.
104 Safier, *Measuring the New World*, pp. 59–60.

selected from third-party reports (thus contradicting the supposed primacy of seeing over hearing tell), once enshrined as true, was often presented to European audiences by these traveler-intellectuals as the product of personal observation. La Condamine did this frequently,[105] defending a number of geographical conformations of the American continent that were no more than echoes of ancient myths—and presenting them as if they were his direct observations. Since d'Anville used him as the principal source of information in drawing up the Amazonas region, the geography that La Condamine considered "true," be it from his personal observations or heard from third parties, wound up appearing on the map. In the case of the inclusion of Lake Parima, another considerable influence was the fact that the most powerful minister of the Spanish state had provided him with the base cartographical document.

During the years d'Anville and La Condamine worked together in Paris, however, Maldonado and Bouguer were already starting to fall out with the latter, accusing him of errors and a certain sleight of hand in the attribution of his not-always-trustworthy sources. Aware of these quarrels, d'Anville continued trusting and privileging the information La Condamine brought him over the sources of the other two.[106] The personal friendship between the two, as well as the respect La Condamine enjoyed in the community of Enlightenment scholars at the time—as opposed to Maldonado, whom d'Anville considered a simple practical navigator—was certainly a determining factor in the geographer's decision to trust his partner's information fully and transport it to his maps.

Finally, we must call attention to the perhaps less evident geopolitical aspect of Parima's inclusion after 1760. In the plate 1 manuscript version of the *Amérique méridionale* (Fig. 11) we can see a few penciled changes made by the geographer after the fact, in which he corrects or adds geographical elements not present in the first manuscript version. We cannot know exactly when these annotations in pencil were made, or if they were done at the same time; they are indications, however, of the alterations made on the map's copperplates in the period following the first printing.

One of these corrections refers precisely to the inclusion of Lake Parima. On the manuscript map, the position of this lake appears in pencil exactly over the dividing line proposed between Brazil and Spanish America—just below the mountain range which would divide the two territories naturally, that is. However, after the second print version from 1760 (Fig. 4), we see that the lake has been moved slightly to the north, not on top of the dividing line

105 Safier, "Como era ardiloso o meu francês," pp. 97–102.
106 Safier, "Como era ardiloso o meu francês," p. 106; *Measuring the New World*, pp. 123–65.

but in Spanish territory. We also see that the inclusion of Parima obliged the cartographer to extend the space between Amucu and the mountain range just to the north; this shifted the mountains away from the proposed dividing line, which itself is moved farther to the south.

What seems to be a simple geographical correction, with no other apparent intention than refining the map, introduces at least two important geopolitical issues. Firstly, the geographer abandons the idea of using natural landmarks—as he had in the 1748 map—to establish the limits between Portugal and Spain in the region, taking advantage of the hefty mountain range. Secondly, the network of rivers connecting the two lakes and allowing for easy transit between Portuguese and Spanish territories, shifts the issue of the demarcation of the border to the diplomatic sphere; the frontier, once tied to a natural element of the terrain, is now apparently fluid and indistinct. The appearance of Parima—which had always been linked to the Spanish, not the Portuguese—thus implies control of the area by the Spanish crown (a claim present in the mythology created in the wake of Pizarro's expedition), while the choice of the toponym for the river most commonly used around Belém—Amazonas, rather than Maranhão or Orellana—placed it in the Portuguese sphere of influence.

It becomes clear, then, that implications of a geopolitical nature, detrimental to Portuguese interests, result from the inclusion of Parima. But why was the lake omitted from the 1754 version and finally included in 1760 if d'Anville had had the Spanish map since at least 1749? It was surely at the request of D. Luis da Cunha and the Duke of Orleans (though the former had died in late 1749 and the latter in 1752) that d'Anville, even with the Spanish map in hand, did not include Lake Parima and would omit it again in 1754, depicting Amucu alone and placing it in Portuguese territory. With no connection to the myth of El Dorado, it was difficult to justify Portuguese control over this part of the territory, especially given that the mountain chain linked to the legend of the Amazonas (who, for their part, would lend their name to the river under Portuguese control) would serve as a natural landmark for the dividing line. These were the echoes of loyalty to an old friend and protector, and the geopolitical injunctions which were to interfere in the *Amérique méridionale* once again.

In general terms, the first given cartographical form of *Amérique méridionale* in 1748 mirrored D. Luis's geopolitical vision for this part of the Amazonas, which was achieved in the Treaty of Madrid in 1750. According to it,

> the frontier will remain through the middle of the Japurá River, and the other rivers that join it and approach the north, until the *peak of the mountain range between the Orinoco and Amazon* or Marañon; and will

> follow the top of these mountains to the east until reaching the dominion of another monarchy.[107]

The year 1760 inaugurated a new era in the Spanish and Portuguese disputes. The bilateral groups that began demarcating the border after the signing of the Treaty of Madrid and taking topographical measurements questioned the terms of the agreement. The Portuguese still smarted at the loss of the Sacramento colony, and the Spanish, for their part, felt they had given up too much. In this sense, 1760 was a prime opportunity to draw up a new edition of the map. And indeed, on February 12 of the next year, the Treaty of El Pardo was signed and canceled the terms of Madrid. In the new version of the map, not only does Lake Parima appear, but it is also clearly positioned in Spanish territory, assuring the latter all rights over the area's promised riches. Portuguese interference in the *Amérique méridionale* was eclipsed by the Spanish influence from La Condamine and Maldonado's collaboration. State motives and geopolitical concerns resounded in Paradise!

---

107 http://www.info.lncc.br/madri.html (accessed July 7, 2009) (italics are the author's).

EPILOGUE

A Lost Treasure

1 Diplomacy and Cartography

Since 1746, after the death of King Filipe V (July 9) of Spain, who was pro-French and openly hostile to reaching an agreement with Portugal, the Portuguese crown made efforts to establish negotiations with Spain on more favorable bases. It was the apex of the strategy to approximate both crowns, initiated as far back as 1727 to 1729 with the double marriage of the crown princes and princesses, descendants of both reigning houses. After the French crown backed off from marrying Louis XV to the Spanish Infanta Maria Ana Vitória (1718–1781), Portuguese diplomacy made efforts for her to marry the future King D. José I, the heir to the Portuguese throne. The negotiations also involved the marriage of Maria Barbara, D. João V's daughter, to Fernando, son of Filipe V and heir to the Spanish throne resulting in an exchange of both princesses in the Caia River, situated on the frontier between Spain and Portugal, in January 1729. This strategy aimed at creating a favorable future environment for Portugal to try to resolve the diplomatic impasses with Spain resulting from the War of the Spanish Succession, which were only partially resolved in the Treaty of Utrecht and in the following negotiations, especially those that took place in 1737 after the Colônia do Sacramento siege; a strategy that now came to fruition with the death of the Spanish king. As soon as he received the news of the death and the consequent rise to the throne of Ferdinand VI, D. João V wrote to his daughter and new queen, Maria Barbara, exhorting her to consider her daughterly love and thus collaborate in the renewal of ties with Spain, which would finally allow both nations to negotiate their differences.[1]

In September 1746, soon after Ferdinand VI's rise to the throne, Tomás da Silva Teles, Viscount of Vila Nova de Cerveira, left for Madrid to take his post as ambassador.[27] Once he arrived in Spain, he embarked on the initial negotiations and D. Luis da Cunha was raised to the position of an important advisor to guide the next steps in order to establish the way forward for Portugal in those negotiations. He was to give his opinion freely, create a general plan of the strategies to be formulated, and, through his vast network of informants,

1 AUC, CDLC, 884, f. 1, July 30, 1746.

 | DOI:10.1163/9789004710764_009

determine the position of the other European powers regarding the differences between Portugal and Spain.[2]

It is at this time that an intense exchange of information begins between the Viscount of Vila Nova de Cerveira in Madrid, the members of the Portuguese court (especially Marco António de Azevedo Coutinho—Minister of Foreign Affairs—and Alexandre de Gusmão—the king's private secretary) in Lisbon, and D. Luis da Cunha in Paris, in order to define actions for the negotiations.[3] D. Luis was in charge of providing the necessary information to the ambassador in Madrid, because since Utrecht he had participated in all attempts to negotiate with Spain, and was therefore perfectly informed of the differences between the two crowns, especially in regards to the Colônia do Sacramento. In the first years that followed, in his role as the king's "oracle" and following D. João V's decisions, the elderly ambassador tried to advise the authorities on what should be done.

In the years leading up to the signing of the Treaty of Madrid in 1750, maps came and went between Lisbon, Madrid, and Paris, "for information as to the borders indicated in the plan" drawn up to resolve the issues at hand.[4] Cartography's importance meant that it was constantly invoked during this process. While maps revealed the territories that could be exchanged in America, or revealed the impossibility of trading them, they also served to illustrate where the line of demarcation ought to go. For this reason, each government set about "draw[ing] with a red line" or a line of another color, "or fixing a few silk threads"[5] on the map to demarcate their territorial claims, including d'Anville's 1748 *Amérique méridionale*, which in the end was not shown to the Spanish ambassador.

Despite Alexandre de Gusmão's initial disagreements, to a great extent the Treaty of Madrid, due especially to the Portuguese initiative, ended up confirming most of the premises advocated by D. Luis da Cunha on how the negotiations on boundaries should be conducted. The issue was not limited to Colônia do Sacramento, or only to the Americas, but also included the Moluccas Islands, located in the Pacific Ocean. Both were the great bargaining chips offered by the Portuguese in exchange for the possessions in the center-west and in the Amazonas region. In the latter area, Portugal ensured its dominion of both banks of the Amazonas River, and hence the monopoly of its

2 AUC, CDLC, 884, f. 1, July 30, 1746, f. 2v.

3 On the differences between Gusmão's and D. Luis's ways of conducting the negotiations in 1736 and in Madrid in 1750, see Furtado, "Portuguese America under Foreign Threat."

4 ANTT, MNE, caixa 612, February 8, 1749.

5 Cortesão, *Alexandre de Gusmão e o Tratado de Madrid*, vol. 4, p. 279.

navigation. No equivalent territory was defined in Europe. Territorial borders were established using as guiding principles, first, natural divisions and second *Uti possidetis*, in other words, each power was to keep the areas it effectively colonized. In this way, the entire *sertão* interior of South America passed to the Portuguese.

Under the inspiration of D. Luis da Cunha, this resignification of *Uti possidetis* appeared in the 1750 Treaty of Madrid. *De facto Uti possidetis* is configured in its Preamble. While in the territories that were known the natural divisions of the land would be used to trace out the frontier, as a secondary recourse it was determined "that each party had to keep what they still possessed, with the exception of mutual ceding of territory."[6] Article XIV stipulated that the dominion of territories disputed by Spain and Portugal on the frontiers of Brazil was to be assured for those who effectively colonized the area. From then on colonization became a judicial condition in itself and the Roman principle of *Uti possidetis* stopped referring to possession based on prior rights, instead becoming based on *de facto* possession. A reflection of this new meaning attributed to the act of colonization is the fact that, progressively, in the second half of the eighteenth century European overseas possessions were increasingly called colonies, and their antagonism with European metropoles and the relationship between them came to be understood from a perspective of exploitation, which became key for the independence movements in the Americas.

To better understand why *Amérique méridionale* was not shown to the Spanish diplomats in Madrid, despite D. Luis da Cunha's wishes, it is necessary to understand how were Portuguese frontiers in South America configured in d'Anville's cartography in the 1737, *c.*1742, and 1748 versions of *Amérique méridionale*, and in its later revisions (Figs. 1, 2, 3, and 4). To answer the question of the extent to which these configurations corresponded to the geopolitical vision that D. Luis came to formulate for America, one must juxtapose the shapes that Brazil takes on in various versions of the *Amérique méridionale*, as well as the various maps presented by the Portuguese at the negotiating table in Madrid in 1750; only thus will we understand how the differing geopolitical conceptions of D. Luis da Cunha and Alexandre de Gusmão were reflected in differing cartographies.

6 Cortesão, *Alexandre de Gusmão e o Tratado de Madrid*, pt. 2, p. 365.

2 The Rio de la Plata Basin

It is clear that the borders d'Anville traced over the course of his cartographical works representing America varied over time. Let us begin with the Rio de la Plata basin, and the maps used to configure the area. D'Anville had samples of both Portuguese and Spanish cartography—not counting, of course, the maps produced by the Jesuits. He had a total of 12 maps of the Rio de la Plata and the mouth of the Paraguay River. Excluding two of his own and one by the royal hydrographer, Jacques-Nicolas Bellin, which was produced after 1748 and based on his own map, he had nine maps of the region. D'Anville also had memoirs and reports from Jesuit missionaries in the region, which he had used to put together his map *Le Paraguay* for the *Lettres edifiantes et curieuses* in 1733.

It is striking to observe the way in which d'Anville represents the region of the Rio de la Plata basin across his maps, in which a political slant is evident, and how he makes use of this base cartography to configure his own maps. The representation would change over time, albeit constantly reflecting that which D. Luis da Cunha believed to be in Portuguese interests. His first map of the region, produced for the Jesuits and entitled *Le Paraguay* (Fig. 9), established no line of demarcation, extending Brazilian territory all the way to the northern bank of the Rio de la Plata, including the Colônia do Sacramento. He justified this in the memoir on the subject by writing, "the sea-coast is uncontested Brazilian land until the Rio de la Plata, where the Portuguese have the Colônia do Sacramento, near the small islands of São Gabriel." Given the "continuation of Brazil" all the way to the river's northern bank,[7] d'Anville drew no borders around this area. However, despite the absence of a line, the mouth of the Rio de la Plata and its confluence with the Paraná and the Paraguay, all the way to Lake Xarais—and above the lake, a small river slightly to the east—configured a natural frontier for the territory, separating the possessions of Spain and Portugal. This map appeared quite in tune with the arguments of the Portuguese at the time, as they "believe that they are entitled to all the coast running from the district of São Vicente (or São Paulo captaincy) to the banks of the Rio de la Plata."[8] The Spanish vision, for its part, was radically opposed, with its negotiators arguing that "the Consejo de Indias has seemingly incontrovertible evidence that the extension of the demarcation of the Portuguese coasts from north to south is not more than 370 leagues,"[9] much less than their demands.

---

7 Anville, "Observations géographiques sur la *Carte du Paraguai*," vol. 8, p. 265.
8 ANTT, MNE, caixa 99, doc. 104, November 30, 1714.
9 ANTT, MNE, caixa 926, doc. 99, November 30, 1714 (italics are the author's).

Though around this period this configuration of Brazil—stretching without borders to the northern bank of the Rio de la Plata—was no longer what d'Anville was representing on his maps, nor what D. Luis da Cunha supported. The strategy suited Portugal's needs well, as it presented the territory as a tabula rasa (in the absence of clearly marked borders), obviating the lengthy disputes between the two crowns since the creation of the Colônia do Sacramento in 1680. In his 1737 *Amérique méridionale* (Fig. 1), d'Anville had already shifted the border of the region—since 1734, after all, he had begun to argue for the need to alter the shape of Paraguay. He declared it would be necessary to extend Brazilian territory up to the Portuguese possession on the northern bank of the Rio de la Plata, the Colônia do Sacramento, but with an interruption in the Montevideo region where the Spanish had settled.[10] We see that on the 1737 map, in the Brazilian south, the border corresponds precisely to this new perception of the geopolitical situation. The line, perpendicular to the coast and not too far from it, falls roughly along where the Tordesillas Meridian would pass. Brazilian lands are thus extended to the mouth of the Rio de la Plata, but they do not take in the whole of the northern bank. Indeed, the Portuguese territory assumes the final form of a fork, or an inverted Y. One of the ends corresponds to the Colônia do Sacramento and the other to the Point of Santa Maria. Between them would fall Montevideo, a Spanish possession secured and fortified since 1736. The dividing line does not follow the Paraguay River, nor does it use Lake Xarais as a natural element to divide the terrain; thus this cannot have been the map D. Luis had on hand in 1737 when negotiating with the French authorities about the Great Siege established by Spain around the Colônia do Sacramento and which he referenced in his *Instruções políticas*. D'Anville draws Paraguay as an autonomous province, lying between Portuguese possessions to the east and the Paraguay River to the west, in a territory mainly occupied by the Jesuit missions.

The 1737 map was a precise reflection of the political context of the region at the time (Fig. 1). Most of all, it reflected D. Luis's interest in showing the French authorities how the territories of both nations in the area were mutually hampered and fostered an explosive situation. The continuation of the Brazilian territory, apparently more or less hugging the Tordesillas Meridian, made the Spanish presence in Montevideo an uncomfortable intrusion on Portuguese claims. The image on the map clearly legitimated the Colônia do Sacramento to the inquisitive eye of the French, whom the ambassador was trying to convince to mediate the negotiations. It is quite likely that the map was drawn

---

10 NL, VB, Ayer MS 1918, doc. 549, f. 4, March 21, 1734.

up alongside the memoir that d'Anville turned out in 1737 on the Tordesillas Meridian, revealing Portuguese interests along the lines supported by D. Luis da Cunha. The closeness between the borders proposed by d'Anville in the *Discussion géographique*,[11] and in this map, when combined with the coincidence in the dates, is enough to safely establish a relationship between the two. The other version of this map, perhaps from around 1742 (Fig. 2), shows the same dividing line; it cannot therefore be the same one that D. Luis sent to Madrid, as it no longer reflected his geopolitical vision for Brazil's borders by the mid eighteenth century.

Beyond a shadow of a doubt, the 1748 version of the *Amérique méridionale* is the most polished presentation of the vision that D. Luis upheld for border negotiations with Spain around the end of the decade (Fig. 38 and detail in 49). Here, the Colônia do Sacramento is no longer connected to Brazil, appearing as a Portuguese enclave on the banks of the Rio de la Plata. This presentation of Sacramento as an isolated portion was more visually helpful in convincing both sides of the advantages of a possible territorial swap, as he encouraged. One had only to glance at the map to conclude that this would be in the interest of the Spanish as well, freeing them from the isolated, inconvenient Portuguese presence in the river delta. Though small, the territory had grave political and economic weight. Not only did Luso-Brazilian merchants drain Peruvian silver through it, but the Colônia do Sacramento had also been a source of constant tension between the two crowns since its foundation in 1680. In freeing itself of the land, Spain could complete its hegemony over the northern bank of the Rio de la Plata.

However, more than convincing Portugal's adversaries, this configuration would help convince the Portuguese themselves, for whom ceding the territory was not a moot point, that they were surrendering an insignificant territory in exchange for significant stretches of land in the midwest and Amazonia. In fact, the trade of the Sacramento—finally settled upon in Madrid—was the source of much contention in Portugal. Opponents of the idea, especially the Society of Jesus, launched a defamatory campaign against Alexandre de Gusmão and blamed him personally for the outcome. The criticisms were hardly justified; Gusmão had resisted the measure as long as he could, giving in only in the last moments of the negotiation. D. Luis, on the other hand, had defended the policy ferociously,[12] and the *Amérique méridionale* served as a visual contribution to diminish the impact of what was referred to in Portugal as the "handover" of the Colônia do Sacramento.

11 NL, VB, Ayer MS 1918, doc. 552.
12 Monteiro, ed., *Meu pai e meu senhor muito do meu coração*, pp. 156–57.

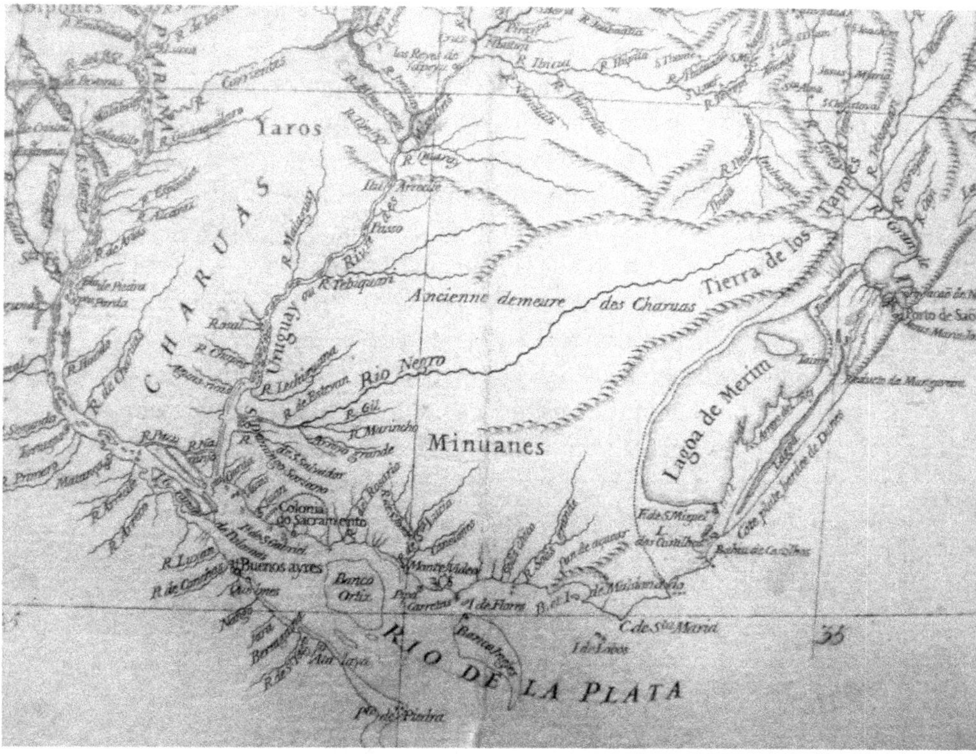

FIGURE 49 Detail of *Amérique méridionale*, manuscript, J.B.B. d'Anville (Paris, by the author at the Louvre galleries, manuscript, ink on paper, 1748). BNF, DCP, GE C-10657 (1)
REPRODUCED COURTESY OF BNF

After a long discussion between Alexandre de Gusmão and D. Luis da Cunha, which can be evidenced in the various versions of the *Dissertation* from the 1730s, Gusmão finally consolidated Portugal's position in a document that became known as the *Grande instrução* (Great Instructions), dated October 1736, addressed to D. Luis and the Secretary of State Marco António de Azevedo Coutinho to guide the negotiations they were leading in France, a traditional ally of Spain.[13] The manuscripts of that document were carefully kept by the ambassador during his life.[14] On the northern bank of the Rio de La Plata, for example, Gusmão doubted that *Uti possidetis* was the ideal argument to use in the dispute, since the continuous confrontations that took place at the time between Spanish and Portuguese forces in the region made it difficult to predict who would be in possession of what when negotiations took place

13 Cortesão, *Alexandre de Gusmão e o Tratado de Madrid*, vol. 3, pp. 420–54.
14 ANTT, MNE, caixa 564, doc. 356, 1756.

as the region was at war, invaded by Spanish military forces. In other words, for Gusmão *Uti possidetis* was an occasional strategy, which could be beneficial or harmful to Luso-Brazilian interests, depending on the current context. D. Luis da Cunha, in turn, had already made by that time *Uti possidetis* the central theme of his arguments with respect to the negotiation of frontiers in the Americas and he had been arguing with the French that the Portuguese possessions had to return to the pre-war position, in other words the entire northwestern bank of the Rio de la Plata was assured by the Treaty of Utrecht, since Portuguese law on this territory was never contested, either before or after the Treaty of Utrecht,[15] and was reinforced by *de facto Uti possidetis*.

Generally speaking, the line drawn on the *Amérique méridionale* map along the southern portion of the territory guaranteed possession of a narrow strip of land along the coast in Rio Grande do Sul, colonized by Brigadier José da Silva Paes with the 1737 establishment of the fort of Jesus Maria José, and the map supported the proposal

> made in 1714, during the negotiations of the Treaty of Utrecht (...) that admitted the extension of Brazil solely to the mouth of the Rio de la Plata, via a tongue of land 10 leagues wide, starting at the district of São Paulo.[16]

From the southernmost Brazilian territory, at the lower portion of the Merim Lagoon, the line turns to the west and from there, using mountains and rivers as landmarks, plunges ever farther inland to the northwest until it hits the Paraguay River; from there it goes north, hugging the river until Lake Xarais. The 1754, 1760, and 1765 versions see no change in the basins of the Rio de la Plata and the Paraguay, and only in 1779 (Fig. 4) does d'Anville add a northern part of the course of the Paraguay River at the upper portion of Lake Xarais, including two tributaries, the Supoaiva (?) and the Jauru (whose mouth had been used in the *Mapa das cortes*—Figure 51—as a border). This edition also modified the frontier line in the Merim Lagoon region (Fig. 50), reflecting what was agreed in the Treaty of Santo Ildefonso in 1777. This artifice allowed the dividing line, which had followed the Paraguay River for a small stretch along the lower part of the lake, to continue following the river as a natural frontier on the upper portion of the map.

15 Cortesão, *Alexandre de Gusmão e o Tratado de Madrid*, vol. 3, p. 303.
16 Cortesão, *Alexandre de Gusmão e o Tratado de Madrid*, vol. 3, p. 754.

FIGURE 50 Detail of Page 2 (south of Brazil) of *Amérique méridionale*, J.B.B. d'Anville (Paris, printed by Gravelot, 1779). Furtado Lins Collection
REPRODUCED COURTESY OF FURTADO LINS COLLECTION

3 The Midwest and Amazonia

On all the versions of the map, starting at the Paraguay River, the line moves farther and farther to the northwest until it hits the Amazonas between São Pedro and Nuestra Señora de Guadalupe. These two are precisely the last Portuguese and the first Spanish settlement established along the Amazonas River. Before hitting this point, the line passes through a territory that—with the exception of the Madeira River—is completely unknown. The course of the Madeira itself and the surrounding tributaries came from information provided by Ignacio dos Reys (Figs. 35, 44, and 45) and from the Portuguese soldier Francisco de Melo Palheta (1670–c.1750), whose upriver expedition was undertaken in 1722–1723. Using this information, d'Anville fixed the configuration of the river network around the mouth of the Madeira (Fig. 3).

Without solid geographical information between Lake Xarais and the Amazonas (the upper course of the Purus, in the absence of any information, is indicated by a dashed line), the dividing line is determined in this area by pure geometry. This way, a rising line moves from southwest increasingly farther to the northwest, cutting through the Madeira in an area with little geographical detail. Around this region, a text by the cartographer states that 22 waterfalls (according to Palheta's diary and map) break up the course of the river between Portuguese territory and the Spanish missions. This constitutes a natural element separating the dominions of the two crowns, justifying the author's move to draw a line without a secure referent. Without known geographical elements, the line runs straight on without approaching rivers or mountains (which in other areas would be possible natural landmarks), until it hits the Amazonas. This line certainly guaranteed Portuguese possession of the whole of the mining regions of the midwest: Minas Gerais, Mato Grosso, and Goiás captaincies, the last strategically created in 1748.[17] This also meant "preserving the fluvial line of the monsoons, connecting [São Paulo] to Cuiabá and ultimately linking Mato Grosso to the territories of Amazonia."[18]

From the Amazonas River, a semi-artificial line curving to the east separates the Portuguese Jesuit missions from those of the Spanish, cutting perpendicularly through several rivers in the region (Figs. 3 and 11). One must highlight the fact that, around this region, d'Anville represents the basins of the Negro and the Solimões Rivers as connected, trusting Gumilla and Acuña and following La Condamine's lead, though the Frenchman never made it to the upper portion of the river.[19] The stretch around the Amucu was established using information from the Prussian traveler Horstman (Fig. 47), and the later inclusion of Lake Parima was due, he reported, to unnamed Spanish sources. While the inclusion of Lake Amucu favored Portuguese interests, the inclusion of Parima in versions after 1760 strengthened Spanish claims in the region (Fig. 4). In the

---

17 Furtado and Gabriel Paquette, "Imperial Competition in the Eighteenth-Century Americas," in Fernando Bouza, Pedro Cardim and Antonio Feros, eds., *The Iberian World* (New York/London: Routledge, 2019), pp. 519–40.

18 Cortesão, *Alexandre de Gusmão e o Tratado de Madrid*, vol. 3, p. 754.

19 RBC, 229, doc. 539 (2), f. 10.

Amazonas region, one can see the steady growth of the Portuguese territory as compared to the two simpler versions of the *Amérique méridionale* from 1737 and *c*.1742, which only take in the lower portion of the river. In the 1748 map, the frontier takes in almost the whole western part of the bay and the northern part of the plain around the river's course, stretching north to Dutch Guiana and west to the viceroyalty of Peru, far beyond the Tordesillas Meridian, which fell roughly across the river's mouth, near Marajó Island.

4 Cabo do Norte and the Guianas

When it came to the Guianas, d'Anville made use of the measurements taken by the French mathematician Jean Richer at Cayenne and later by La Condamine at Cayenne and the fort of Oyapock, respectively, that were used to determine the Cayenne Meridian through comparison with the Paris Meridian. La Condamine blamed the Treaty of Utrecht for taking the Oyapock River as the Vicente Pinzón. He claimed that he had discovered a second mouth of the Araguari River and this was in fact the river of Vicente Pinzón, located in the Bay of Vicente Pinzón. According to him, it was there that the frontier line should be positioned, which would give 50 leagues of territory to the French.[20] D'Anville followed his information and drew the border line from this cape, called Cabo do Norte, running more or less perpendicular until the region of Lake Amucu. He seems to merge French and Portuguese interests with some advantage to the first. Since the late 1720s, after all, French cartography had pushed for a distinction between the Oyapock and Vicente Pinzón Rivers with an eye to reducing Portuguese territories around Cabo do Norte.[21]

Relying heavily on the observations of French travelers who visited the coast of the region—principally La Condamine, who, on his way from Belém to Cayenne, drew sketches as he went and took measurements of the places he passed—d'Anville maintained La Condamine's stratagem. The place chosen in the *Amérique méridionale* manuscript map for the dividing line is the north mouth of the Araguari River, positioning it at the north of the Cabo do Norte (Fig. 11). But one can see that he drew a second possible line a little southern, cutting the Cabo do Norte in the middle. It was this second position that he chose for the printed version of the map in 1748, which gave the Portuguese

20 La Condamine, *Abrégée d'un Voyage fait dans l'Intérieur de L'Amérique méridionale*, pp. 194–95.

21 Max Justo Guedes, "Os limites Territoriais do Brasil a Norte e Nordeste," in Luis de Albuquerque, ed., *Portugal no mundo*, vol. 5 (Lisbon: Publicações Alfa, 1989), p. 211.

a little more territory. Though a cursory glance at the map seems to confirm an exclusive expression of French interests, the Portuguese had in fact kept what was most important for them. This division gave Portuguese America all of its old forts at the mouth of the Amazonas River, constructed along what it had considered to be its territory by right; and, more importantly still, it guaranteed control over the estuary and the whole course of the Amazonas, making it possible to ban foreign trade on the river. In this sense, the map reflected the culmination of a slow process of constructing Portuguese hegemony along the course of the Amazonas basin, which D. Luis da Cunha had pressed for since Utrecht, and which was finally achieved with the 1750 Treaty of Madrid,[22] although it didn't follow exactly what the ambassador desired for the Cabo do Norte.

In conclusion, it thus becomes evident that the representation of the Brazilian south, west, and north in d'Anville's 1748 *Amérique méridionale*, and not in the 1737 and *c.*1742 versions, illustrated the ideas D. Luis da Cunha upheld for the configuration of the Brazilian territory around this period. To the south, from São Paulo onward, Brazil stretches out in a narrow strip of coastline along the Patos and Merim Lagoons, ending just before the mouth of the Rio de la Plata—thus stopping before the Colônia do Sacramento and the Sete Povos das Missões, in Paraguay, evidently positioned beyond the Tordesillas Meridian but still under Portuguese control. The map clearly outlines the possibilities for trading between these two territories. To the west, the borderline puts all of Minas Gerais, Mato Grosso, and Goiás in the Portuguese zone, guaranteeing control over the mineral heart of America and connecting it to Amazonia, as suggested by D. Luis da Cunha who saw in the Brazilian mines "the old lady's goods." In the north, though much of this area was also positioned outside Tordesillas, the Portuguese had won total control over the Amazonas and even many of its northern tributaries, extending their territory all the way to the border with the Guianas and the viceroyalty of Peru. In this way, the map was not merely the result of the process of revealing the continent's geography, made possible by Europeans' progressive push inland (as d'Anville would have it), but also brought to light the geopolitical plan that D. Luis da Cunha believed to be the most favorable to Portuguese interests in America. This was not, then, a neutral document, as Enlightenment thinkers believed, but one loaded with intentionality; though this did not mean distorting the territory in question, but rather configuring it in the most realist form possible.

22 BNL, R, cód. 11.209, 312, April 13, 1713.

5 The Treaty of Madrid and *Uti possidetis*

In the year 1749, the months leading up to the Portuguese decision to cede the Colônia do Sacramento were marked by unprecedented tension. The Portuguese were insisting that the Spanish indicate an equivalent to be exchanged and dug in their heels on the claim that their possession of the enclave was legitimate. In this context, any information leak could be extremely prejudicial. This was no longer a matter of simply choosing an equivalent in land that could be traded between the two nations. As D. Luis da Cunha himself had pointed out for years, the issue had taken on larger dimensions, implying the definition of a broad new line of demarcation throughout all the American territories and also dragging in the disputed regions in the Pacific Ocean—the Moluccas and the Philippines. This was a negotiation of global dimensions, and no precautions could be spared.

The negotiation was based on the invocation of treaties, authorities, and historical documents from the process of the colonization of America, including historic and contemporary maps that might be used to justify each side's claims. In first place, it was established "that the line be preferentially guided by the most conspicuous and notable landmarks, large mountains or rivers," and secondly, the right of *Uti possidetis*, in which "each would preserve what it had occupied."[23] The idea of *Uti possidetis* was first and foremost sketched out by D. Luis da Cunha and not by Alexandre de Gusmão as Jaime Cortesão claimed and is usually repeated in Brazilian historiography.[24]

The use of *de facto* occupation as a mechanism to guide the drawing of the frontiers of South America was presented by the Portuguese to the Spanish shortly after the negotiations began in Madrid. It was on July 9, 1747, during the first meetings with D. José de Carvajal, Spanish Secretary of Foreign Affairs, that the Portuguese ambassador, the Viscount of Vila Nova de Cerveira, verbalized for the first time the concept of *Uti possidetis*, though without naming it, certainly inspired by D. Luis da Cunha's correspondence that was sent from Paris to instruct him. Questioned by the former about the difficulties involved in defining frontiers in the Americas, he proposed that the new treaty be very succinct "and that the main part would be to give ownership to what each had." Cautiously, the Castilian promised him "to put this clause at the end of the memoir [which he had written about the Castilian position], in the case that if

23 ANTT, MNE, caixa 612, May 16, 1749.
24 Furtado, "Portuguese America under Foreign Threat."

it could not be put into execution the terms would be adjusted *a quo* to remove the dividing line."[25]

On July 10, 1748, almost a year later, as the negotiations advanced, Viscount of Vila Nova de Cerveira, who headed the negotiations in Madrid, wrote to Portugal forwarding and commenting on the letters which D. Luis da Cunha had sent him from Paris to guide his actions. He stated that he was not aware that the Spanish had intervened to "take possession of everything we occupy from the Maranhão [or Amazonas] River to the Rio de la Plata," since what it was intended to prevent was the advance of the Portuguese beyond these points, which could threaten the Spanish monopoly on silver reserves. He warned that in this and "in everything else in the opinion of D. Luis da Cunha, I am satisfied with what he says." He guaranteed that he took care to ensure that the articles of the new treaty showed

> with greater clarity all the land we occupied and possessed, so that they not only recognize that we legitimately possess them, but that, if by some accident it comes to be understood that they belonged to the Castilians, they will cede them to us.[26]

As can be observed in the comments of the viscount, D. Luis insisted that *Uti possidetis* be used as the principal negotiation mechanism, that the terms of the treaty had to be clear in relation to the Portuguese-held territories past the Tordesillas line and for this he insisted that d'Anville's map *Amérique méridionale*, which he had sent to Madrid the previous year, had to be used as the base for the representation of this territory, since it showed the area in question as precisely as possible and defended that Colônia do Sacramento be used as a means of exchange for mining territories in the center-west and the Amazonas.

Documents passed between Portugal and Madrid and from the Portuguese ambassador in the city to the Spanish Ministry of Foreign Affairs. On September 7, 1748, Cerveira made his second commentary on the second memoir he had presented to the Spanish, where he listed the main points defended by the Portuguese which had to appear in the treaty. Afterwards Alexandre de Gusmão made his observations on the two documents and listed eight principal guidelines which Cerveira should follow in the negotiations. As a third point he argued that it was "preferable for Spain to compensate some occupied areas with others," referring to territory beyond the Tordesillas line which the Spanish held in the Pacific Ocean, and that "current possession would

25 Cortesão, *Alexandre de Gusmão e o Tratado de Madrid*, vol. 2, p. 282.
26 Cortesão, *Alexandre de Gusmão e o Tratado de Madrid*, vol. 4, p. 166.

continue to serve as legitimate, fixed, and unalterable title of ownership of both parties."[27]

In the same month, Portugal finally first proposed trading the territory of Colônia do Sacramento for an equivalent in land, as D. Luis da Cunha had always advocated, but instructed that this land should be in Europe and not in the Americas as the ambassador had wished. It was therefore not a mechanism to exchange possessions between the two crowns in the New World, nor a question of considering the borders in the Americas as a whole, nor a way of guaranteeing Portugal's possession of large extensions of land, for example, in Amazonia or the center-west. Gusmão wrote to Cerveira that it was better to let the Spanish propose the equivalent they were willing to cede in Europe, advising them that Portugal would consider the proposal, but that it would also be of interest if they also presented a general proposal defining where they considered that the demarcation line should go through. In other words, at that point he let the adversaries take the lead in the negotiations.[28]

The *Primeira minuta do tratado* (First Draft of the Treaty), sent from Portugal also in September 1748, was the first official Portuguese document to determine that demarcation would

> use as its first basis the determination of preserving as much as possible, and where this does not offer significant inconveniences to the contrary, current possession and occupation, and choosing everywhere, insofar as the situation allows this, the natural divides of mountains and rivers.

Just after, in another letter, dated May 16, 1749 the Secretary of State Marco António de Azevedo Coutinho stated that since the return to diplomatic negotiations Portugal had adopted "as preliminary maxims, in first place that in the lands already occupied by any of the parties, each would keep what they had occupied," highlighting the importance of *Uti possidetis* as a strategic pillar.[29]

However, while since the *Primeira minuta* of 1748 the preliminary maximum proposed by the Portuguese to regulate the limits was *Uti possidetis*, followed by the natural divisions of the terrain, advocated by Spanish authorities, in the final text of the Treaty of Madrid their precedence and hierarchy were inverted, which constituted a victory for Carvajal.[30] In its Preamble, in first place it was established that in lands that were known the natural divisions

27 Cortesão, *Alexandre de Gusmão e o Tratado de Madrid*, vol. 4, p. 173.
28 ANTT, Ministério dos Negócios Estrangeiros, caixa 612, September 11, 1748.
29 Cortesão, *Alexandre de Gusmão e o Tratado de Madrid*, vol. 4, p. 312.
30 Cortesão, *Alexandre de Gusmão e o Tratado de Madrid*, vol. 2, p. 375.

of the terrain would be used and, as a secondary recourse, it was determined "that each party has to keep what it currently possesses, with the exception of mutual cessions."[31]

The Spanish and Portuguese then embarked on an episode without precedent in the long diplomatic battle over the definition of Brazil's borders. The 1750 treaty did not put an end to the disputes between the two nations; squabbles over borders in South America would continue throughout the eighteenth, nineteenth, and twenty centuries, going on through the post-independence period and involving the young American nations. However, the negotiation did represent a landmark in giving Portuguese America a very similar shape to that of modern-day Brazil, inventing a Brazilian territory that was then not truly possessed by the Portuguese. The agreement overrode the Treaty of Tordesillas; and, for the first time, instead of an imaginary line of demarcation, the treaty established limits using concrete experience, either through effective colonization or the natural borders of the land, though in several cases both were produced by human invention.

This dispute was a war of cartographical images.

6 The *Amérique méridionale* versus the *Mapa das cortes*

Getting ahead of the Spanish, the Portuguese were able to elaborate a fundamental cartographical document, presented by the Viscount of Vila Nova de Cerveira—entitled *Mapa dos confins do Brazil com as terras da coroa de Espanha na América meridional*, it came to be known as the *Mapa das cortes* (Map of the Courts—Fig. 51), as it ultimately served to settle the Treaty of Madrid boundaries in South America, signed by the two courts in 1750. This document reflected the stance that Portugal stuck to in terms of the format of both nations' territories in America. Since the Andes wound up forming a natural barrier to the Spanish push inland from the western coast, the vast *sertões* of the central highlands—though promised to Spain by the Treaty of Tordesillas—were more easily reached by Luso-Brazilian explorers. Through their clever use of the network of rivers in the region, the Portuguese wound up colonizing the midwest; and as its explorers penetrated the interior, the crown's geographical knowledge of these territories increased. This precedence, combined with Spain's relative ignorance of the true dimensions of the American interior, allowed Portugal to gain the upper hand in the negotiations

31 Cortesão, *Alexandre de Gusmão e o Tratado de Madrid*, vol. 2, p. 365.

FIGURE 51 *Mapa dos confins do Brasil com as terras da coroa de Espanha na América, ou Mapa das cortes* (Rio de Janeiro, manuscript, facsimile, ink and color, of the Spanish original, July 12, 1750). Furtado Lins Collection
REPRODUCED COURTESY OF FURTADO LINS COLLECTION

and guarantee at Madrid that the line of demarcation was dictated by the *Mapa das cortes*.

The different geopolitical visions of Alexandre de Gusmão, the great architect of the Treaty of Madrid, and D. Luis da Cunha are directly reflected in the *Mapa das cortes* and d'Anville's *Amérique méridionale*. The differences suggest that, although D. Luis's political thought was largely followed through on by the negotiators, under Gusmão's influence, the cartographical representation of the territory and the way the negotiations were carried out diverged. In order to serve the ends that both men believed to be in the interest of the Portuguese monarchy, the *Mapa das cortes* distorts the dimensions of America, while the *Amérique méridionale* sought to represent its true dimensions through a series of clear principles.

Since the first quarter of the eighteenth century, in large part due to the insistence of D. Luis da Cunha, the Portuguese government spared no effort in collecting a serious cartographical record of Brazil, looking to arm its ambassadors in European negotiations after the War of Spanish Succession. In 1719, the *Conselho Ultramarino* recommended that the government sponsor the production of maps of Brazil.[32] After nearly 30 years of local and regional geodesic surveys, using many reports and maps from frontiersmen, Gomes Freire de Andrade (1688–1763), then viceroy of Brazil, oversaw the production of the map entitled *Descripçãm do continente da América Meridional* (Fig. 52) in 1746. Executed in Brazil, this map consolidated all cartographical knowledge in and of the Brazilian territory, as well as Portugal's position in relation to the South American continent; the document sketched a dividing line between the two nations' possessions, although still respecting the borders imposed by the Treaty of Tordesillas. This map was then sent to Gusmão to serve as the basis for the maps to be drawn up during the negotiations. From Lisbon, a copy was sent to Madrid[33] and remained in the archives of the Portuguese ambassador in Madrid as late as 1778–1779, when Portugal negotiated a new treaty with Spain, that of Santo Ildefonso, regarding their frontiers in South America.[34]

In December 1748, Cerveira received instructions about the maps he was permitted to display. Initially, the list included the map by Diogo Soares, for the Rio Grande do Sul region; d'Anville's *Le Paraguay* (Fig. 9), for the Rio de la Plata basin; and Gomes Freire's for the central region of Brazil, including Goiás and Mato Grosso (Fig. 52). From there to the Amazonas, he was instructed to use the *Carta das missões dos Moxos*, which appeared in the 12th volume of the *Lettres édifiantes*, and for the Amazonas, the map by La Condamine and d'Anville (Fig. 24). At the time, the Portuguese resigned themselves to using "the maps at hand [or] the conclusion of the treaty would be postponed indefinitely."[35] There is no mention of d'Anville's *Amérique méridionale*.

32 Almeida, *A formação do espaço brasileiro*.

33 ANTT, MNE, livro 826, f. 112v, December 14, 1748.

34 Furtado, "Cartography in Dispute: The Frontiers of Brazil in Abbé Raynal's *Histoire des deux Indes*", *Culture & History*, 10, 2 (2021), p. 11.

35 ANTT, MNE, livro 826, f. 112v, December 14, 1748.

FIGURE 52 *Descripção do continente da América Meridional que nos pertence com os rios, e montes, que os certanejos mais experimentados, dizem ter encontrado, cuja divisão se faz, mandado fazer por Gomes Freire de Andrade* (Rio de Janeiro, manuscript in ink and color, December 5, 1746). Biblioteca Guita e José Mindlin

REPRODUCED COURTESY OF BIBLIOTECA GUITA E JOSÉ MINDLIN/USP

Despite this first order, on December 28, the crown instructed Cerveira that, "as for the map sent by Gomes Freire de Andrade, it would not be wise to present it as it stands." The reason for this decision was that "the governor indicated the line that seemed best to him, and as he knew nothing of our position, the territory is much smaller than that which we should like to claim."[36] The proposed border hewed fairly close to that of the 1737 *Amérique méridionale* (Fig. 1), taking in a small part of Amazonia and generally following the Tordesillas Meridian. The orders dictated that, if it should be necessary

> to make use of that map, on the way from São Paulo to Cuiabá, and to Mato Grosso (that being the part that it best presents), you shall have only that part copied, or the whole map with the exception of the future dividing line.

Cerveira was then warned that "the general map that I have promised you, and which will suffice to demonstrate everything, is not yet finished."[37] Between December 14 and 28 of 1748, however, we see that the Portuguese crown abandoned the idea of presenting only fragments or copies of maps already in the ambassador's hands and resolved to produce a new map with the dividing line it proposed to claim.

In 1749, the new map was finally finished and sent to Spain. Drawn up under the supervision of Alexandre de Gusmão, the *Mapa das cortes* would indeed become the key cartographical piece in establishing the borders of the Treaty of Madrid.[38] Nevertheless, and despite what the Portuguese authorities had affirmed, the differences between this document and Gomes Freire's map were not limited to the matter of the dividing line. The *Mapa das cortes* was marked by serious imprecisions:[39] "several important urban centers in the interior of Brazil were pushed to the east; the city of Cuiabá was moved east by nearly 5°, situating it much closer to the line of the Tordesillas Meridian, and the distance between the meridians of Belém and the Colônia do Sacramento was set at just 7°20'." These and other shifts were the result of the map's fixation on longitudes, although only latitudes were numbered on the document. With this stratagem, all of Portuguese America was shifted east, and "the extra-Tordesillas area of Brazil was (…) significantly diminished, which gave the

36 ANTT, MNE, livro 826, f. 114, December 28, 1748.

37 ANTT, MNE, livro 826, f. 114, December 28, 1748.

38 Ferreira, *O Tratado de Madrid e o Brasil Meridional*; Costa, Furtado, Renger, Santos, *Cartografia das Minas Gerais: da capitania à província* (Belo Horizonte: Ed. UFMG, 2002), p. 15.

39 Ferreira, "O *Mapa das cortes* e o Tratado de Madrid."

impression of there being few territorial gains for the Portuguese to the west of this meridian." When the distortions of the map are corrected, positioning the various geographical features correctly in relation to the meridians, the territorial gains that the Portuguese had hoped to swindle out of the Spanish come to light.[40] Alexandre de Gusmão's geopolitical project, then, was based on a cartographical illusion—a deformed mirror of Portuguese America—"drawn more to deceive than to inform."[41]

For its part, D'Anville's *Amérique méridionale*, which mirrored D. Luis da Cunha's geopolitical vision for the region, represented the Portuguese territory in terms much closer to those believed to be real, using information provided by the ambassador, La Condamine, and other scholars and travelers, and bringing them together via a method of geographical critique. This was what the ambassador had hoped: that a single accurate map could put an end to nearly a half-century of territorial disputes. Indeed, even with incomplete information on many areas, one does not see any shift to the east here. This bears witness to the maxims that D. Luis believed should guide the negotiations: with accurate maps and no sleight of hand being played with geographical reality, the negotiations ought to follow what had been settled in previous treaties, respecting *Uti possidetis* and the natural features of the terrain.

The discrepancies between the *Amérique méridionale* signal that the two men's geopolitical visions and the ways in which they envisioned conducting the negotiations differed considerably. The experienced ambassador saw it as a matter of presenting the territory with the greatest possible accuracy, sharing the information that the Portuguese had gathered with the Spanish, and attempting to resolve these conflicts diplomatically while making use of previous accords. In this sense, d'Anville's map was a faithful mirror of the Brazil that the Portuguese had just invented. Gusmão, meanwhile, saw a need to disguise the territory's true shape and conduct negotiations by taking advantage of the Spaniards' relative ignorance of significant portions of land in the center of South America, doing so in such a way that their adversaries would accept Portugal's claims.[42]

Paradoxically, the two maps opposed two old allies who, while they had had discussions and differences in the past, had always come to an agreement: Alexandre de Gusmão and D. Luis da Cunha. With the shared intention of

---

40 Costa, Furtado, Renger, Santos, *Cartografia das Minas Gerais*, pp. 30–31; Furtado and Cintra, "Bourguignon D'Anville's *Carte de l'Amérique méridionale.*"

41 Jordana Dym, "'Mais calculado para enganar do que para informar': os viajantes e o mapeamento da América central (1821–1945)," *Varia Historia*, 23, 37 (2007), pp. 81–109.

42 Ferreira, *O Tratado de Madrid e o Brasil Meridional*, p. 87.

serving the interests they understood to be those of the Portuguese government, each invented a new Brazil, but using different means: the *Mapa das cortes* distorted its dimensions, while the *Amérique méridionale* sought to reestablish accurate spatiality in the interior of the American continent, even as many mythological conformations of the continent stubbornly remained. In terms of the Treaty of Madrid, Alexandre de Gusmão was victorious; despite d'Anville's later efforts to defend his territorial conformation of the continent, it was the *Mapa das cortes* that served as the basis for the establishment of the treaty, and would come to be considered the map that had invented this new Brazil. But the treaty also stipulated that bilateral expeditions ought to establish stone markers to demarcate these borders in America. When these groups began to take local measurements, the distortions of the *Mapa das cortes* quickly became evident.

D'Anville's 1748 *Amérique méridionale* was for a while considered a lost treasure. Its primacy and importance in the construction of Brazilian territory, extending it out in a triangular shape far to the west of the Tordesillas Meridian, were long forgotten. This was due to the fact that the map was intentionally hidden from the Spanish during negotiations, even though D. Luis da Cunha had sent a copy to Cerveira. The *Mapa das cortes* was chosen instead. Soon afterwards, even the quality of the *Amérique méridionale* would be called into question. In 1750, Francisco Tosi Colombina, a Genovese engineer hired by Portugal at the recommendation of d'Anville himself to take part in the border-establishing expedition in Brazil after the Treaty of Madrid,[43] would study the map at the orders of the D. Marcos José de Noronha e Brito (1712–1768), 6th Count of Arcos, then governor of Goiás captaincy, and conclude that "one cannot base oneself on this map of d'Anville's (...) and as for the interior of America, which belongs to this crown, it is quite wrong and cannot even serve to accurately divide its districts and bishoprics."[44] Colombina, who had been in Brazil for nearly a year, traveling around some parts of Goiás, had knowledge of other districts by way of local geodesic surveying, especially the 1730s reports of the Priest Mathematicians. Around this time, the Portuguese had fairly accurate longitude and latitude measurements, allowing Colombina to conduct a rigorous critique. Most of the divergences he points out in the *Amérique méridionale* are on the order of a degree of latitude or longitude—insignificant, especially given the dimensions of the map, the fact that it was a product of armchair geography, the fact that d'Anville had no

43 Monteiro, ed., *Meu pai e meu senhor muito do meu coração*, p. 163.
44 Cortesão, *Alexandre de Gusmão e o Tratado de Madrid*, vol. 3, pp. 139–44.

information for many regions, and that, given the imprecision of the instruments and methods of the period, the very measurements he was working with were imperfect.

The continued negotiations between Portugal and Spain after the Treaty of Madrid (the Treaties of Pardo, 1761, and of San Ildefonso, 1777)[45] seemed to provide an occasion for d'Anville to rise in defense of his *Amérique méridionale*; to this end, he wrote two new memoirs in 1779, in which, with a few variations, he clarified the sources he had used in drawing up the map and his profitable collaboration with D. Luis da Cunha.[46] However, this was not sufficient to turn the tide against the relative oblivion settling over the map. Interestingly enough, in all the border negotiations involving the Minister of Foreign Affairs in Brazil from the late nineteenth century to the early twentieth century, d'Anville's maps were invoked as documentary proof of Brazil's right over the territories in dispute. In negotiations with Suriname in 1897, for example, the famous Brazilian diplomat, the Baron of Rio Branco (1845–1912), José da Silva Paranhos, affirmed that "d'Anville drew the frontiers of Dutch Guiana in accordance with official surveys from Holland," and used the map to support Brazilian's territorial claims. He went on to affirm that the British government had used the same map in its favor during negotiations with Venezuela, in 1896.[47] This stratagem went against the context in which the map had been produced. Having invented Brazil and serving neither Dutch nor Spanish aims, the *Amérique méridionale* was subsequently invoked by the diplomats of several countries as a historical document proving a previously assured right over one of the parts of the territories in question. That is to say, the fact that the map had preceded the space it was inventing had been forgotten; now it was used as proof that the borders drawn there already existed at the time.

This book hopes to ensure that the 1748 *Amérique méridionale* is no longer seen as a lost treasure. It brought together two great *savants* of its time: D. Luis da Cunha and Jean Baptiste Bourguignon d'Anville. It was thanks to the two of them, as they fused politics, diplomacy, geography, and cartography and together produced a map revealing the borders the Portuguese desired for their American territories, that a continental Brazil was invented and used later on

45 For the use of maps by Portuguese and Castilian ambassadors in the context of San Ildefonso Treaty and as political propaganda, see Furtado, "Cartography in Dispute: The Frontiers of Brazil in Abbé Raynal's *Histoire des deux Indes*," *Culture & History*, 10, 2 (2021), pp. 1–25.

46 RBC, 229, doc. 539 (1) (2).

47 José Maria da Silva Paranhos Júnior, *Mémoire sur la question des limites entre les États-Unis du Brésil et la Guyane Britannique* (Rio de Janeiro: Instituto Rio Branco, 1897), f. 2.

as a diplomatic strategy.[48] This invention was configured as a future plan, first put into practice with the 1750 Treaty of Madrid and later in the treaties negotiated by Portugal and Brazil already as an independent country. In this sense, while the Portuguese ambassador and the French geographer are considered oracles of Enlightenment geopolitics, the *Amérique méridionale* can be seen as a valuable oracle of the Brazilian territory, because the map was essential for its continental dimensions to be built. The map invented Brazil!

48 For the use of d'Anville's map and especially its toponomy as a strategy for later diplomatic negotiations see Kantor, "L'appropriation des cartes de d'Anville dans le monde luso-brésilien," pp. 287–303.

Bibliography

Primary Sources

Acuña, Cristóbal. *Nuevo Descubrimento del Gran Rio de las Amazonas.* Madrid: Iberoamericana, 2009.

Alegrete, Manuel Teles da Silva. *Colecçam dos Documentos Estatutos e Memorias da Academia Real da Historia.* Lisboa: Officina de José António da Sylva, 1727.

Alembert, M. d'. "Preface," in *Éloges Lus dans les Séances Publiques de l'Académie Françoise.* Paris: Panckoucke, 1779, aiij–xxxiv.

Alembert, M. d'. *Journal et Mémoires du Marquis D'Argenson.* Vol. 2. Paris: Veuve de J. Renouard, 1860.

Almeida, Francisco José de Lacerda e. *Travessia da África.* Introduction by Manuel Múrias. Lisboa: Agência Geral das Colónias, 1936.

Anson, George. *A Voyage Round the World, in the Years 1740, 41, 42, 43, 44.* London: John and Paul Knapton, 1749.

Antonil, André João. *Cultura e Opulência do Brasil por suas Drogas e Minas.* Introduction by Andrée M. Diniz Silva. Lisboa: Comissão Nacional para as Comemorações dos Descobrimentos Portugueses, 2001.

Anville, Jean-Baptiste Bourguigon d'. "Ethésiens," in *L'Encyclopedie ou Dictionnaire Raisonée des Sciences, des Arts et des Métiers,* ed. Denis Diderot and D'Alembert. Vol. VI. Paris, 1756, 50–51.

Anville, Jean-Baptiste Bourguigon d'. *Analyse Geographique de l'Italie.* Paris: Veuve Estienne et Fils, 1744.

Anville, Jean-Baptiste Bourguigon d'. *Considération Générales sur l'Étude et les Connoissances que Demande la Composition des Ouvrages de Géographie.* Paris: Imprierie de Lambert, 1777.

Anville, Jean-Baptiste Bourguigon d'. *Dissertation sur l'Étendue de l'Ancienne Jérusalem et de son Semple, et sur les Mesures Hébraïques de Longueur.* Paris: Prault Fils, 1747.

Anville, Jean-Baptiste Bourguigon d'. *Eclaircissemens Géographiques sur l'Ancienne Gaule, Précédés D'un Traité des Mesures Itinéraires des Romains et de la Lieue Gauloise.* Paris: Veuve Estienne, 1741.

Anville, Jean-Baptiste Bourguigon d'. *Éclaircissemens Géographiques sur la Carte de l'Inde.* Paris: l'Imprimerie Royale, 1753.

Anville, Jean-Baptiste Bourguigon d'. *États Formés en Europe Après la Chute de l'Empire Romain en Occident.* Paris: l'Imprimerie Royale, 1771.

Anville, Jean-Baptiste Bourguigon d'. *Géographie Ancienne Abrégée.* 3 vols. Paris: Chez Merlin, Libraire, 1768.

Anville, Jean-Baptiste Bourguigon d'. *Géographie Ancienne Abrégée.* Vol. 1, contenant Europe. Paris: Chez P. Théophile Barrois, 1782.

Anville, Jean-Baptiste Bourguigon d'. *Géographie Ancienne et Historique, Composé D'Aprés les Cartes de D'Anville par L.B.D.M.* Paris: Egron, 1807.

Anville, Jean-Baptiste Bourguigon d'. *l'Empire Turc Considéré dans son Établissement et dans ses Accroissemens Successifs.* Paris: l'Imprimerie Royale, 1772.

Anville, Jean-Baptiste Bourguigon d'. "Lettre de M. d'Anville, à M. Folkes, Président de la Société Royale de Londres, sur une Copie Faite à Londres de la *Carte de l'Amérique Septentrionale,*" *Mercure de France,* Paris (15 de novembro de 1750), 151–167.

Anville, Jean-Baptiste Bourguigon d'. "Lettre de M. d'Anville à l'Auteur du *Mercure* sur la publication d'une nouvelle *Carte de l'Afrique,*" *Mercure de France* Paris (Mars 1751), 157–161.

Anville, Jean-Baptiste Bourguigon d'. "Lettre de M. d'Anville, à M. Remond de Sainte Albine, sur une nouvelle *Carte d'Amérique méridionale,*" *Mercure de France* Paris (17 November 1749), 169.

Anville, Jean-Baptiste Bourguigon d'. *L'Euphrate et le Tigre.* Paris: l'Imprimerie Royale, 1779.

Anville, Jean-Baptiste Bourguigon d'. *Mémoire Instructif pour Dresser sur les Lieux des Cartes d'un Canton de Pays Renfermant Dix ou Douze Paroisses ...* Juillet 1743.

Anville, Jean-Baptiste Bourguigon d'. *Mesure Conjecturale de la Terre sur l'Équateur, en Conséquence de l'Étendue de la Mer du Sud.* Paris: Chaubert, 1736.

Anville, Jean-Baptiste Bourguigon d'. *Notice de l'Ancienne Gaule Tirée des Monumens Romains ...* Paris: Desaint & Saillant, 1760.

Anville, Jean-Baptiste Bourguigon d'. "Premiere Lettre de Monsieur D'Anville, à Messieurs du Journal des Sçavans, sur une Carte de l'Amérique Méridionale qu'Il Vient de Publier," *Journal des Sçavans* (March 1750), 522–533.

Anville, Jean-Baptiste Bourguigon d'. "Seconde Lettre de Monsieur D'Anville, à Messieurs du Journal des Sçavans, sur la Carte qu'Il a Publiée de l'Amérique Méridionale," *Journal des Sçavans* (Avril, 1750), 625–673.

Anville, Jean-Baptiste Bourguigon d'. *Proposition d'une Mesure de la Terre dont il Résulte une Diminution Considérable dans sa Circonférence sur les Parallèles.* Paris: Chaubert, 1735.

Anville, Jean-Baptiste Bourguigon d'. *Recueil Factive des Cartes de D'Anville, Suit une Première Page Intitulée 'Liste Générale des Cartes de D'Anville'.* Paris, 1765–1771.

Anville, Jean-Baptiste Bourguigon d'. *Réponse de M. D'Anville au Mémoire Envoyé à l'Académie Royale des Sciences, Contre la Mesure Conjecturale des Degrez de l'Équateur, en Conséquence de l'Étendue de la Mer du Sud,* 1738.

Anville, Jean-Baptiste Bourguigon d'. *Traité des Mesures Itinéraires Anciennes et Modernes.* Paris: l'Imprimerie Royale, 1769.

Anville, Jean-Baptiste Bourguigon d'. "Histoire de l'Art en France. Gravelot," *L'Artiste* (1er Avril 1853), 69–70.

Anville, Jean-Baptiste Bourguigon d'. "Observations Géographiques sur la Carte du Paraguai," *Lettres Édifiantes et Curieuses, Écrites des Missions Étrangéres*, Toulouse 8 (1810–1811), 254–266.

Anville, Jean-Baptiste Bourguigon d'. "Eloge de Monieur Gravelot," in *Le Nécrologe des Hommes Célèbres de France (1772–1775)*. Vol. 9. Paris: l'Imprimerie de Moreau, 1974, 132–145.

Argenson, René-Louis de Voyer d'. *Journal et Mémoires du Marquis D'Argenson*. Vol. 1. Paris: Veuve de J. Renouard, 1859.

Azevedo, Pedro, ed. *Instruções Inéditas a Marco António de Azevedo Coutinho*. Coimbra: Imprensa da Universidade de Coimbra, 1929.

Barleo, Gaspar. *Rerum per Octennium in Brasilia, et Alibi Nuoer Gestarum*. Amsterdam: Ioannis Blaeu, 1647.

Beauchamp, Alphonse de. *Histoire du Brèsil, Depuis sa Découverte en 1500 Jusqu'en 1810*. 3 vols. Paris: A La Librairie d'Éducation et de Jurisprudence D'Alexis Eymery, 1815.

Bellin, Jacques-Nicolas. "Remarques de M. Bellin … sur les Cartes et Plans qu'Il a Eté Chargé de Dresser. Dissertation Préliminaire, sur l'Origine des Amériquains," in Pierre-François-Xavier de Charlevoix, ed., *Histoire et Description Générale de la Nouvelle France avec le Journal Historique d'un Voyage Fait par Ordre du Roi dans l'Amérique Septentrionnale*. Vol. 3. Paris: Chez Pierre-François Giffart, 1744, i–xii.

Bestrman, Theodore, ed. "Correspondence de Voltaire," in *Les Œuvres Complètes de Voltaire*. Genève: Institut et Musée Voltaire, 1971.

Bestrman, Theodore, ed. *Correspondance de Voltaire*. 13 vols. Paris: Gallimard, c.1977.

Blém, Antonio. *Escola do Mundo, ou Instrucção de um Pae para seu Filho, Pertencente ao Modo com que se Deve Viver no Mundo, Dividida em Diálogos, por Le Noble na Lingua Francesa e Traduzida na Portuguesa*. 2 vols. Lisboa: Oficina da Música, 1722–1724.

Bluteau, Raphael. *Vocabulário Português e Latin* … 8 vols. Coimbra: No Collegio das Artes da Companhia de Jesus, 1728.

Bluteau, Raphael. *Dicionário da Língua Portuguesa*. Extended by Antônio de Morais. Lisboa: Oficina de Thadeo Ferreira, 1739.

Bluteau, Raphael. *Suplemento ao Vocabulário Português e Latino que Acabou de Sair à Luz, Ano 1721*. 2 vols. Lisboa: Patriarcal Officina da Musica, 1727–1728.

Bougainville, Louis Antoine de. *Voyage Autour du Monde*. Genève: Éditions de Crémille, 1973.

Carvajal, Gaspar de. *Descobrimento do Rio Orellana*. São Paulo: Companhia Editora Nacional, 1941.

Cassini, Jean Domenique. *Mémoires pour Servir a l'Histoire des Sciences et a Celle de l'Observatoire Royal de Paris, Suivis de la Vie de J.D. Cassini Écrite par Lui-Même et des Elóges de Plusieurs Académiciens Morts Pendant la Revolution*. Paris: Bleuet, 1810.

Catalogue des Livres de feu M. Sanchès. Paris: Chez Bure, 1783.

Catalogue des Livres de la Bibliothèque de feu Mr. de Couvay, Chevalier des Ordres du Roi de Portugal. Paris: Chez Damonneyville, 1755.

Chevalier, Nicolas. *Recherche Curieuse d'Antiquités Venuës d'Italie, de la Grèce, d'Egypte, & Trouvées à Nimegue, à Santen, au Château de Wiltenburg Proche d'Utrecht, dans le Château de Britten Proche de Leyde, & a Tongres*. Utrecht: N. Chevalier, 1707.

Chevalier, Nicolas. *Relation des Fêtes que son Excellence Monseigneur le Comte de Tarouca a Donnés au Sujet des Naisssances des Deux Princes de Portugal, et de Plusiers Autres Fêtes Aissi Donnés par son Excellence en Differentes Occasions*. Utrecht: N. Chevalier, 1714.

Choiseul-Gouffier, Marie-Gabriel-Florent-Auguste. *Discours Préliminaire du Voyage Pittoresque de la Grèce*. Paris: Imprimerie de Ph. D. Pierres, 1783.

Comte, P. Louis le. "Nouveaux Memoires Sur l'État Present de la Chine. Par le P. Louis le Comte, de la Compagnie de Jesus, Mathematicien du Roy," *Le Journal des Savans* (28 Janvier 1697), 37–44.

Condorcet, Jean-Antoine-Nicolas de Caritat (marquis de). "Éloge de M. D'Anville," in *Œuvres de Condorcet*. Vol. 2, Arthur Condorcet O'Connor and M. F Arago, eds. Paris: Firmin Didot Frères, 1847, 528–540.

Cortesão, Jaime. *Alexandre de Gusmão e o Tratado de Madrid*. 10 vols. Rio de Janeiro: Ministério das Relações Exteriores, 1950–1963.

Cunha, Luis da. "Carta de D. Luis da Cunha em Resposta do Aviso que o Secretario da Academia lhe Fez de Estar Nomeado Académico Supranumerário," in *Collecçam dos Documentos Estatutos e Memorias da Academia Real da Historia Portugueza: Que neste Anno de 1723 se Compuzeraõ, e se Imprimiraõ por Ordem dos seus Censores Dedicada a El Rey Nosso Senhor pelo Marquez de Alegrete Manoel Telles da Sylva, Secretario da Mesma Academia*, Marques de Alegre, ed. Lisboa Occidental: Officina de Pascoal da Silva, Impressor de Sua Magestade, e da Academia Real, 1723, 85–90.

Cunha, Luis da. "Carta de Instruções a Luís da Cunha Manuel," in *Instruções Políticas*, Abílio Diniz Silva, ed. Lisboa: Comissão Nacional para as Comemorações dos Descobrimentos Portugueses, 2001, 173–179.

Cunha, Luis da. "Carta de Instruções a Marco Antônio de Azevedo Coutinho," in *Instruções Políticas*, Abílio Diniz Silva, ed. Lisboa: Comissão Nacional para as Comemorações dos Descobrimentos Portugueses, 2001, 181–373.

Cunha, Luis da. "Carta de 1725," in *Instruções Políticas*, Abílio Diniz Silva, ed. Lisboa: Comissão Nacional para as Comemorações dos Descobrimentos Portugueses, 2001, 375–378.

Cunha, Luis da. *Instruções Políticas*, Abílio Diniz Silva, ed. Lisboa: Comissão Nacional para as Comemorações dos Descobrimentos Portugueses, 2001.

Cunha, Luis da. *Testamento Político* (1748). São Paulo: Alfa–Omega, 1976.

Dacier, Bon-Joseph. "Éloge de M. D'Anville," in *Notice des Ouvrages de M. D'Anville Précédée de son Éloge, Premier Géographe du Roi, Membre de l'Académie des Inscriptions et Belles-Lettres …*, Louis Charles Joseph de Manne, Jean-Denis Barbié du Bocage and Bon-Joseph Dacier, eds. Paris: Imprimerie de Delance, 1802, 17–44.

Dacier, Bon-Joseph. "Eloge de M. D'Anville," *Académie des Inscriptions et Belles-Lettres. Histoire, 1770–1784* 45 (1793), 160–174.

De Sauvigny. "Précis Historique sur la vie de M. Sanchès," in *Catalogue des Livres de feu M. Sanchès*. Paris: Chez Bure, 1783, 5–28.

Delisle, M. "Determination Geographique de la Situation et de l'Etendue des Diferentes Parties de la Terre," *Des Sciences*, Paris, Académie des Sciences de Paris (1722), 365–384.

Diderot, D. "Philosophe," in *Œuvres Complètes*. Vol. XVI. Paris: A.T. edition, s.d.

Du Bocage, Jean-Denis Barbié. "Avant Propos," in *Notice des Ouvrages de M. D'Anville Précédée de son Éloge, Premier Géographe du Roi, Membre de l'Académie des Inscriptions et Belles-Lettres …*, J.C.J. de Manne and Jean-Denis Barbié du Bocage, eds. Paris: Imprimerie de Delance, 1802, 5–15.

Du Halde, Jean-Baptiste. *Description Géographique, Historique, Chronologique, Politique et Physique de l'Empire de la Chine et de la Tartarie Chinoise, Enrichie des Cartes Generales & Particulières de ces Pays, de la Carte Générale & des Cartes Particulières du Thibet & de la Corée & Ornée d'un Grand Nombre de Figures & de Vignettes Gravées en Taille-Douce*. Paris: P.G. Lemercier, 1735.

Du Halde, Jean-Baptiste and J.-B.B. d'Anville. "Cartes de la Chine," in *Recueil de Mémoires Orientaux: Textes et Traductions Publiés par les Professeurs de l'École Spéciale des Langues Orientales Vivantes à l'Occasion du XVIe Congrès International des Orientalistes Réuni à Alger, Avril 1905*. Paris: E. Leroux, 1905, 392–400.

Du Halde, Jean-Baptiste, ed. *Lettres Édifiantes et Curieuses, Écrites des Missions Étrangères*. Nouvelle édition in 10 vols. Toulouse: Noel-Etienne Sens/Auguste Gaude, 1810–1811.

Du Retail, Boutillier. *Documentation sur Jean Baptiste Bourguignon D'Anville*. Paris: Charavay, 1880. BNF. banque DOSS < FOL–LN1–232 (600).

Duvivier, Benjamin. *Portrait de Jean-Baptiste Bourguignon D'Anville*. BNF–MFILM SG PORTRAIT–1382.

Ferreiro, J.A. Pinto, ed. *Correspondência de D. João V e D. Bárbara de Bragança, Rainha de Espanha* (1746–1747). Coimbra: Livraria Gonçalves [Deposit.], 1945.

François, P.J. *La Science de la Géographie*. Rennes: J. Hardy, 1652.

Frézier, Amédée. *Relation du Voyage de la Mer du Sud aux Côtes du Chili, du Pérou et de Brésil, Fait Pendant les Années 1712, 1713 1714. Ouvrage Enrichi de Quantité de Planches en Taille-Douce*. Paris: Pierre Humbert, 1717.

Green, John. *Remarks in Support of the New Chart of North and South America, in Six Sheets*. London: J. Jefferys, 1753.

Gibbon, Edward. *The History of the Decline and Fall of the Roman Empire* [*1776–1788*]. 3 vols. London: Penguin, 1976–1994.

Gumilla, Joseph (S.J., Le P.). *El Orinoco Ilustrado, Historia Natural, Civil y Geographica, de este Gran Rio, y de sus Caudalosas Vertientes: Govierno, Usos, y Costumbres de los Indios sus Habitadores, con Nuevas y Utiles Noticias de Animales, Arboles, Frutos, Aceytes, Resinas, Yervas, y Raìces Medicinales: Y Sobre Todo, se Hallaràn Conversiones Muy Singulares à Nuestra Fé, y Casos de Mucha Edification*. Madrid: Manuel Fernandez, 1741.

Gusmão, Alexandre de. "Dissertation qui Determine Tant Geografiquement que par les Traités Faits Entre la Courone de Portugal et Celle D'Espagne quels sont les Limites de Leurs Dominations en Amerique c'est a Dire du Coté de la Riviere de la Plate. (Com a Colaboração de Dom Luís da Cunha)," in *Alexandre de Gusmão e o Tratado de Madrid*, ed. Jaime Cortesão. Part 3, vol. 1. Rio de Janeiro, Ministério dos Negócios Exteriores, 1951, 367–417.

Guzmán, Ruy Dias de. *La Argentina*. Madrid: Historia 16, 1986.

Hassenfratz, Jean-Henri. *Encyclopédie Méthodique. Physique*. Vol. 2. Paris: Panckoucke, 1793.

Humboldt, Alexander Von. *Relation Historique du Voyage aux Régions Équinoxiales du Nouveau Continent fait en 1799, 1800, 1802, 1803 et 1804*. 3 vols. Stuttgart: F.A. Brockhaus, 1970.

La Condamine, Charles-Marie de. *Journal du Voyage Fait par Ordre du Roi a l'Equateur, Servant d'Introduction Historique a la Mesure des Trois Premiers Degrés du Méridien*. Paris: l'Imprimerie Royale, 1751.

La Condamine, Charles-Marie de. *Relation Abrégée d'un Voyage Fait dans l'Interieur de l'Amérique méridionale, Depuis la Côte de la Mer du Sud, Jusqu'aux Côtes du Brésil et de la Guiane*. Paris: Vve Pisot, 1745.

La Condamine, Charles-Marie de. *Viagem na América Meridional Descendo o Rio Amazonas* [1745]. Brasília: Editora do Senado Federal, 1984.

La Condamine, Charles-Marie de. *Voyage sur l'Amazone* [1745]. Paris: La Découverte, 2004. Introduction, notes et choix de textes par Hélène Minguet.

La Condamine, Charles-Marie de. [1745] *A succinct Abridgment of a Voyage Made Within the Inland Parts of South-America*. London: E. Whiters, 1747.

La Condamine, Charles-Marie de. *Abrégée d'un Voyage fait dans l'Intérieur de l'Amérique Méridionale, depuis la Côte de la Mer du Sud, Jusqu'aux Côtes du Brésil & de la Guyane*. Maestricht: Chez Jean-Edme Dufour & Philippe Roux, 1778.

Le Cabinet d'Un Grand Amateur, P.J. Mariette, 1694–1774. Paris: Musée du Louvre, 1967.

Le Nécrologe des Hommes Célèbres de France. 17 vols. Paris: l'Imprimerie de Moreau, 1767–1782.

Leme, Pedro Taques de A.P. *Notícias das Minas de São Paulo e dos Sertões da mesma Capitania*. Belo Horizonte/São Paulo: Itatiaia/EDUSP, 1980.

Les Commentaires de César. Paris: Veuve J. Camusat et P. Le Petit, 1650.

Longuerue, Louis Du Four (abbé). *Description Historique et Géographique de la France Ancienne et Moderne*. 2 parts in 1 vol. Paris: J.-H. Pralard, 1719.

Longuerue, Louis Du Four (abbé). *Description Historique et Géographique de la France Ancienne et Moderne, Enrichie de Plusieurs Cartes Géographiques*. 2 parts in 1 vol. Paris, 1722.

Lozano, P.P. *Description Chorographica del Terreno, Rios, Arboles y Animals de las Provincias del Gran Chaco, Gualamba y de los Ritos & de las Naciones Barbaras que le Habitan* (in 4°. na Biblioteca de Ribeiro Sanches, 83). Cordoba: El Colegio de La Assumpcion, 1733.

Luynes, Duc de. (Charles Philippe D'Albert). *Mémoires du Duc de Luynes, sur la Cour de Louis XV. (1735–1758)*. Vols. 1, 2, 4 and 8. Paris: Firmin Didot Frères, 1862.

Malte-Brun. *Précis de la Geographie Universelle ou Description de Tout Parties du Monde, d'Après les Grandes Divisions Naturelles du Globe*. Paris: Aimé André, 1836.

Mandroux-França, Marie-Thérèse and Maxime Préaud, eds. *Catalogues de la Collection d'Estampes de Jean V, Roi de Portugal, par Pierre-Jean Mariette*. 3 vols. Lisbonne/Paris: Fundação Calouste Gulbenkian/Bibliothèque Nationale de France/Fundação da Casa de Bragança, 2003.

Manne, Louis Charles Joseph de. *Catalogue des Livres de feu M. de Manne ... Suivi des Manuscrits, Lettres Autographes, et Autres Documents Provenant du Cabinet de M. D'Anville*. Paris: François, 1863.

Manne, Louis Charles Joseph de. *Œuvres de D'Anville*. Paris: F.-G. Levrault, 1834.

Manne, Louis Charles Joseph de, Jean-Denis Barbié du Bocage and Bon-Joseph Dacier. *Notice des Ouvrages de M. D'Anville Précédée de son Éloge, Premier Géographe du Roi, Membre de l'Académie des Inscriptions et Belles-Lettres ...* Paris: Imprimerie de Delance, 1802.

Marcel, M. Gabriel. "Lettres Inédites du Cardinal Passionei à Danville," *Bulletin de Géographie Historique et Descriptive* 3 (1904), 418–438.

Marcel, M. Gabriel. "Lettres Inédites de Michel Hennin à Danville," *Bulletin de Géographie Section de Geographie. Comité de Travaux Historiques et Scientifiques* 3 (1904), 446–448.

Mascart, Jean. "La Correspondance de D'Anville," *Ciel et Terre—Revue Populaire d'Astronomie, de Météorologie et de Physique du Globe* 33 (1912), 8–13.

Monteiro, Nuno Gonçalo, ed. *Meu Pai e meu Senhor Muito do Meu Coração: Correspondência do Conde de Assumar, Para seu Pai, o Marquês De Alorna*. Lisboa: ICS/Quetzal Editores, 2000.

Novinsky, Anita. *Inquisição, Rol dos Culpados: Fontes Para a História do Brasil, Século XVII*. Rio de Janeiro: Expressão e Cultura, 1992.

Oliveira, Francisco Xavier de. *Memórias das Viagens de Francisco Xavier de Oliveira.* Vol. 1. Amsterdam, 1741.

Ozanan, Didier. *La Diplomacia de Fernando VI: Correspondencia Reservada entre D. Jose de Carvajal y el Duque de Huescar, 1746–1749.* Madrid: Escuela de Historia Moderna, 1975.

Perrin, Antoine. *Almanach de la Librarie.* Paris: Aubel, [1781] 1984.

Pimentel, Manoel. *Arte de Navegar: Em que se Ensinam as Regras Práticas e os Modos de Cartear e Graduar a Balestrilha por Via de Números e Muitos Problemas Úteis à Navegação e Roteiro das Viagens e Costas Marítimas de Guiné, Angola, Brasil, Índias e Ilhas Ocidentais e Orientais.* Lisboa: Oficina de Francisco da Silva, 1746.

Piso, Willem, Georg Marggraf and Johannes Laet. *Historia naturalis Brasiliae. Auspicio et Beneficio Illustriss. I. Mauritii Com Nassau.* Leiden: Franciscum Hackium/ Amsterdam: Ludovicus Elzevirium, 1648.

Raleigh, Walter (Sir). *The Experience of the Large, Rich and Bewtiful Empire of Guiana, with a Relation of the Great and Golden Citie of Manoa (Which the Spaniards Call Eldorado)* [1596]. Manchester: Manchester University Press, 1997.

Rathery, E.J.B., ed. *Journal et Mémoires du Marquis D'Argenson.* 2 vols. Paris: Jules Renouard, 1859.

Rau, Virginia and Maria Fernanda Gomes da Silva. *Os Manuscritos do Arquivo da Casa de Cadaval Respeitantes ao Brasil.* 2 vols. Coimbra: Universidade de Coimbra, 1955–1958.

Ripauld, Louis-Madeleine. *Report of the Commission of Arts to the First Consul Bonaparte on the Antiquities of Upper Egypt and the Present State of All the Temples* ... London: J. Debrett, 1800.

Rojas, Alonso de. *Relação do Descobrimento do Rio das Amazonas.* São Paulo: Companhia Editora Nacional, 1941.

Rollin, Charles. *Abrégé de l'Histoire Ancienne, de Monsieur Rollin, par M. L'Abbé Tailhié, Prêtre. Nouvelle Edition, Revue, Corrigé e Augmentée par l'Auteur, avec Figures.* Paris: Barrois\Savoye, 1782.

Rollin, Charles. *Abrégé de l'Histoire Romaine de Rollin, ou l'on Suivi le Plan de cet Auteur. Par Ant. C**.* Paris: Brunot–Labbe, 1818.

Saint-Martin, Vivien de. *De l'État Actuel de la Cartographie en Europe, Extrait du Bulletin de la Société de Géographie.* Paris: Imprimerie de L. Martinet, 1855.

Saint-Martin, Vivien de. *Histoire de la Géographie et des Découvertes Géographiques Depuis les Temps les plus Reculés Jusqu'à Nous Jours.* Paris: Hachette, 1873.

Silva, Manoel Telles da. "Prática do Conde de Assumar, Depois da Declaração do Diretor em 18 de Março de 1721," in *História da Academia Real da História Portuguesa.* Lisboa Ocidental: Oficina de Joseph Antonio da Sylva, 1726.

"Sobre o Descobrimento dos Diamantes na Comarca do Serro do Frio. Primeiras Administrações," *Revista do Arquivo Público Mineiro* 8 (1902), 251–263.

Sousa, Antônio Caetano de *História Genealógica da Casa Real Portugueza: Desde a sua Origem até o Presente, com as Familias Illustres, que Procedem dos Reys, e dos Serenissimos Duques de Bragança.* Vol. VIII. Lisboa Occidental: na Officina de Joseph Antonio da Sylva, 1741.

Soulié, E. and M.M. Dussieux. *Mémoires du Duc de Luynes sur la Cour de Louis XV* (1735–1758). Paris: Firmin Didot Fréres, 1862.

Turpin de Crisse, M. le Cte. *Commentaires de César, Avec des Notes Historiques, Critiques et Militaires.* 3 vols. Amsterdam, Leide: Les Libraires Associés, 1787.

Vaugoundy, M. Robert de. *Essai Sur l'Histoire de la Géographie ou Sur son Origine, ses Progrès et son État Actuel.* Paris: Antoine Boudet, 1755.

Vaugoundy, M. Robert de. "Géographie," in *L'Encyclopedie ou Dictionnaire Raisonée des Sciences, des Arts et des Métiers*, ed. Denis Diderot & D'Alembert. Vol. 7. Paris: Briasson, 1757, 608–626.

Vaugoundy, M. Robert de. *Mémoire sur Une Question de Géographie Pratique* ... Paris: Antoine Boudet, 1775.

Vayrac, Jean de (abade). *Etat Présent de l'Espagne où l'on Voit Une Géographie Historique du Pays, l'Établissement de la Monarchie, ses Révolutions, sa Décadence, son Rétablissement et ses Accroissemens.* 3 vols. Amsterdam: Steenhouwer et Uytwerf, 1719.

Voltaire. *Correspondence and Related Documents*, ed. Theodore Besterman. Genève: Institut et Musée Voltaire/Toronto: Toronto University Press, 1971.

Secondary Literature

Abbattista, Guido. "Establishing the 'Order of Time and Place': 'Rational Geography', French Erudition and the Emplacement of History in Gibbon's Mind," *Studies on Voltaire Studies on Voltaire & the Eighteenth Century—Edward Gibbon: Bicentenary Essays.* Edited by David Womersley 355 (1997), 45–72.

Abou-el-Haj, Rifaat A. "The Formal Closure of the Ottoman Frontier in Europe: 1699–1703," *Journal of the American Oriental Society*, 89, 3 (July–September 1969), 467–475.

Adonias, Isa. *Mapa: Imagens da Formação Territorial Brasileira.* Rio de Janeiro: Fundação Emílio Odebrecht, 1993.

Adonias, Isa. *Mapas e Planos Manuscritos Relativos ao Brasil Colonial.* 2 vols. Rio de Janeiro: Ministério Relações Exteriores, 1960.

Aguillar, José. *Historia de la Cartografia: La Tierra de Papel.* Buenos Aires: Codex, 1968.

Albuquerque, Martim de. *Estudos de Cultura Portuguesa.* Vol. 1. Lisboa: Imprensa Nacional/Casa da Moeda, 1984.

Albuquerque, Martim de, Inácio Guerreiro, Feliciano Novoa Portela and Elena Postigos Castellanos, eds. *Encontros e Desencontros Ibéricos: Tratados Hispano-Portugueses desde a Idade Média*. Lisboa: Chaves Ferreira Publicações, 2006.

Alder, Ken. *A Medida de Todas as Coisas: A Odisséia de Sete Anos e o Erro Encoberto que Transformaram o Mundo*. Rio de Janeiro: Objetiva, 2003.

Aldridge, A.O. *The Ibero-American Enlightenment*. Urbana: University of Illinois Press, 1971.

Alegria, Maria Fernanda and João Carlos Garcia. "Aspectos da Evolução da Cartografia Portuguesa (Séculos XV–XIX)," in *Os Mapas em Portugal: Aos Novos Rumos da Cartografia*, Maria Helena Dias, ed. Lisboa: Cosmos, 1995, 27–84.

Almeida, André Ferrand de. *A Formação do Espaço Brasileiro e o Projecto do Novo Atlas da América Portuguesa*. Lisboa: Comissão Nacional para a Comemoração dos Descobrimentos Portugueses, 1991.

Almeida, André Ferrand de. "Os Jesuítas Matemáticos e os Mapas da América Portuguesa (1720–1748)," *Oceanos* 40 (1999), 79–92.

Almeida, André Ferrand de. "Samuel Fritz and the Mapping of the Amazon". *Imago Mundi* 55 (2003), 113–119.

Almeida, Luís Ferrand de. "A Autenticidade do Testamento Político de D. Luís da Cunha". *Anais da Academia Portuguesa de História* 17 (1968), 81–114.

Almeida, Luís Ferrand de. "A Autenticidade do Testamento Político de D. Luís da Cunha," in *Páginas Dispersas: Estudos de História Moderna de Portugal*. Coimbra: Instituto de História Econômica e Social/Faculdade de Letras da Universidade de Coimbra, 1995, 37–58.

Almeida, Luís Ferrand de. "A Colónia do Sacramento e a Formação do Sul do Brasil," in *Páginas Dispersas: Estudos de História Moderna de Portugal*. Coimbra: Instituto de História Econômica e Social/Faculdade de Letras da Universidade de Coimbra, 1995, 163–182.

Almeida, Luís Ferrand de. *A Colônia do Sacramento na Época da Sucessão de Espanha*. Coimbra: Instituto Histórico Dr. António de Vasconcelos, Faculdade de Letras da Universidade de Coimbra, 1957.

Almeida, Luís Ferrand de. "A Colônia do Sacramento nos Princípios do Século XVIII, uma Fonte Importante para seu Estudo," *Revista Portuguesa de História* XVI (1976), 333–341.

Almeida, Luís Ferrand de. "Dom João V e a Biblioteca Real," in *Páginas Dispersas: Estudos de História Moderna de Portugal*. Coimbra: Instituto de História Econômica e Social/Faculdade de Letras da Universidade de Coimbra, 1995, 209–228.

Almeida, Luís Ferrand de. "Memórias e Outros Escritos de D. Luís da Cunha," *Separata do Arquivo de Bibliografia Portuguesa* 7, n. 25–26 (1961), 1–13.

Almeida, Luís Ferrand de. "O Absolutismo de Dom João v," in *Páginas Dispersas: Estudos de História Moderna de Portugal.* Coimbra: Instituto de História Econômica e Social/Faculdade de Letras da Universidade de Coimbra, 1995, 183–207.

Almeida, Luís Ferrand de. "O Naturalista Merveilleux em Portugal," *Revista Portuguesa de História* 24 (1988), 273–292.

Almeida, Luís Ferrand de. *Alexandre de Gusmão, o Brasil e o Tratado de Madrid* (1735–1750). Coimbra: INIC/Universidade de Coimbra, 1990.

Amzalak, Moses Bensabat. *A Economia Política em Portugal. Dom Luis da Cunha, como Economista.* Lisboa, 1922.

Anastasia, Carla M.J. and Junia Ferreira Furtado. "A Estrada Real na História de Minas Gerais," *História e Perspectivas* 16–17 (1999), 36–37.

Andrade, António Alberto Banha. "Antecedentes da Travessia da África," *Anais da Academia Portuguesa da História* 2, n. 27 (1982), 323–354.

Andrade, Francisco Eduardo. *A Invenção das Minas Gerais: Empresas, Descobrimentos e Entradas nos Sertões do Ouro da América Portuguesa.* Belo Horizonte: Autênica/PUC–MG, 2008.

Anton, Nigel. "Edward Gibbon," in *Encyclopedia of the Enlightenment.* Vol. 2, Alan Charles Kors, ed. Oxford: Orford University Press, 2003.

Araújo, Ana Cristina. *A Cultura das Luzes em Portugal: Temas e Problemas.* Lisboa: Livros Horizontes, 2003.

Araújo, Ana Cristina. "Livros de uma Vida: Critérios e Modalidades de uma Livraria Particular no Século XVIII," *História das Idéias* 20 (1999), 149–185.

Archier, Edwige. "Anville, Jean-Baptiste Bourguignon d'," in Dörflinger, Kretschmer, and Wawrik, eds., *Lexikon zur Geschichte der Kartographie.* Wien: Franz Duticke, 1986, 18–21.

Azevedo, J. Lúcio. *Épocas de Portugal Econômico: Esboços de História.* Lisboa: Livraria Clássica Editora, 1973.

Bacou, Roseline. "Mariette: la Collection de Dessins," in *Le Cabinet D'un Grand Amateur, P.J. Mariette, 1694–1774.* Paris: Musée du Louvre, 1967, 17–23.

Bagrow, Leo and R.A. Skelton. *History of Cartography.* London: C.A. Watts & Co., 1964.

Baker, J.N.L. *A History of Geographical Discovery and Exploration.* London: G. Harrap, 1932.

Banha, António Rodrigues. *Vernei e a Cultura de seu Tempo.* Coimbra: Editora da Universidade de Coimbra, 1966.

Barber, Peter, ed. *The Map Book.* London: Weidenfeld & Nicolson, 2005.

Barber, Peter. *Monarchs, Ministers and Maps: The Emergence of Cartography as a Tool of Government in Early Modern Europe,* David Buisseret, ed. Chicago: University of Chicago Press, 1992.

Baumer, Franklin L. *O Pensamento Europeu Moderno: Séculos XVII e XVIII*. Vol. 1. Lisboa: Edições, 1990.

Bély, Lucien. *Espions et Ambassadeurs au Temps de Louis XIV*. Paris: Fayard, 1990.

Bernardino, Teresa. *Sociedades e Atitudes Mentais em Portugal (1777–1810)*. Lisboa: Imprensa Nacional/Casa da Moeda, 1986.

Black, Jeremy. *European International Relations, 1648–1815*. Basingstoke: Palgrave, 2002.

Black, Jeremy. *Italy and the Grand Tour*. New Haven/London: Yale University Press, 2003.

Black, Jeremy. *Mapas e História: Construindo Imagens do Passado*. Baur: Edusc, 2005.

Black, Jeremy. *Maps and Politics*. Chicago: The University of Chicago Press, 1997.

Black, Jeremy. *Visions of the World: A History of Maps*. Great Britan: Octopus Publishing Group, 2003.

Blais, Hélène and Isabelle Laboulais. *Géographies Plurielles*. Paris: L'Harmattan, 2006.

Blakemore, M.J. and J.B. Harley. "Concepts in the History of Cartography: A Review and Perspective," *Cartographica* 17, Monograph 26 (1980), 1–120.

Blond, Stéphane and Lucile Haguet. "Anville, Jean-Baptiste Bourguignon d'," in *Une Carrière de Géographe au Siècle des Lumières: Jean-Baptiste d'Anville*, Lucile Haguet and Catherine Hofmann, eds. Chicago: Chicago University Press, 2020, 111–115.

Bödeker, Hans Erich. "Academias," in *Diccionario Histórico de la Ilustración*, ed. Vicenzo Ferrone and Daniel Roche, 220–227. Madrid: Alianza Editorial, 1998.

Bodenstein, Wulf. "D'Anville's 1752 Map of India, His Commentary and English Translation," *Maps in History* 47 (September 2013), 19–25.

Bord, Jean-Paul and Pierre Robert Baduel. *Les Cartes de la Connaissance*. Paris/Tours: Karthala/Urbama, 2004.

Bord, Jean-Paul. "La Carte et la Construction des Savoir en Géographie et dans les Sciences Sociales," in *Les Cartes de la Connaissance*. Paris/Tours: Karthala/Urbama, 2004, 17–35.

Boureau, Alain. "La Norme Épistolaire, une Invention Médiévale," in *La Correspondance: Les Usages de la Lettre au XIX*ᵉ*. Siecle*, Roger Chartier, ed. Paris: Fayard, 1991, 127–159.

Borges, João Vieira. *Conquista de Madrid: Portugal faz Aclamar Rei de Espanha o Arquiduque Carlos de Habsburgo (1706)*. Lisboa: Tribuna, 2003 (Coleção Batalhas de Portugal).

Bourguet, Noëlle and Christian Licoppe. "Voyages, Measures et Instruments: Une Nouvelle Expérience du Monde au Siècle des Lumières," *Annales: Économies, Sociétés, Civilisations* 5 (1997), 1115–1151.

Bourguet, Noëlle, Christian Licoppe and H. Otto Sibum, eds. *Itineraries of Precision from the Seventeenth to the Twentieth Century*. London/New York: Routledge, 2002.

Boxer, Charles R. *A Idade de Ouro do Brasil*. Rio de Janeiro: Nova Fronteira, 2000.

Brandão, Fernando de Castro. *De D. João V a Dona Maria I, 1707–1799: Uma Cronologia*. Odivelas: Europress, 1994.

Brandão, Fernando de Castro. *História Diplomática de Portugal: Uma Cronologia*. Lisboa: Livros Horizonte, 2002.

Brigola, João Carlos. *Coleções, Gabinetes e Museus em Portugal no Século XVIII*. Lisboa: Fundação Calouste Gulbenkian, 2003.

Brigola, João Carlos. "Introdução," in *Colecionismo no Século XVIII—Textos e Documentos*. Vol—XXIV. Porto: Porto Editora, 2009.

Broc, Numa. *La Geographie des Philosophes: Géographes et Voyages Français au XVIIIe. Siècle*. Lille: University Paul Valéry de Montpellier, 1972.

Broc, Numa. "'Deux Géographes 'Revolutionnaires'—Philippe Buache et Jean Nicolas Buache de la Neuville," *Cartes, Cartographes et Géographes: Actes du 114ᵉ Congrès National des Sociétés Savants*. Paris: CTHS, 1990, 6–13.

Broc, Numa. *La Géographie de la Renaissance—1420–1620*. Paris: CTHS, 1986.

Broc, Numa. *La Géographie des Philosophes, Géographie et Voyagers Français au XVIIIᵉ*. Paris: Edition Ophrys, 1975.

Bourguet, Noëlle; Licoppe, Christian. "Voyages, Mesures et Instruments: Une Nouvelle Expérience du Monde au Siècle des Lumières," *Annales: Économies, Sociétés, Civilisations* 5 (1997), 1115–1151.

Bross, Mag. *America and the East: Early Geographies*. London: Mag Bross, 2008.

Brown, Lloyd A. *The Story of Maps*. Boston: Little, Brown and Company, 1949.

Bruckner, Martin. *The Geographic Revolution in Early America: Maps, Literacy and National Identity*. Virginia: Omohundro Institute of Early American History and Culture, 2006.

Bueno, Beatriz Piccolotto Siqueira. *Desenho e Desígnio: O Brasil dos Engenheiros Militares (1500–1822)*. São Paulo: EDUSP/FAPESP, 2011.

Bueno, Beatriz Piccolotto Siqueira. "O Engenheiro Artista: As Aquarelas e as Tintas nos Mapas do Novo Mundo," in *Sons, Formas, Cores e Movimentos na Modernidade Atlântica: Europa, Américas e África*, Junia Ferreira Furtado, ed. São Paulo: Annablume, 2008, 375–383.

Bueno, Beatriz Piccolotto Siqueira. "Decifrando Mapas: Sobre o Conceito de Território e suas Vinculações com a Cartografia," *Anais do Museu Paulista* 12 (jan/dez 2004), 193–236.

Buisseret, David, ed. *Monarchs, Ministers and Maps: The Emergence of Cartography as a Tool of Goverment in Early Modern Europe*. Chicago: Chicago University Press, 1992.

Bulletin de la Section de Géographie. Comité des Travaux Historiques et Scientifiques 41 (1926), 93–145.

Burnet, D. Graham. *Masters of All They Surveyed: Exploration, Geography, and a British El Dorado*. Chicago: University Of Chicago Press, 2001.

Cabria, Juan Vicente. *Cartografia Manuscrita de Brasil em las Colecciones Españolas* (1500–1822). Salamanca: Universidad de Salamanca/Centro de Estúdios Brasileños, 2008.

Caetano, Joaquim Oliveira, ed. *Gravura e Conhecimento do Mundo: O Livro Impresso Ilustrado nas Coleções da Biblioteca Nacional.* Lisboa: Biblioteca Nacional, 1998.

Camporesi, Piero. *Hedonismo e Exotismo: A Arte de Viver na Época das Luzes*. São Paulo: Unesp, 1996.

Cañizares-Esguerra, Jorge. *How to Write the History of the New World: Histories, Epistemologies, and Identities in the Eighteenth-Century Atlantic World.* Stanford: Stanford University Press, 2001.

Cardim, Pedro. "A Prática Diplomática em Europa do Antigo Regime," in *História e Relações Internacionais: Temas e Debates,* Luís Nuno Rodrigues and Fernando Martins, ed. Lisboa: Edições Colibri, 2004, 11–53.

Cardim, Pedro. "La Jurisdicción Real y su Afirmación en la Corona Portuguesa y Sus Territórios Ultramarinos (Siglo XVI–XVIII), Reflexiones sobre la Historiografia," in *De Re Publica Hispaniae: Uma Vindicación de la Cultura Política em los Reinos Ibéricos em la Primera Modernidade,* Francisco José Aranda Pérez e José Damião Rodrigues, ed. Madrid: Sílex, 2008, 349–388.

Cardoso, José Luís et al. *O Tratado de Methuen (1703), Diplomacia, Guerra, Política e Economia.* Lisboa: Livros Horizonte, 2003.

Cardoso, José Luís. *Pensar a Economia em Portugal: Digressões Históricas*. Lisboa: Difel, 1997.

Cardozo, Manoel. "The Internationalism of the Portuguese Enlightenment: The Role of the Estrangeirado," in *The Ibero-American Enlightenment,* A.O. Aldridge, ed. Urbana: University of Illinois Press, 1971, 153–160.

Carneiro, Ana, Ana Simões, and Maria Paula Diogo. "The *Estrangeirados* and their Communication Networks," *Social Studies of Science* 30, n. 4 (August 2000), 591–619.

Carvalho, Augusto da Silva. "Um Agente de Portugal em França, Francisco Mendes de Góis," *Anais da Academia Portuguesa da História* 2, 2 (1949), 211–240.

Carvalho, Rômulo. *Relações entre Portugal e a Rússia no séc. XVIII*. Lisboa: Sá da Costa, 1979.

Carvalho, Rômulo. *Relações Científicas do Astrônomo Francês Joseph-Nicolas de l'Isle com Portugal.* Coimbra: Atlântida Editora, 1967.

Carvalho, Rômulo. *A Astronomia em Portugal no Século XVIII*. Lisboa: ICLP, 1985.

Castro, Zília Osório de, ed. *Diplomatas e Diplomacia: Retratos, Cerimônias e Práticas.* Lisboa: Livros Horizontes, 2004.

Cavalcante, Paulo. *Negócios de Trapaça: Caminhos e Descaminhos na América Portuguesa (1700–1750)*. São Paulo: Hucitec, 2006.

Chartier, Roger. "The Man of Letters," in *Enlightenment Portraits*, Michel Volvelle, ed. Chicago: Chicago University Press, 1997, 142–89.

Chartier, Roger. "O Homem de Letras," *O Homem do Iluminismo*, ed. Michel Volvelle. Lisboa: Editorial Presença, 1997, 117–54.

Chartier, Roger. *A Ordem dos Livros: Leitores, Autores e Bibliotecas na Europa entre os Séculos XIV e XVIII*. Brasília: UNB, 1998.

Choiseul-Gouffier, Marie-Gabriel-Florent-Auguste. *Discours Préliminaire du Voyage Pittoresque de la Grèce*. Paris: Pierres, 1783.

Cintra, J.P. and J.C. Freitas. "Sailing down the Amazon River: La Condamine's Map," *Survey Review* 43, 323 (2011), 550–566.

Cluny, Isabel and Paulo J.S. Barata. "A Propósito de um Documento da Política Cultural Joanina," *Leituras—Revista da Biblioteca Nacional de Lisboa* 3, n. 3 (abr/out 1998), 129–140.

Cluny, Isabel. *D. Luís da Cunha e a Ideia de Diplomacia em Portugal*. Lisboa: Livros Horizontes, 1999.

Cluny, Isabel. "Os Diplomatas de Negociação e de Representação em Portugal no Século XVIII," in *Diplomatas e Diplomacia: Retratos, Cerimônias e Práticas*, Zília Osório de Castro, ed. Lisboa: Livros Horizontes, 2004, 53–68.

Cluny, Isabel. "A Correspondência do Conde de Tarouca," in *Correspondências: Usos da Carta no Século XVIII*, Nuno Gonçalo Monteiro, Teresa Sousa Almeida and Vanda Anastácio, eds. Lisboa: Colibri, 2005, 103–120.

Cluny, Isabel. *O Conde de Tarouca e a Diplomacia na Época Moderna*. Lisboa: Livros Horizonte, 2006.

Compère, Marie-Madeleine. *Les Collèges Français 16e–18e Siècle*, Répertoire 3. Paris: Institut National de Recherche Pédagogique, 2002.

Cortesão, Armando. *History of Portuguese Cartography, With Two Chapters on History of Astronomical Navigation by Luís de Albuquerque*. 2 vols. Coimbra: Junta de Investigações do Ultramar, 1971.

Cortesão, Jaime. *Alexandre de Gusmão e o Tratado de Madrid*. 10 vols. Rio de Janeiro: Instituto Rio Branco, 1950–1963.

Cortesão, Jaime. *Alexandre de Gusmão e o Tratado de Madrid*. 4 vols. Lisboa: Livros Horizonte, 1984.

Cortesão, Jaime. *História da Expansão Portuguesa*. Lisboa: Imprensa Nacional/Casa da Moeda, 1993.

Cortesão, Jaime. *História do Brasil nos Velhos Mapas*. 2 vols. Rio de Janeiro: Instituto Rio Branco, 1965/1971.

Cortesão, Jaime. *O Tratado de Madrid*. 2 vols. Brasília: Senado Federal, 2001.

Cosgrove, Denis, ed. *Mappings*. London: Reaktion Books, 1999.

Cosgrove, Denis, ed. *Apollo's Eye: A Cartographic Genealogy of the Earth in the Eastern Imagination*. Baltimore: Johns Hopkins, 2001.

Costa, Fernando Marques, Francisco Contente Domingues and Nuno Gonçalo Monteiro, eds. *Do Antigo Regime ao Liberalismo* (1750–1850). Lisboa: Veja, 1979.

Costa, Gilberto Antônio, Junia Ferreira Furtado, Friedrich E. Renger and Márcia Maria D. Santos. *Cartografia das Minas Gerais: Da Capitania à Província*. Belo Horizonte: Editora UFMG, 2002.

Costa, Gilberto Antônio, ed., Junia Ferreira Furtado, Friedrich E. Renger and Márcia Maria D. Santos. *Cartografia da Conquista das Minas*. Lisboa: Kappa/Belo Horizonte: Editora da UFMG, 2004.

Costa, Gilberto, ed. *Os Caminhos do Ouro e a Estrada Real*. Lisboa: Kappa/Belo Horizonte: Editora da UFMG, 2005.

Costa, Maria de Fátima. *História de Um País Inexistente: O Pantanal entre os Séculos XVI e XVIII*. São Paulo: Estação Liberdade/Kosmos, 1999.

Crone, C.R. *Maps and Their Makers: An Introduction to the History of Cartography*. 5th ed. Kent/England: Dawson/Archon Books, 1978.

Cunha, Norberto A.F.G. "A Ilustração Científica de D. Francisco Xavier de Meneses, 4°. Conde da Ericeira (1685–1720)," *Diacrítica*, Braga 1 (1989), 167–187.

Cunha, Norberto Ferreira da. *Elites e Acadêmicos na Cultura Portuguesa Setecentista*. Lisboa: Imprensa Nacional/Casa da Moeda, 2006.

Curto, Diogo Ramada. *O Discurso Político em Portugal (1600–1650)*. Lisboa: Projeto Universidade Aberta, 1988.

Curvelo, Alexandra. "O Poder dos Mapas," in *O Olhar do Viajante: Dos Navegadores aos Exploradores*, Fernando Cristóvão, ed. Coimbra: Almedina, 2003, 106–119.

Dainville, François de (S.J.). *Le Langage des Géographes: Termes, Signes, Couleurs des Cartes Anciennes, 1500–1800*. Paris: Éditions Picard, 1964.

Dainville, François de (S.J.). *L'Éducation des Jésuites (XVIe–XVIIIe siècles)*. Paris: Les Éditions de Minuit, 1978.

Dawson, Nelson Martin. *L'Atelier Delisle: L'Amérique du Nord sur la Table à Dessin*. Québec: Editions du Septentrion, 2000.

Delaforce, Angela. *Art and Patronage in Eighteenth-Century Portugal*. Cambridge: Harvard University Press, 2002.

Delano-Smith, Catherine. "Smoothed Lines and Empty Spaces: The Changing Face of the Exegetical Map Before 1600," in *Combler les Blancs de la Carte, Modalities et Enjeux de la Construction des Savoirs Géographiques (XVIIe–XXe)*. Strasbourg, Presses Universitaires de Strasbourg, 2004, 29–31.

De Vos, Paula. "Enlightenment—Spanish America," in *Iberia and the Americas: Culture, Politics and History*, Michael J. Francis ed. Santa Barbara: ABC–Clio, 2005, 450–453.

Delvaux, Marcelo Motta. *As Minas Imaginárias: o Maravilhoso Geográfico nas Representações Sobre o Sertão da América Portuguesa—Séculos XVI a XIX*. Master Dissertation, History. Belo Horizonte: Universidade Federal de Minas Gerais, 2009.

Dew, Nicholas. "Scientific Travel in the Atlantic World: the French expedition to Gorée and the Antilles, 1681–1683," *British Journal of the History of Science (BJHS)* 43, n. 1 (March 2010), 1–17.

Dias, Maria Helena. *Os Mapas em Portugal: Da Tradição aos Novos Rumos da Cartografia*. Lisboa: Cosmos, 1995.

Dias, Maria Odila Leite da Silva. "A Interiorização da Metrópole," in *A interiorização da Metrópole e Outros Estudos*. São Paulo: Alameda, 2005, 7–37.

Dion, Roger. *Les Frontières de la France*. Paris: Hachette, 1947.

Domingues, Francisco Contente and Luís Filipe Barreto. *A Abertura do Mundo: Estudos de História dos Descobrimentos Europeus*. 2 vols. Lisboa: Editorial Presença, 1986.

Du Bus, C. "Les Collections D'Anville à la Biblioteque Nationale," *Comité des Travaux Historiques et Scientifiques: Bulletin de la Section de Géographie* 41 (1926), 93–145.

Duarte, Paulo Araújo. *Fundamentos de Cartografia*. Florianópolis: Editora UFSC, 2002.

Ducène, Jean-Charles. "L'Utilisation des Sources Orientales par Jean-Baptiste d'Anville," in *Une Carrière de Géographe au Siècle des Lumières: Jean-Baptiste d'Anville*, Lucile Haguet and Catherine Hofmann, eds. Oxford/Paris: Voltaire Foundation/Bibliothèque Nationale de France, 2018, 179–202.

Duranton, Henri. "La Recherche Historique à l'Académie des Inscriptions: l'Exemple de l'Histoire de France," *Colloque Historique Franco-Allemand*. Bonn: L. Röhrscheid, 1976, 207–235.

Duzer, Chet Van. *Frames that Speak*. Leiden: Brill, 2023.

Dym, Jordana. "Mais Calculado para Enganar do que Para Informar: Os Viajantes e o Mapeamento da América Central (1821–1945)," *Varia Historia* 23, n. 37 (jan/jun 2007), 81–109.

Edney, Matthew H. "A História da Publicação do Mapa da América do Norte de John Mitchell de 1755," *Varia Historia* 23, n. 37 (jan.jun.2007), 31–52.

Edney, Mathew H. and Mary S. Pedley, eds. *The History of Cartography: Cartography in the European Enlightenment*. Vol. 4. Chicago: Chicago University Press, 2020.

Ehrenberg, Ralph E., ed. *Mapping the World: An Illustrated History of Cartography*. Washington: National Geographic, 2006.

Ehrenberg, Ralph E. and Seymour I. Schwartz. *The Mapping of America*. New Jersey: Wellfleet Press, 2001.

Estrelas de Papel: Livros de Astronomia dos Séculos XIV a XVIII. Lisboa: Biblioteca Nacional de Portugal, 2009.

Faria, Ana Leal. *Arquitetos da Paz: A Diplomacia Portuguesa de 1640 a 1815*. Lisboa: Tribuna, 2008.

Fernandes, Mário Gonçalves, ed. *Manoel Azevedo Fortes: Cartografia, Cultura e Urbanismo (1660–1749)*. Porto: GEDES, 2006.

Fernandes, Neusa. *A Inquisição em Minas Gerais*. Rio de Janeiro: Eduerj, 2000.

Ferreira, Mário Clemente. "Uma Idéia de Brasil num Mapa Inédito de 1746," *Oceanos* 43 (2000), 184–195.

Ferreira, Mário Clemente. *O Tratado de Madrid e o Brasil Meridional*. Lisboa: Comissão Nacional para as Comemorações dos Descobrimentos Portugueses, 2001.

Ferreira, Mário Clemente. "O Mapa das Cortes e o Tratado de Madrid: A Cartografia a Serviço da Diplomacia," *Varia Historia* 23, n. 37 (2007), 53–69.

Ferreira, Mário Clemente. "O Conhecimento da Área de Fronteira entre Mato Grosso e a América Espanhola no Século XVIII: A Procura de Informações Geográficas e Cartográficas por Portugueses e Castelhanos," in *Mapas de la Mitad del Mundo: La Cartografía y la Construcción Territorial de los Espacios Americanos Siglos XVI al XIX*, Francisco Roque de Oliveira and Héctor Mendoza Vargas, eds. Lisboa/México: Centro de Estudos Geográficos da Universidade de Lisboa/Universidade do México, 2010, 267–297.

Ferrone, Vicenzo and Daniel Roche, eds. *Diccionario Histórico de la Ilustración*. Madrid: Alianza editorial, 1998.

Findley, Paula. *Possessing Nature: Museus, Colleting, and Scientific Culture in Early Modern Italy*. Berkeley: University of California Press, 1996.

Fisher, H.E.S. *De Methuen a Pombal: O Comércio Anglo-português de 1700 a 1770*. Lisboa: Gradiva, 1984.

Flores, Kátia Maia. *Caminhos que Andam: o Rio Tocantins e a Navegação Fluvial dos Sertões*. Goiânia: Ed. UCG, 2009.

Fontaine, Joëlle and Arksan Simaan. *A Imagem do Mundo: Dos Babilônios a Newton*. São Paulo: Companhia das Letras, 2003.

Fordham, Sir H.G. *Some Notable Surveyors and Map-Makers of the Sixteenth, Seventeenth and Eighteenth Centuries and Their Work: A Study on History of Cartography*. Cambridge: Cambridge University Press, 1929.

Franzen, Beatriz Vasconcelos. "A Presença Ibérica no Extremo-Sul do Brasil (séculos XVII–XVIII)," *Mare Liberum* 4 (dez.1992), 255–259.

Freitas, Maria do Carmo Telles de and Susana Madalena Oliveira, eds. *Os Espaços de um Império*. 2 vols. Lisboa: Comissão Nacional para as Comemorações dos Descobrimentos Portugueses, 1999.

Frigo, Daniela. "Ambassadeurs et Diplomatie à l'Époque Moderne: Modèles et Pratiques entre l'Italie et l'Europe," in *Diplomatas e Diplomacia: Retratos, Cerimônias e Práticas*, Zília Osório de Castro, ed. Lisboa: Livros Horizontes, 2004, 29–52.

Fumaroli, Marc. *Diplomatie de l'Espirit. Colloque l'Europe des Traites de Westphalie*. Paris, 1968.

Fumaroli, Marc. *La Diplomatie de l'Esprit. De Montaigne à La Fontaine*. Paris: Hermann, 1994.

Furtado, Junia Ferreira. "Chuva de Estrelas na Terra: O Paraíso e a Busca dos Diamantes nas Minas Setencentistas," in *História e Meio Ambiente. O Impacto da Extensão Européia*. Funchal: Editora CEHA Centro de Estudos de História do Atlântico, 1999, 445–457 (Coleção Memórias 26).

Furtado, Junia Ferreira. *Homens de Negócio: A Interiorização da Metrópole e do Comércio nas Minas Setecentistas*. São Paulo: Hucitec, 1999.

Furtado, Junia Ferreira. *O Mundo Francês em Minas*. Belo Horizonte: BDMG Cultural, 2000.

Furtado, Junia Ferreira. "As Índias do Conhecimento ou a Geografia Imaginária da Conquista do Ouro". *Anais de História de Além-Mar* 4 (2003), 155–212.

Furtado, Junia Ferreira. "José Rodrigues Abreu e a Geografia Imaginária Emboaba da Conquista do Ouro," in *Modos de Governar: Idéias e Práticas Políticas no Império Português (séc. XVI a XIX)*, Maria Fernanda Bicalho and Vera Lúcia Amaral Ferlini, eds. São Paulo: Alameda, 2005, 277–295.

Furtado, Junia Ferreira. "Uma Correspondência de Negócios nas Minas Gerais Setecentistas: Possibilidades de Leitura," in *Cultura Letrada no Brasil: Objetos e Práticas*, Márcia Abreu and Nelson Schapochnik, eds. Campinas: Mercado de Letras, 2005, 135–157.

Furtado, Junia Ferreira. "Le Merveilleux, le Monstrueux et le Singulier dans la Cosmographie de la Terre D'Andre Thevet," in *Naturalia, Mirabilia & Monstruosa en los Imperios Ibéricos (siglos XVI–XIX)*, Johan Verberckmoes and Thomas Werner, eds. Leuven: University Press, 2006, 193–212 (Avisos de Flandes, 12).

Furtado, Junia Ferreira. "Um Cartógrafo Rebelde? José Joaquim da Rocha e a Cartografia de Minas Gerais," *Anais do Museu Paulista* 17, 2 (jul.–dez. 2009), 155–187.

Furtado, Junia Ferreira. "Collectionisme en Smaak. De Portugese Aaankopen van Boeken en Prenten in de Zuidelijke Nederlanden," in *Een Wereld op Papier. Zuid-Nederlandse Boeken, Prenten en Kaarten in Het Spaanse en Portugese Wereldrijk (16de–18de Eeuw)*, Werner Thomas and Eddy Stols, eds. Leuven: Acco, 2009, 411–425.

Furtado, Junia Ferreira. "O Oráculo que Sua Majestade foi Buscar: Dom Luís da Cunha e a Geopolítica do Novo Império Luso-Brasileiro," in *Na Trama das Redes: Política e Negócios no Império Português, Séculos XVI–XVIII*, João Fragoso and Maria de Fátima Gouvêa, eds. Rio de Janeiro: Ed. Civilização Brasileira, 2010, 373–400.

Furtado, Junia Ferreira. "Cartographic Independence," in *Mapping Latin America: A Cartographic Reader*, Karl Offen and Jordana Dym, eds. Chicago/London: Chicago University Press, 2011, 114–120.

Furtado, Junia Ferreira. "Guerra, Diplomacia e Mapas: a Guerra da Sucessão Espanhola e a América Portuguesa na Cartografia de D'Anville," *Topoi* 12, 23 (Jul.–Dec. 2011), 66–83.

Furtado, Junia Ferreira. *Oráculos da Geografia Iluminista: Dom Luís da Cunha e Jean Baptiste Bourguignon d'Anville na Construção da Cartografia do Brasil.* Belo Horizonte: Ed. UFMG, 2012.

Furtado, Junia Ferreira. "História da Engenharia," in *Engenharia: História em Construção,* Heloísa M. Murguel Starling and Lígia Beatriz de P. Germano, eds. Belo Horizonte: Ed. UFMG, 2012, 21–69.

Furtado, Junia Ferreira. *O Mapa que Inventou o Brasil.* São Paulo/Rio de Janeiro: Odebrecht/Versal Editores, 2013.

Furtado, Junia Ferreira. "Trajetórias Carto—Geográficas de uma Família de Cristãos—Novos dos Sertões das Gerais aos Cárceres da Inquisição: O Caso dos Irmãos Nunes," in *Travessias Inquisitoriais das Minas Gerais aos Cárceres do Santo Ofício: Diálogos e Trânsitos Religiosos no Império Luso-Brasileiro* (*Sécs. XVI–XVIII*), Junia F. Furtado and Maria Leônia Chaves de Resende, eds. Belo Horizonte: Fino Traço, 2013, 189–238.

Furtado, Junia Ferreira. "Bosque de Minerva: Artefatos Científicos no Colecionismo Joanino," in *Formas do Império: Ciência, Tecnologia e Política em Portugal e no Brasil. Séculos XVI ao XIX*, Heloisa Meireles Gesteira, Luís Miguel Carolino and Pedro Marinho, eds. Rio de Janeiro: Paz e Terra, 2014, 229–273.

Furtado, Junia Ferreira. "Dom João V e a Década de 1720: Novas Perspectivas na Ordenação do Espaço Mundial," in *O Brasil Colonial* (*1720–1821*). Vol. 3, João Luís Ribeiro Fragoso and Maria de Fátima Gouvêa, eds. Rio de Janeiro: Editora Civilização Brasileira, 2014, 61–110.

Furtado, Junia Ferreira. "Between Angola and Mozambique: The Portuguese Ambassador Dom Luís da Cunha, the French Geographer Jean-Baptiste Bourguignon D'Anville and the Discovery of the South Africa's Geography (XVIII century)," in *Век Просвещения/Le Siècle des Lumières* (La Géographie à L'Époque des Lumières: Entre Imaginaire et Réalité). Vol. V, Sergueï Karp, ed. Москва: Naouka, 2015, 168–185.

Furtado, Junia Ferreira. "El Embajador Portugués, el Cartógrafo Francés y la Biblioteca Real de Don João V na Construcción de la Cartografía de la América," in *Dibujar y Pintar el Mundo: Arte, Cartografía y Política*, Mauricio Nieto Olarte and Sebastián Díaz Ángel, eds. Bogotá: Uniandes, 2016, 131–136.

Furtado, Junia Ferreira. "Metamorphoses of Colonization: The Tocantins River and the Expansion to the West in Maps and Reports (18th century)," *Tempo* 22, n. 40 (May–Aug. 2016), 367–399.

Furtado, Junia Ferreira. "Evolving Ideas: J.B. D'Anville's Maps of Southern Africa, 1725/1749," *Imago Mundi—The International Journal for the History of Cartography*, 69 (2017), 202–215.

Furtado, Junia Ferreira. "Uma República de Letras Luso-Brasileira: a Academia Real da História Portuguesa," in *El Ocaso del Antiguo Régimen en los Impérios Ibéricos*, Scarlett O'Phelan Godoy and Margarita Eva Rodríguez García, eds. Lima/Lisboa: Fondo Editorial/Cham, 2017, 345–369.

Furtado, Junia Ferreira. "Elargir ses Réseaux, Diversifier ses Commandes: Les Travaux de D'Anville pour la Couronne Portugaise," in *Une Carrière de Géographe au Siècle des Lumières: Jean-Baptiste d'Anville*, Lucile Haguet and Catherine Hofmann, eds. Oxford/Paris: Voltaire Foundation/Bibliothèque Nationale de France, 2018, 53–83.

Furtado, Junia Ferreira. "Portuguese America Under Foreign Threat and the Creation of the Concept of *Uti Possidetis* in the First Half of the 18th Century," *Espacio, Tiempo y Forma. Série IV, Historia Moderna* 34 (2021), 45–74.

Furtado, Junia Ferreira. "Cartography in Dispute: The Frontiers of Brazil in Abbé Raynal's *Histoire des Deux Indes*," *Culture & History* 10, n. 2 (2021), 1–25.

Furtado, Junia Ferreira. *Quebra-Cabeça Africano: Como um Embaixador Português, um Geógrafo Francês, um Escritor Inglês e um Pirata Imaginário Transformaram a Cartografia da África no Século XVIII*. Belo Horizonte: Odisseia/Miguilim, 2021.

Furtado, Junia Ferreira. "A Expansão para o Oeste: a *Carte Manuscrite de la Navigation de la Rivière des Tocantins*, c. 1742-3," in *História do Brasil em 25 Mapas*, Junia F. Furtado, and Andréa Doré, eds. São Paulo: Companhia das Letras, 2022, 119–133.

Furtado, Junia Ferreira. "Da Conquista à Dicotomia Colônia—Metrópole: Uma Cartografia dos Vocábulos Políticos no Contexto da Independência," in *Uma Cartografia dos Brasis: Poderes, Disputas e Sociabilidades na Independência*, Furtado and Andréa Slemian, eds. Belo Horizonte: Fino Traço, 2022, 41–68.

Furtado, Junia Ferreira. "Portuguese Jesuits, the French Cartographer J.-B. B. d'Anville, and the Representation of the Sources of the Nile on European Maps (Fifteenth to Eighteenth Centuries)", forthcoming.

Furtado, Junia Ferreira. "O Tratado de Utrecht e a Construção da Cartografia como Instrumento da Diplomacia Portuguesa," in *Da Restauração às Revoluções Atlânticas: Estratégias Políticas e Práticas Culturais da Diplomacia na Época Moderna (sécs. XVII–XIX). Estudos Oferecidos a Ana Leal de Faria*, José Damião Rodrigues and Sónia Borges, eds. Lisboa: Editora do Centro de História da Universidade de Lisboa, forthcoming.

Furtado, Junia Ferreira. "Embates na Terra de Papel: As Longitudes e a Cartografia na Negociação das Fronteiras da América Portuguesa na Primeira Metade do Século XVIII," in *As Matemáticas e as Ciências Afins na Construção dos Territórios*, Heloisa Meireles Gesteira, Fernando Figueiredo, Luís Miguel Carolino and Thomás Haddad, eds. Coimbra: Editora da Universidade de Coimbra, forthcoming.

Furtado, Junia Ferreira and Gabriel Paquette. "Imperial Competition in the Eighteenth-Century Americas," in *The Iberian World*, Fernando Bouza, Pedro Cardim and Antonio Feros, eds. New York/London: Routledge, 2019, 519–540.

Furtado, Junia Ferreira and Jorge Pimentel Cintra. "Bourguignon D'Anville's *Carte de l'Amérique Méridionale*: A Comparative Amazonian Cartography in Perspective". *Revista Brasileira de História* 31, 62 (2011), 273–316.

Furtado, Junia Ferreira and Neil Safier. "O Sertão das Minas como Espaço Vivido: Luís da Cunha e Jean-Baptiste Bourguignon D'Anville na Construção da Cartografia Européia sobre o Brasil," in *Brasil-Portugal: Sociedades, Culturas e Formas de Governar no Mundo Português (Séculos XVI–XVIII)*, Eduardo França Paiva, ed. São Paulo: Editora Annablume, 2006, 263–277.

Furtado, Junia Ferreira and Neil Safier. "Indigenous Peoples and European Cartography," in *The History of Cartography: Cartography in the European Enlightenment*. Vol. 4, part 1, Mathew H. Edney and Mary S. Pedley, eds. Chicago: Chicago University Press, 2020, 663–672.

Gandía, Enrique de. *História Crítica de los Mitos y Leyendas de la Conquista Americana.* Buenos Aires: Centro Difusor del Libro, 1946.

Garcia, João Carlos, Joaquim Romero Magalhães and Jorge Manuel Flores, eds. *Lugares e Regiões em Mapas Antigos.* Lisboa: Comissão Nacional para a Comemoração dos Descobrimentos Portugueses, 1997.

Garcia, João Carlos, Joaquim Romero Magalhães and Jorge Manuel Flores, eds. *Cartografia e Diplomacia no Brasil do Século XVIII.* Lisboa: Comissão Nacional para a Comemoração dos Descobrimentos Portugueses, 1997.

Garcia, João Carlos, Joaquim Romero Magalhães and Jorge Manuel Flores, eds. *Tesouros da Cartografia Portuguesa.* Lisboa: Comissão Nacional para a Comemoração dos Descobrimentos Portugueses, 1997.

Garcia, João Carlos, Joaquim Romero Magalhães and Jorge Manuel Flores, eds. *O 2º. Visconde de Santarém e a História da Cartografia.* Lisboa: Biblioteca Nacional, 2006.

Godlewska, Anne Marie Claire. *Geography Unbound: French Geographic Science from Cassini to Humboldt.* Chicago: Chicago University Press, 1999.

Goffart, Walter. *Historical Atlases: The First Hundred Years, 1570–1870.* Chicago: Chicago University Press, 2003.

Goes Filho, Synesio Sampaio, ed. *Navegantes, Bandeirantes, Diplomatas: Um Ensaio sobre a Formação das Fronteiras do Brasil.* Brasília: FUNAG, 2015.

Gomes, Plínio Freire. "O Amazonas e o Prata na Mitogeografia da América," *Topoi* 3 (2001), 41–61.

Grell, Chantal. "De l'Antiquité aux Temps Modernes: l'Histoire Utile aux Princes de France, au Siècle des Lumières," in *Parallèle des Anciens et des Modernes: Réthorique, Histoire et Esthétique,* M.A. Bernier, ed. Québec: Presses de l'Université Laval, 2006, 81–106.

Gruzinski, Serge. *Les Quatre Parties du Monde*: *Histoire d'une Mondialisation.* Paris: La Matinière, 2004.

Guedes, Max Justo. "A Cartografia Holandesa do Brasil," in *O Brasil e os Holandeses (1630–1654),* Paulo Herkenhoff, ed. Rio de Janeiro: GMT Editores, 1999, 64–85.

Guedes, Max Justo. "Os Limites Territoriais do Brasil a Norte e Nordeste," in *Portugal no Mundo.* Vol. 5, Luis de Albuquerque, ed. Lisboa: Publicações Alfa, 1989, 202–228.

Gunn, Geoffrey C. *First Globalization: The Eurasian Exchange 1500–1800.* Lanham/Boulder/New York/Toronto/Oxford: Rowman Littlefied Publishers, 2003.

Habermas, J. *Mudança Estrutural da Esfera Pública.* Rio de Janeiro: Tempo Brasileiro, 1984.

Haguet, Lucile. *Aegyptus, de l'Égypte de l'Occident: Concept et Représentation de l'Égypte dans la Cartographie Occidentale du XV*e *au XVIII*e *Siècle.* PhD Dissertation. Paris: Université de Paris—IV Sorbonne, 2007.

Haguet, Lucile. "J.-B. d'Anville as Armchair Mapmaker: The Impact of Production Contexts on His Work," *Imago Mundi* 63, n.1 (2011), 88–105.

Haguet, Lucile. "Specifying Ignorance in Eighteenth-Century Cartography, a Powerful Way to Promote the Geographer's Work: The Example of Jean-Baptiste d'Anville," in *The Dark Side of Knowledge*, Cornel Zwierlein, ed. Leiden/Boston: Brill, 2016, 358–384.

Haguet, Lucile and Catherine Hofmann, eds. *Une Carrière de Géographe au Siècle des Lumières: Jean-Baptiste d'Anville.* Oxford/Paris: Voltaire Foundation/Bibliothèque Nationale de France, 2018 (Oxford University Studies in the Enlightenment).

Haguet, Lucile and Catherine Hofmann. "Introduction," in *Une Carrière de Géographe au Siècle des Lumières: Jean-Baptiste d'Anville*, Lucile Haguet and Catherine Hofmann, eds. Oxford/Paris: Voltaire Foundation/Bibliothèque Nationale de France, 2018, 113–116.

Haguet, Lucile. "La Bibliothèque Cartographique, Outil de Travail du Géographe de Cabinet," in *Une Carrière de Géographe au Siècle des Lumières: Jean-Baptiste d'Anville*, Lucile Haguet and Catherine Hofmann, eds. Oxford/Paris: Voltaire Foundation/Bibliothèque Nationale de France, 2018, 117–154.

Haguet, Lucile. "Splendeur et Décadence d'un 'Grand Homme': Réception et Postérité de d'Anville et Son Œuvre," in *Une Carrière de Géographe au Siècle des Lumières Jean-Baptiste d'Anville*, Lucile Haguet and Catherine Hofmann, eds. Oxford/Paris: Voltaire Foundation/Bibliothèque Nationale de France, 2018, 311–360.

Haguet and Stéphane Bond. "Anville, Jean-Baptiste Bourguignon d'", in *The History of Cartography: Cartography in the European Enlightenment.* Vol. 4, Mathew H. Edney and Mary S. Pedley, eds. Chicago: Chicago University Press, 2020, 111–115.

Hampson, Norman. *The Enlightenment: An Evaluation of its Assumptions, Attitudes and Values.* London: Penguin Books, 1990.

Harley, John Brian. "Maps, Knowledge and Power," in *The Iconography of Landscape: Essays on the Symbolic Representation, Design, and Use of Past Environments*, Denis Cosgrove and Stephen Daniels, eds. New York: Cambridge University Press, 1988, 277–312.

Harley, John Brian. *Conference on the History of Cartography: Early Maps as Historical Evidence.* London: International Conference on the History of Cartography, 1967.

Harley, John Brian. *Le Pouvoir des Cartes.* Paris: Éd. Anthropos, 1995.

Harley, John Brian. "Relire les Cartes de Redécouverte de Colomb," in *Le Pouvoir des Cartes.* Paris: Éd. Anthropos, 1995, 97–107.

Harley, John Brian. *The New Nature of Maps: Essays in the History of Cartography.* Baltimore/London: The Johns Hopkins University Press, 2001.

Harley, John Brian. "The Map and the Development of the History of Cartography," in *History of Cartography: Cartography in Prehistoric, Ancient, and Medieval Europe and Mediterranean.* Vol. 1, John Harley and David Woodward, eds. Chicago: Chicago University Press, 1987, 1–42.

Harley, John Brian and David Woodward, eds. *The History of Cartography*. 3 books, 6 vols. Chicago/London: The University of Chicago Press, 1987–2007.

Harris, C.A. and J.A.J. de Villiers. *Storm van's Gravesande, the Rise of British Guiana*. Vol 1. London: The Hakluyt Society, 1911.

Harris, Nathaniel. *Mapping the World: Maps and Their History*. San Diego: Thunder Bay Press, 2002.

Hazard, Paul. *O Pensamento Europeu no Século XVIII*. Lisboa: Editorial Presença, 1989.

Hélie, Jérôme. *Les Relations Internationales dans l'Europe Moderne. Conflits et Equilibres Européens—1453–1789*. Paris: Armand Colin, 2008.

Herkenhoff, Paulo, ed. *O Brasil e os Holandeses (1630–1654)*. Rio de Janeiro: GMT Editores, 1999.

Hespanha, António Manuel. *História de Portugal: O Antigo Regime*. Vol. 4, José Mattoso, ed. Lisboa: Editorial Estampa, 1993.

Heywood, Colin and Ivan Parvev, eds. *The Treaties of Carlowitz (1699): Antecedents, Course and Consequences*. Leiden: Brill, 2019.

Hodgkiss, Alan. *Understanding Maps: A Systematic History of Their Use and Development*. Folkstone: Dawson, 1981.

Hoff, Ulrich Im. *A Europa no Século das Luzes*. Lisboa: Editorial Presença, 1995.

Hofmann, Catherine. "La Genèse de l'Atlas Historique em France (1630–1800)," *Bibliothèque de l'École des Charts* 158 (2000), 97–128.

Hofmann, Catherine. "La 'Collection d'Anville' au Ministère des Affaires Étrangeres (1772–1828), Modalités et Enjeux D'une Approriation," in *Une Carrière de Géographe au Siècle des Lumières: Jean-Baptiste d'Anville*, Lucile Haguet and Catherine Hofmann, eds. Oxford/Paris: Voltaire Foundation/Bibliothèque Nationale de France, 2018, 361–396.

Hofmann, Catherine. "D'Anville et la Bibliothèque Royale/Nationale: Forces et Ambiguïtés d'un Héritage," in *Une Carrière de Géographe au Siècle des Lumières: Jean-Baptiste d'Anville*, ed. Lucile Haguet and Catherine Hofmann. Oxford/Paris: Voltaire Foundation/Bibliothèque Nationale de France, 2018, 397–431.

Holanda, Sérgio Buarque de. "Um Mito Geopolítico: A Ilha Brasil," in *Tentativas de Mitologia*. São Paulo: Prespectiva, 1979, 61–84.

Holanda, Sérgio Buarque de. *O Extremo Oeste*. São Paulo: Brasiliense, 1986.

Holanda, Sérgio Buarque de. *Monções*. 3rd ed. São Paulo: Brasiliense, 1990.

Holanda, Sérgio Buarque de. *Caminhos e Fronteiras*. 3rd ed. São Paulo: Companhia das Letras, 1994.

Holanda, Sérgio Buarque de. *Visão do Paraíso*. 6a ed. Brasiliense: São Paulo, 1994.

Hontanilla, Ana. *El Gusto de la Razón: Debates de Arte y Moral en el Siglo XVIII Español*. Madrid: Iberoamericana, 2010.

Horne, Herbert P. *The Binding of Books: An Essay in the History of Gold-Tooled Bindings*. New York: The Grolier Club, 1884.

Hsu, Mei-Ling. "The Role of Projections in Modern Map Design," *Cartographica* 27, 18, n. 2 (1981), 151–169.

Iby, Elfriede. "L'Éducation Artistique des Enfants Impériaux," in *Marie-Antoinette*, 54–69. Paris: Éditions de la Réunion des Musées Nationaux, 2008.

Jacob, Christian. *L'Empire des Cartes: Approche Théorique de la Cartografphie à Travers l'Histoire*. Paris: Albin Michel, 1992.

Jacob, Christian. *The Sovereign Map: Theoretical Approaches in Cartography Throughout History*. Chicago: Chicago University Press, 2007.

Jacob, Margaret C. *Living the Enlightenment*. Oxford: Oxford University Press, 1991.

James, Preston Everett. *All Possible Worlds: A History of Geographical Ideas*. New York: J. Wiley & Sons, 1981.

Kafker, Frank Arthur and Serena Kafker. "Anville, Jean-Baptiste Bourguignons D'," in *The Encyclopedists as Individuals: A Biographical Dictionary of the Authors of the "Encyclopédie."* Oxford: Voltaire Foundation, 1988, 11–12.

Kafker, Frank Arthur, ed. *The Encyclopedists as a Group: A Collective Biography of the Authors of the "Encyclopédie."* Oxford: Voltaire Foudation, 1996.

Kantor, Íris. *Esquecidos e Renascidos: Historiografia Acadêmica Luso-Americana (1724–1759)*. São Paulo: Hucitec, 2004.

Kantor, Íris. "Cartografia e Diplomacia: Usos Geopolíticos da Informação Toponímica (1750–1850)," *Anais do Museu Paulista* 17, n. 2 (jul.–dez. 2009), 39–61.

Kantor, Íris. "L'Appropriation des Cartes de d'Anville dans le Monde Luso-Brésilien: Mémoire Toponymique et Stratégie Diplomatique dans la Région Amazonienne, 1798 et 1904," in *Une Carrière de Géographe au Siècle des Lumières: Jean-Baptiste d'Anville*, Lucile Haguet and Catherine Hofmann, eds. Oxford/Paris: Voltaire Foundation/Bibliothèque Nationale de France, 2018, 287–303.

Keynes, Geoffrey, ed. *The Library of Edward Gibbon: A Catalogue of His Books*. London: The Bibliography Society, 1950.

King, Geoff. *Mapping Reality: An Exploration of Cultural Cartographies*. London: Macmillan, 1996.

Klinghoffer, Arthur Jay. *The Power of Projections: How Maps Reflect Global Politics and History*. Westport: Praeger, 2006.

Koeman, Cornelis. "Cartas Manuscritas do Estreito de Magalhães de Gerard Van Keulen (1711–1724)," in *A Abertura do Mundo: Estudos de História dos Descobrimentos Europeus*. Vol. 2, Francisco Contente Domingues and Luís Filipe Barreto, eds. Lisboa: Editorial Presença, 1987, 219–226.

Kok, Glória. *O Sertão Itinerante: Expedições da Capitania de São Paulo no Século XVIII*. São Paulo: Hucitec, 2004.

Konvitz, Josef. *Cartography in France, 1660–1848: Science, Engineering and Statecraft*. Chicago/London: University of Chicago Press, 1987.

Konvitz, Josef. "The Nation-State, Paris and Cartography in Eighteenth and Nineteenth-Century France," *Journal of Historical Geography* 16, n. 1 (1990), 3–16.

Koppel, Susanne. *Biblioteca Brasiliana da Robert Bosch GMBH*. Rio de Janeiro: Livraria Kosmos Editora, 1992.

Kriegel, Blandine. *Les Académies de l'Histoire*. Paris: Presses Universitaires de France, 1988.

Kuhn. Fábio. "Uma Fronteira do Império: O Sul da América Portuguesa na Primeira Metade do Século XVIII," *Anais do Além-Mar* 8 (2007), 103–121.

Laboulais, Isabelle. "La Géographie dans les Arbres Encyclopédique," in *Géographies Plurielles: Les Sciences Géographiques au Moment de l'Émergence des Sciences Humaines, 1750–1850*, Hélène Blais and Isabelle Laboulais, eds. Paris/Budapest/Kinshasa: l'Harmattan/DL2006/14–Condé-sur-Noireau/Impr. Corlet, 2006, 63–93.

Labourdette, Jean-François. "L'Ambassade de Monsieur de Chavigny à Lisbonne (1740–1743)," *Bulletin du Centre d'Histoire des Espaces Atlantiques* 1 (1983), 27–80.

Labourdette, Jean-François. *La Nation Française a Lisbonne de 1669 a 1790: Entre Colbertisme et Liberalisme*. Paris: Fondation Calouste Gulbenkian, 1988.

Lazzarotto, Danilo. *História do Rio Grande do Sul*. 7th ed. Ijuí: Editora Unijuí, 2001.

Le fait divers. Paris: Musée National des Arts et Traditions Populaires/Ministère de la Culture, 1982.

Le Gentil, Georges. *Découverte du Monde*. Paris: Presses Universitaires de France, 1954.

Le Gof, Jacques. "Documento/Monumento," in *Enciclоédia Einaudi. Memória—História*. Vol. 1. Lisboa: Imprensa Nacional/Casa da Moeda, 1997, 95–106.

Leclerc, Éric and Denis Retaillé. "Autorité ou Pouvoir des Cartes," in *Les Cartes de la Connaissance*, Jean-Paul Bord and Pierre Robert Baduel, eds. Paris/Tours: Urbama, 2004, 165–172.

Lefebvre, Henri. *La Production de l'Espace*. Paris: Anthropos, 1974.

Leoni, Sylviane and Réal Ouellet, eds. *Mythes et Géographies des Mers du Sud*. Dijon: Editions Universitaires de Dijon, 2006.

Lepape, Pierre. *Voltaire: Nascimento dos Intelectuais no Século das Luzes*. Rio de Janeiro: Zahar, 1995.

Lestringant, Frank. *L'Atelier du Cosmographe ou l'Image du Monde à la Renaissance*. Paris: Albin Michel, 1991.

Levasseur, E. *Le Brésil*. Paris: H. Lamirault et cie., 1889.

Licoppe, Christian. *La Formation de la Pratique Scientifique: Le Discurs de l'Experience en France et en Angleterre, 1630–1820*. Paris: Ed. La Découverte, 1996.

Lima, Sérgio Eduardo Moreira. "Apresentação," in *Navegantes, Bandeirantes, Diplomatas: Um Ensaio sobre a Formação das Fronteiras do Brasil*, Synesio Sampaio Goes Filho, ed. Brasília: FUNAG, 2015, 5–8.

Livingstone, David N. and Charles Withers. *Geography and Enlightenment*. Chicago: University of Chicago Press, 1999.

Lynch, John. *Bourbon Spain*. Oxford: B. Blackwell, 1989.
Macedo, Jorge Borges de. "Estrangeirados, um Conceito a Rever," *Revista da Faculdade de Letras de Lisboa* 3, n. 4 (1960), 79–97.
Macedo, Jorge Borges de. *História Diplomática Portuguesa: Constantes e Linhas de Força*. Vol. 1. Lisboa: Tribuna, 2006.
Maderuelo, Rafael Díaz, ed. *La Aventura del Amazonas*. Madrid: Historia 16, 2012.
Magalhães, Basílio de. *Expansão Geográfica do Brasil Colonial*. 4th ed. São Paulo: Compahia Editora Nacional, 1978.
Magalhães, Joaquim Romero, ed. *História de Portugal: No Alvorecer da Modernidade*. Vol. 3, José Matoso. Lisboa: Editorial Estampa, 1993.
Magalhães, José Calvet, eds. "A Acção Diplomática no Pensamento dos Diplomatas Portugueses dos séculos XVII e XVIII," in *A Diplomacia na História de Portugal: Actas do Colóquio*. Lisboa: Academia Portuguesa da História, 1990, 14–27.
Magnoli, Demétrio. *O Corpo da Pátria: Imaginação Geográfica e Política Externa no Brasil (1808–1912)*. São Paulo: Moderna/Edunesp, 1997.
Makham, Clements. *Major James Rennell and the Rise of Modern English Geography*. London: Cassell, 1895
Mandroux-França, Marie-Thérèse and Maxime Préaud, eds. *Catalogues de la Collection d'Estampes de Jean V, Roi de Portugal, par Pierre-Jean Mariette*. Lisbonne/Paris: Fundação Calouste Gulbenkian/Bibliothèque Nationale de France/Fundação da Casa de Bragança, 2003.
Mankin, Robert. "D'Anville, Gibbon et L'Espace des Empires: La Réception Britannique du Géographe Français à Traves l'Exemple de l'Historien Anglais," in *Une Carrière de Géographe au Siècle des Lumières: Jean-Baptiste d'Anville*. Lucile Haguet and Catherine Hofmann, eds. Oxford/Paris: Voltaire Foundation/Bibliothèque Nationale de France, 2018, 261–286.
Manne, J.C.J. de. *Oeuvres de D'Anville*. Vols. I–XVII. Paris, 1834.
Marino, John. "Administrative Mapping in the Italian States," in *Monarchs, Ministers and Maps: The Emergence of Cartography as a Tool of Goverment in Early Modern Europe*. David Buisseret, ed. Chicago: Chicago University Press, 1992, 5–25.
Marques, Alfredo Pinheiro. *A Cartografia dos Descobrimentos*. Lisboa: Elo, 1994.
Marques, Alfredo Pinheiro. "O Imperialismo Português no Brasil nos Séculos XVIII–XIX: O Papel dos Cartógrafos e dos Engenheiros Militares na Fixação dos Limites do Brasil," in *Portugal no Mundo*, Luis de Albuquerque, ed. Vol. 5. Lisboa: Publicações Alfa, 1989, 180–190.
Marques, Alfredo Pinheiro. *Origem e Desenvolvimento da Cartografia Portuguesa na Época dos Descobrimentos*. Lisboa: Imprensa Nacional/Casa da Moeda, 1987.
Marques, Miguel da Silva. *Cartografia Antiga: Tabela de Equivalências de Medidas. Cálculo de Escalas e Conversão de Valores de Coordenadas Geográficas*. Lisboa: Biblioteca Nacional, 2001.

Marques, Oliveira A.H. de and João José Alves Dias. *Atlas Histórico de Portugal e do Ultramar Português*. Lisboa: Centro de Estudos Históricos, 2003.

Martins Filho, Enéas. "Os Três Caminhos para as Minas Gerais," *Anais do Congresso Comemorativo do Bicentenário da Transferência da Sede do Governo do Brasil da Cidade do Salvador para o Rio de Janeiro* 1 (1963), 171–211.

Martins, Elze Vonk. *As Academias Literárias na Época Joanina*. Lisboa: Claro-Escuro, 1989.

Martins, Rocha. "Os Serviços e o Espírito de Alexandre de Gusmão," *Congresso do Mundo Português* 9 (1940), 129.

Mascart, Jean. "La Correspondance de D'Anville," *Ciel et Terre—Bulletin de la Societè Belge D'Astronomie, Bruxelles* 1 (1912), 1–7.

Masseau, Didier. *L'Invention de l'Intellectuel dans l'Europe du XVIIIe Siècle*. Paris: PUF, 1994.

Mauro, Frédéric. *O Império Luso-Brasileiro (1620–1750)*. Lisboa: Estampa, 1991.

Maxwell, Kenneth. *Marquês de Pombal: Paradoxo do Iluminismo*. São Paulo: Paz e Terra, 1996.

Mayhew, Robert John. *Landscape, Literature and English Religious Culture, 1660–1800*. Basingstoke/New York: Palgrave Macmillan, 2004.

Mayhew, Robert John. *Enlightenment Geography: The Political Languages of British Geography, 1650–1850*. Basingstoke: Macmillan, 2000.

Mayhew, Robert John. "On the Cusp of Modern Geography: Fieldwork and Textuality in the Career of James Rennell, 1764–1830," in *Enlightenment Geography: The Political Languages of British Geography, 1650–1850*. Basingstoke: Macmillan, 2000, 193–206.

Mayhew, Robert John. "Introduction," in *The Grand Tour: A Journey Through Netherlands, Germany, Italy and France*, Thomas Nugent, ed. Vol. 1. London: Ganesha, 2004, v–xxi.

Mello, Francisco Manuel de. *The Government of a Wife, or Wholsom and Pleasant Advice for Married Men*, translated by John Stevens. London: Printed for Jacob Tonson and R. Knaplock, 1697.

Mendes, H. Gabriel. "Subsídios para a História da Cartografia e Engenharia Portuguesas no Século XIX," *Revista do Instituto Geográfico Cadastral*, 1 (1981), 25–74.

Mendes, H. Gabriel. *Cartografia Portuguesa do Marquês de Pombal a Filipe Folque (1750–1900)*. Lisboa: Fundação Calouste Gulbenkiam, 1982.

Miceli, Paulo, ed. *O Tesouro dos Mapas: A Cartografia na Formação do Brasil*. São Paulo: Instituto Cultural Banco Santos, 2002.

Michaud, Louis-Gabriel. "Anville (Jean-Baptiste Bourguignon D')," in *Biographie Universalle Ancienne et Moderne: Histoire par ordre Alphabétique de la Vie Publique et Privée de Tous les Hommes* ... Vol. 2. Paris: C. Desplayes, 1858, 97–98.

Michaud, Louis-Gabriel. "D'Anville (Jean-Baptiste B.)," in *Biographie Universelle Ancienne et Moderne*. Vol. 2. Paris, 1811, 296–298.

Michaud, Louis-Gabriel. "Longuerue (Louis Defour, abbé de)," in *Biographie Universelle Ancienne et Moderne*. Vol. 25. Paris: C. Desplayes, 1858, 78–81.

Michaud, Louis-Gabriel. "Longuerue (Louis Defour, abbé de)," in *Biographie Universalle Ancienne et Moderne*. Vol. 25. Paris, 1858, 79–80.

Minguet, Charles. *Alexandre de Humboldt Historien et Géographe de l'Amerique Espagnole, 1799–1804*. Paris: L'Harmattan, 1997.

Minguet, Hélène. "Introduction," in *Voyage sur l'Amazone, de Charles-Marie de La Condamine*. Paris: F. Maspero, 1981, 5–27.

Miranda, João Manuel da Silva Alves. *As Relações Diplomáticas entre Portugal e a Rússia na Época de Pedro, O Grande (1700–1725)*. Lisboa: Fundação Calouste Gulbenkian, 2005.

Monmonier, Mark. *Drawing the Line: Tales of Maps and Cartocontroversy*. New York: Henry Holt and Co., 1995.

Monteiro, Nuno Gonçalo. "Introdução," in *Meu Pai e Meu Senhor Muito do Meu Coração: Correspondência do Conde de Assumar, para Seu Pai, o Marquês de Alorna*. Nuno Gonçalo Monteiro, ed. Lisboa: ICS/Quetzal Editores, 2000, 9–22.

Monteiro, Nuno Gonçalo, Teresa Sousa Almeida and Vanda Anastácio. *Correspondências: Usos da Carta no Século XVIII*. Lisboa: Colibri, 2005.

Monteiro, Nuno Gonçalo and Pedro Cardim. "La Diplomacia Portuguesa Durante el Antiguo Régimen. Perfil Sociológico y Trayectorias," *Cuadernos de Historia Moderna* 30 (2005), 7–40.

Monteiro, Rodrigo Bentes. *Uma Cidadela no Império: A Colônia do Sacramento e os Dilemas da Colonização no Sul da América Portuguesa. (1680–1750)*. Master Dissertation, History. Rio de Janeiro: Universidade Federal Fluminense, 2005.

Moravia, Sergio. "The Enlightenment and the Sciences of Man," *History of Science* 17 (1989), 247–268.

Mormiche, Pascale. *Devenir Prince: L'École du Povoir en France*. Paris: CNRS Éditions, 2015.

Mormiche, Pascale. "Participer à l'Éducation des Princes: D'Anville et son Élève Louis XV (1718–1730)," in *Une Carrière de Géographe au Siècle des Lumières: Jean-Baptiste d'Anville*, Lucile Haguet and Catherine Hofmann, eds. Oxford/Paris: Voltaire Foundation/Bibliothèque Nationale de France, 2018, 23–51.

Mota, Avelino Teixeira. "Dom Luis da Cunha e a Carta da África Meridional de Bourguignon D'Anville (1725)". *Separata da Revista Portuguesa de História* 10 (1962), 1–16.

Mota, Avelino Teixeira. "Os Portugueses e a Cartografia Antiga da África Central (1550–1800)," *Separata do Boletim da Academia das Ciências de Lisboa* 35 (1963), 13–27.

Mota, Avelino Teixeira. *A Cartografia Antiga da África Central e a Travessia entre Angola e Moçambique* (1500–1860). Lourenço Marques: Sociedade de Estudos de Moçambique, 1964.

Mota, Isabel Ferreira da. *A Academia Real da História: Os Intelectuais, o Poder Cultural e o Poder Monárquico no Século XVIII*. Coimbra: Minerva, 2003.

Mota, Isabel Ferreira da. "Os Historiadores, o Mecenato e o Clientelismo: Autonomia e Dependência (1700–1750)," *Revista de História das Ideias* 19 (1977), 471–493.

Novinsky, Anita. "Ser Marrano em Minas Colonial," *Revista Brasileira de História* 21, n. 40 (2001), 168.

Nowell, Charles E. *The Rose-Colored Map: Portugal's Attempt to Build an African Empire from the Atlantic to the Indian Ocean.* Lisboa: Junta de Investigações Científicas do Ultramar, 1982.

Oceanos—A Construção do Brasil Urbano 41 (janeiro/março 2000).

Oldroyd, David. "Geography," in *Encyclopedia of the Enlightenment*, Alan Charles Kors, ed. Vol. 2. Oxford: Orford University Press, 2003, 114–119.

Oliveira Martins. *História de Portugal.* 21th ed. Lisboa: Guimarães Editores, 2004.

Ozanan, Didier. *La Diplomacia de Fernando VI: Correspondencia Reservada entre D. Jose de Carvajal y el Duque de Huescar, 1746–1749.* Madrid: C.S.I.C., 1975.

Palma-Ferreira, João. *Academias Literárias dos Séculos XVII e XVIII.* Lisboa: Biblioteca Nacional de Portugal, 1982.

Palsky, Gilles. "Aux Origins de la Cartographie Thématique: Les Cartes Speciales Avant 1800," in *La Cartografia Francesa. 5è curs, Cicle de Conferènces sobre Historia de la Cartografia.* Catalunya: Institut Cartogràfic de Catalunya, 1992, 129–148.

Pedley, Mary Sponberg. *Bel et Utile: The Work of the Robert de Vaugondy Family of Mapmakers.* Great Britain: Map Collector Publications, 1992.

Pedley, Mary Sponberg. "New Light on an Old Atlas: Documents Concerning the Publication of the 'Atlas Universal' (1757)," *Imago Mundi* 36 (1984), 48–62.

Pedley, Mary Sponberg. "O Comércio de Mapas na França e na Grã Bretanha durante o Século XVIII," *Varia Historia* 23, n. 37 (jan./jun.2007), 14–30.

Pedley, Mary Sponberg. *The Commerce of Cartography: Making and Marketing Maps in Eighteenth-Century France and England.* Chicago: The University of Chicago Press, 2005.

Pedreira, Jorge M. "Brasil, Fronteira de Portugal. Negócio, Emigração e Mobilidade Social (Séculos XVII e XVIII)," in *Do Brasil à Metrópole: Efeitos Sociais (Séculos XVII–XVIII)*, Mafalda Soares da Cunha et al., eds. Évora: Universidade de Évora, 2001, 47–72.

Pelletier, Monique and Henriette Ozanne. *Portraits de la France: Les Cartes, Témoins de l'Histoire.* Paris: Hachette/BNF, 1995.

Pelletier, Monique. *Cartographie de la France et du Monde de la Renaissance au Siècle des Lumiéres.* Paris: Bibliotèque Nationale de France, 2001.

Pelletier, Monique. "Des Cartes Pour Communiquer: De la Localisation des Etapes, a la Figuration du Parcours 17e–18e Siècles," in *La Cartografia Francesa. 5è curs, Cicle de Conferènces sobre Historia de la Cartografia.* Catalunya: Institut Cartogràfic de Catalunya, 1992, 33–48.

Pelletier, Monique. “Des Cartes Pour Partager: Divisions Administratives, Frontières, Plans Terriers du 16e au 18e Siècle,” in *La Cartografia Francesa. 5è curs, Cicle de Conferènces sobre Historia de la Cartografia.* Catalunya: Institut Cartogràfic de Catalunya, 1992, 17–31.

Pelletier, Monique. “Le Monde vu de France: Évolution du 16e au 18e Siècle,” in *La Cartografia Francesa. 5è curs, Cicle de Conferènces sobre Historia de la Cartografia.* Catalunya: Institut Cartogràfic de Catalunya, 1992, 49–58.

Pelletier, Monique. *Les Cartes des Cassini: La Science au Service de l'État et des Régions.* Paris: Éditions du CTHS, 2002.

Pelletier, Monique. “Philippe Buache,” in *The History of Cartography: Cartography in the European Enlightenment.* Vol. 4, Edney, Mathew H. and Mary S. Pedley, eds. Chicago: Chicago University Press, 2020, 228–230.

Pereira, José Esteves. *O Pensamento Político em Portugal no Século XVIII.* Lisboa: Imprensa Nacional/Casa da Moeda, 1983.

Pérez, José Manuel. “Introdução: A Formação Territorial do Brasil Através dos Mapas Manuscritos nos Arquivos Espanhóis: Reflexo de Três Séculos de História Compartida,” in *Cartografia Manuscrita de Brasil em las Colecciones Españolas (1500–1822),* Juan Vicente Cabria, ed. Salamanca: Universidad de Salamanca/Centro de Estúdios Brasileños, 2008, 55–59.

Perla, Georges A. “La Géographie dans l'Encyclopédie,” *Revue de Synthèse* 115 (Juillet–Septembre 1984), 298–310.

Piazza, Walter F. “Um ‘Iluminado’ Oitocenteista: José da Silva Paes,” *Boletim do Arquivo Histórico Militar* 51 (1982), 201–223.

Piazza, Walter F. *O Brigadeiro José da Silva Paes, Estruturador do Brasil Meridional.* Florianópolis: Editora da UFSC, 1988.

Poindron, P. “Les Cartes Géographiques du Ministère des Affaires Étrangères (1780–1789), Jean-Denis Barbié du Bocage et la Collection D'Anville,” *Sources, Etudes, Recherches, Informations des Bibliothèques Nationales de France* 1, 1 (1943), 46–72.

Possamai, Paulo. *A Vida Cotidiana na Colónia do Sacramento: Um Bastião Português em Terras do Futuro Uruguai.* Lisboa: Bertrand, 2006.

Pounds, N.J.G. “The Origin of the Idea of Natural Frontiers in France.” *Annals, Association of American Geographers* 44, n. 1 (1951), 51–62.

Prado, Fabrício. *Colônia do Sacramento: O Extremo Sul da América Portuguesa.* Porto Alegre: F.P. Prado, 2002.

Pratt, Mary Louise. *Os Olhos do Império: Relatos de Viagem e Transculturação.* Bauru: Edusc, 1999.

Ract, Patrice. “La Guerre et les Affaires Étrangères: Réseaux de Sociabilité dans les Départements Ministériels (Seconde Moitié du XVIIIe Siècle),” *Le Monde des Cartes* 187 (mars 2006), 26–32.

Ramos, Luis A. de Oliveira. *Sob o Signo das 'Luzes.'* Lisboa: Imprensa Nacional/Casa da Moeda, 1988.

Rêgo, João de Figueirôa. *Reflexos de um Poder Discreto*. Lisboa: CHAM, 2008.

Reinhartz, Dennis. *The Cartographer and the Literati: Herman Moll and His Intellectual Circle*. Lewiston/NY: Queenston, 1997.

Renger, Friedrich E. "Primórdios da Cartografia das Minas Gerais (1585–1735), Dos Mitos aos Fatos," in *História de Minas Gerais: As Minas Setecentistas*. Vol. 1, Maria Efigênia Lage de Resende and Luiz Carlos Villalta, eds. Belo Horizonte: Autêntica, 2007, 103–126.

Reynaud, Denis. "Le Temps de l'Information dans la Presse Politique au XVIIIe Siècle: Grandeur et Décadence du Modèle Gazette," *Cadernos de Cultura—Gazetas: a Informação nos Finais do Antigo Regime*, João Luís Lisboa, ed., 4 (2001), 13–29.

Richard, Hélène. *Les Globes de Coronelli*. Paris: Bibliothèque Nationale de France, 2006.

Rio-Branco, José Maria da Silva Paranhos Júnior (Barão do). *Mémoire sur la Question des Limites entre les États-Unis du Brésil et la Guyane Britannique*. Rio de Janeiro: Ministério das Relações Exteriores 1897.

Rio-Branco, José Maria da Silva Paranhos Júnior (Barão do). *Questão de limites*. Rio de Janeiro: Ministério das Relações Exteriores, 1945–1947.

Rio-Branco, José Maria da Silva Paranhos Júnior (Barão do). *Memóire sur la Question des Limites entre les Etats-Unis du Brésil et la Guyane Britanique*. Rio de Janeiro: Ministério das Relações Exteriores, sd.

Ricupero, Rubens. *A Diplomacia na Construção do Brasil, 1750–2016*. Rio de Janeiro: Versal Editores, 2016.

Robinson, Arthur H. *Elements of Cartography*. 3rd ed. New York, London, Sydney, J. Wiley and sons, 1969.

Robinson, Arthur H. and Barbara Bartz Petchenick, eds. *The Nature of Maps: Essays Toward Understanding Maps and Mapping*. Chicago: Chicago University Press, 1976.

Rodrigues, Ana Maria, ed. *A Construção do Brasil*. Lisboa: Catálogo de Exposição no Palácio da Ajuda, 2000.

Rodrigues, Eugénia. "D. Luís da Cunha e Jean-Baptiste Bourguignon D'Anville: Das Representações da África Austral aos Projetos de Reconfiguração do Império Português," in *Actas do III Encontro Internacional de História de Angola*. Luanda: Arquivo Nacional de Angola, 2007, 1–31.

Rodrigues, Luís Nuno and Fernando Martins, eds. *História e Relações Internacionais: Temas e Debates*. Lisboa: Edições Colibri, 2004.

Rodrigues, Manuel Augusto. *A Biblioteca de António Nunes Ribeiro Sanches*. Coimbra: Publicações do Arquivo da Universidade de Coimbra, 1986.

Romeiro, Adriana. *Um Visionário na Corte de Dom João V*. Belo Horizonte: Ed. UFMG, 2001.

Rovira, Maria del Carmen. *Ecléticos Portugueses del Siglo XVIII, y Algunas de sus Influencias en América.* México: Universidade Autónoma de México, 1979.

Safier, Neil. "Como era Ardiloso o Meu Francês: Charles Marie de la Condamine e a Amazônia das Luzes," *Revista Brasileira de História* 29, n. 57 (jan.–jun. 2009), 91–114.

Safier, Neil. *Measuring the New World: Enlightenment Science and South America.* Chicago: Chicago University Press, 2008.

Saldanha, Nun. *Poéticas da Imagem.* Lisboa: Caminho Editorial, 1995.

Sanchez, Jean-Pierre. *Mythes et Légendes de la Conquête de l'Amérique.* Rennes: Presses Universitaires de Rennes, 1996.

Santarem, Visconde de. *Quadro Elementar das Relações Políticas e Diplomáticas de Portugal.* Lisboa: Academia Real das Sciencias de Lisboa, 1842.

Santos, Douglas. *A Reinvenção do Espaço: Diálogos em Torno da Construção do Significado de uma Categoria.* São Paulo: Editora UNESP, 2002.

Santos, Maria Emília Madeira. *Viagens de Exploração Terrestre dos Portugueses em África.* Lisboa: Centro de Estudos de História e Cartografia Antiga, 1978.

Schwartz, Seymour I. *The Mismapping of America.* Rochester/New York: The University of Rochester Press, 2003.

Seed, Patrícia. "Cerimônias: Os Teatrais Ritos Franceses de Tomada de Posse Política," in *Cerimônias de Posse na Conquista Européia do Novo Mundo (1492–1640).* São Paulo: Unesp, 1999, 63–99.

Sekolec, Jernej. "Orbis Lusus: A Forgotten 17th Century Geographical Game," in *Maps, Myths and Narratives. Program and Paper Abstracts of 23nd International Conference on the History of Cartography.* Copenhagen: The Royal Library, 2009, 26.

Serrão, Joaquim Veríssimo. *História de Portugal (1640–1750), A Restauração e a Monarquia Absoluta.* Vol. 5. 2nd ed. Lisboa: Verbo, 1982.

Serrão, Joaquim Veríssimo. *História de Portugal (1750–1807), O Despotismo Iluminado.* Vol. 6. 5th ed. Lisboa: Verbo, 1996.

Seruya, Ana Isabel, ed. *Globos Coronelli.* Lisboa: Instituto Português de Conservação e Restauro, 2004.

Shank, J.B. *The Newton Wars & the Beginning of the French Enlightenment.* Chicago: The University of Chicago Press, 2008.

Shapin, Steven and Simon Schaffer. *Leviathan and the Air-Pump: Hobbes, Boyle, and the Experimental Life.* Princeton: Princeton University Press, 1985.

Silbert, Albert. *Do Portugal de Antigo Regime ao Portugal Oitocentista.* Lisboa: Livros Horizonte, 1972.

Silva, Abílio Diniz. "D. Luís da Cunha e o Brasil," in *De Cabral a Pedro I: Aspectos da Colonização Portuguesa no Brasil,* Maria Beatriz Nizza da Silva, ed. Porto: Universidade Portucalense, 2001, 261–276.

Silva, Abílio Diniz. "Introdução," in *Instruções Políticas de D. Luís da Cunha*, Abílio Diniz Silva, ed. Lisboa: Comissão Nacional para as Comemorações dos Descobrimentos Portugueses, 2001, 9–117.

Silva, Cândido Marciano da. "D. João v Patrono do Astrônomo Bianchini," in *Estrelas de Papel: Livros de Astronomia dos Séculos xiv a xviii*. Lisboa: Biblioteca Nacional de Portugal, 2009, 49–65.

Silva, Júlio Joaquim da Costa Rodrigues da. *Ideário Político de uma Elite de Estado: Corpo Diplomático (1777/1793)*. Lisboa: Fundação Calouste Gulbenkian, 2002.

Skelton, Raleigh Ashlin. *Maps: A Historical Survey if Their Study and Collecting*. Chicago: Chicago University Press, 1972.

Skelton, Raleigh Ashlin. *Explorers' Maps: Chapters in the Cartographic Record of Geographical Discovery*. New York: Frederick A. Praeger, 1958.

Skelton, Raleigh Ashlin. *Looking at an Early Map*. Lawrence: University of Kansas libraries, 1965.

Snyder, John P. *Flattening the Earth: Two Thousand Years of Map Projections*. Chicago: Chicago University Press, 1993.

Sobel, Dava. *Longitude: The True Story of a Lone Genius Who Solved the Greatest Scientific Problem of His Time*. New York: Pinguim Books, 1995.

Sørensen, Madeleine Pinault. "Les Voyagers Artistes en Amérique du Sud au xviiie Siècle," in *Les Naturalistes Français en Amérique du Sud, xvi–xix Siècles*. Yves Laissus, ed. Paris: Comité des Travaux Historiques et Scientifiques, 1995, 43–55.

Souza, Laura de Mello e. "D. Pedro de Almeida Portugal e a Revolta de 1720 em Vila Rica," in *Cultura Portuguesa na Terra de Santa* Cruz, Maria Beatriz Nizza da Silva, ed. Lisboa: Editorial Estampa, 1995, 193–213.

Souza, Laura de Mello e. "Teoria e Prática do Governo Colonial: D. Pedro de Almeida, Conde de Assumar," in *O Sol e a Sombra: Política e Administração na América Portuguesa do Século xviii*. São Paulo: Companhia das Letras, 2006, 185–252.

Souza, Néri de Almeida. "Peregrinação, Conquista e Povoamento. Mito e 'Realismo Desencantado' Numa Hagiografia Medieval Portuguesa," *Revista Brasileira de História* 21, n. 40 (2001), 70.

Taunay, Affonso de E. *História das Bandeiras Paulistas*. 2 vols. São Paulo: Edições Melhoramentos, sd.

Tengarrinha, José, ed. *História de Portugal*. Bauru/São Paulo: EDUSC/Editora UNESP, 2001.

Thrower, Norman J.W. *Maps & Civilization: Cartography in Culture and Society*. 2nd ed. Chicago: Chicago University Press, 1999.

Thuillier, Jacques. "Introduction," in *Catalogues de la Collection D'Estampes de Jean v, Roi de Portugal, par Pierre-Jean Mariette*, Marie-Thérèse Mandroux-França and Maxime Préaud, eds. Vol. 1. Lisbonne/Paris: Fundação Calouste Gulbenkian/ Bibliothèque Nationale de France/Fundação da Casa de Bragança, 2003, 19–41.

Thomas, Werner and Eddy Stols, eds. *Un Mundo Sobre Papel. Libros, Grabados y Mapas de Flandres en los Impérios Español y Portugués (siglos XVI–XVIII)*. Lovaina: Acco, 2009.

Todorov, Tzvetan. *A Conquista da América: A Questão do Outro*. São Paulo: Martins Fontes, 1983.

Tolias, Georges. "Une Science de Pure Érudition: La Géographie Critique et Comparée selon Jean-Baptiste D'Anville," in *Une Carrière de Géographe au Siècle des Lumières: Jean-Baptiste d'Anville*, Lucile Haguet and Catherine Hofmann, eds. Oxford/Paris: Voltaire Foundation/Bibliothèque Nationale de France, 2018, 155–178.

Tooley, Ronald Vere. "Anville, Jean Baptiste Bourguignon D'," in *A Dictionary of Mapmakers*. Vol. 1. London: Map Collectors' Circle, 1965, 36–39.

Tooley, Ronald Vere. "Delahaye [De La Haye]," in *A Dictionary of Mapmakers*. Vol. 1. London: Map Collectors' Circle, 1965, 351.

Tooley, Ronald Vere. "Gravelot, [Bourguignon], Hubert-François," in *A Dictionary of Mapmakers*. Vol. 2. London: Map Collectors' Circle, 1965, 204–205.

Török, Zsolt, ed. *Sacred Places on Maps*. Budapest: Pannonhalma, 2005.

Török, Zsolt. "Karlowitz, Treaty of (1699)," in *Encyclopedia of the Ottoman Empire*, Gábor Ágoston and Bruce Masters, eds. Chicago: Chicago University Press, 2008.

Turner, Gerard L'E. *Scientific Instruments* (1500–1900). Berkeley: University of California Press, 1990.

Vedia, Enrique, ed. *Historiadores Primitivos de Indias*. Madrid: Atlas, 1946–1947.

Veiga, Raul. *Catálogo de Documentos do Cartório de dom Luis da Cunha (1709–1749)*. Coimbra: INIC, 1991.

Ventura, Maria das Graças Mateus and Luís Jorge Semedo de Matos. *As Novidades do Mundo: Conhecimento e Representação na Época Moderna*. Lisboa: Edições Colibri, 2003.

Verdier, Nicolas. "Entre Publicité, Débat Scientifique et Vulgarisation: Jean-Baptiste d'Anville dans les Journaux," in *Une Carrière de Géographe au Siècle des Lumières: Jean-Baptiste d'Anville*, Lucile Haguet and Catherine Hofmann, eds. Oxford/Paris: Voltaire Foundation/Bibliothèque Nationale de France, 2018, 237–259.

Verdier, Nicolas and Jean-Marc Besse. "Art and Dsign of Maps," in *The History of Cartography: Cartography in the European Enlightenment*. Vol. 4, Edney, Mathew H. and Mary S. Pedley, eds. Chicago: Chicago University Press, 2020, 117–125.

Vivien de Saint-Martin, Louis. *Histoire de la Géographie et des Découvertes Géographiques Depuis les Temps les Plus Reculés Jusqu'à nos Jours*. Paris: Hachette, 1873.

Voss, Jürgen. "Das Elsaß als Mittler Zwischen Deutscher und Französischer Geschichtswissenschaft im 18. Jahrhundert," *Pariser Historische Studien*, 13 (1976), 334–350.

Washburn, Wilcomb E. "As Ilhas Canárias e a Questão do Meridiano de Referência: A Busca do Rigor na Medição da Terra," in *A Abertura do Mundo: Estudos de História dos Descobrimentos Europeus*, Francisco Contente Domingues and Luís Filipe Barreto, ed. Vol. 1. Lisboa, Editorial Presença, 1986, 213–220.

Wilford, John Noble. *The Mapmakers*. New York: Vintage, 2000.

Wilken, Patrick. *O Império a Deriva: A Corte Portuguesa no Rio de Janeiro*. Rio de Janeiro: Objetiva, 2005.

Wills Jr, John E. *1688: O Início da Era Moderna*. Rio de Janeiro: Campus, 2001.

Withers, Charles W.J. and David N. Livingstone, eds. *Geography and Enlightenment*. Chicago: Chicago University Press, 1999.

Withers, Charles W.J. and David N. Livingstone, eds. "Introduction: On Geography and Enlightenment," in *Geography and Enlightenment*, Withers, Charles W.J. and David N. Livingstone, eds. Chicago: Chicago University Press, 1999, 1–31.

Withers, Charles W.J. "Geography, Enlightenment and the Paradise Question," in *Geography and Enlightenment*, Withers, Charles W.J. and David N. Livingstone, eds. Chicago: Chicago University Press, 1999, 67–92.

Withers, Charles W.J. *Placing the Enlightenment: Thinking Geographically About the Age of Reason*. Chicago: Chicago University Press, 2007.

Wolf, Hans. *America: Early Maps of the New World*. Munich: Prestel, 1992.

Wolkenhauer, W.J.B. "Bourguignon D'Anville," in *Deutsche Rundschau für Geographie und Statistik* 19 (1897), 468–471.

Wolkenhauer, W.J.B. "Bourguignon D'Anville," in *Nouvelle Biographie Générale*. Vol. 2. Paris: M. Firmin–Didot frères, 1859, 868–870.

Wood, Denis. *The Power of Maps*. New York: Guilford Press, 1992.

Woodward, David, ed. *Five Centuries of Map Printing*. Chicago: Chicago University Press, 1975.

Woodward, David, ed. *Art and Cartography: Six Historical Essays*. Chicago/London: University of Chicago Press, 1987.

Woodward, David. *Maps and the Rationalization of Geographic Space*. Mimeo, 1992.

Woodward, David. *Maps as Prints in the Italian Renaissance: Makers, Distributors and Consumers*. London: The British Library, 1996.

Woodward, David. "The Study of History of Cartography: A Suggested Framework," *American Cartographer* 1 (1974), 101–115.

Yolton, John W., ed. *Enlightenment*. 2nd ed. Oxford: Blackwell, 1996.

Index

www.ingramcontent.com/pod-product-compliance
Lightning Source LLC
Chambersburg PA
CBHW071326230926
51849CB00002B/325
* 9 7 8 9 0 0 4 7 0 8 3 1 0 *